Student Solutions Guide

Laurel Carpenter

to accompany

Calculus Concepts:
An Informal Approach to the Mathematics of Change

LaTorre/Kenelly/Fetta/Harris/Carpenter

Third Edition

Houghton Mifflin Company Boston New York

Sponsoring Editor: Lauren Schultz
Editorial Associate: Kasey McGarrigle
Senior Manufacturing Coordinator: Jane Spelman
Senior Marketing Manager: Danielle Potvin

Printed in the U.S.A.

ISBN:0-618-40133-4

1 2 3 4 5 6 7 8 9 –QD - 09 08 07 06 05

Preface

Learning calculus requires time, concentration, and energy. You will need to carefully listen to and do your best to understand your teacher during class, but more than that, you will need to apply yourself diligently and on a daily basis to studying outside of class. Read the text! Not only read the text, but as you read, work through the discussion and examples with a calculator, paper, and pencil so that you can duplicate the results with understanding.

Work on the activities. Don't give in to the temptation to do the minimum work that your teacher will allow you to get away with. Work for a thorough understanding of the concepts underlying the activities. Calculus is not a spectator sport; it takes application and effort. Do your best and you will eventually see the rewards.

We hope that you will enjoy the applications in this book.

I would like to thank Dave George and Jennifer King at Houghton Mifflin Company for their help and patience with this project. Any comments or suggestions concerning this *Guide* can be directed to the publisher.

This work is dedicated to my husband, Dean, and my children Jessica, Travis, Lydia, and Carl. Without their patience and understanding, this work would not have been possible.

Table of Contents

Chapter 1

Section 1.1 Models, Functions, and Graphs

1. Input: weight of letter
Input variable: w
Input units: ounces
Output: first-class domestic postage
Output variable: $R(w)$
Output units: cents
R is a function of w because a letter of one weight cannot have two different domestic first-class postage amounts.

3. Input: day of the week
Input variable: m
Input units: none
Output: amount spent on lunch
Output variable: $A(m)$
Output units: dollars
A is not a function of m unless you always spend the same amount on lunch every Monday, the same amount every Tuesday, etc., or unless the input is the days in only 1 week.

5. The table represents a function because each input (age) corresponds to only one output (percent with flex schedule.

7. The table does not represent a function because some inputs $(5'3''$ and $6'0'')$ are listed more than once in the table and correspond to more than one output.

9. Graphs b and c are functions. Graph a is not a function because vertical lines cutting through the circle touch it at two points.

11. **a.** P(Honolulu, HI) = 295

b. P(Providence, RI) = 137.8

c. P(Portland, OR) = 170.1

13. **a.** In 1988 cotton exports had a value of $1,975,000,000.

b. In 1992 cotton exports had a value of $1,999,000,000.

15.a.

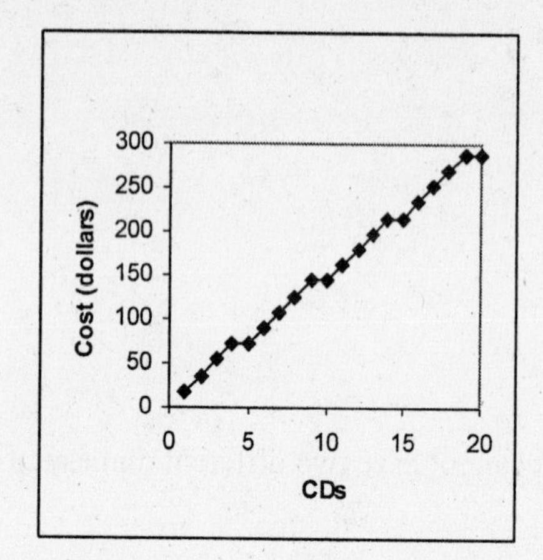

b. The cost of 6 CDs is the cost of 5 CDs at \$18 each, plus 1 free CD: (5 CDs)(\$18 per CD) = \$90.

c. You could buy 2 CDs for \$36. Because \$36/\$18 per CD = 2 CDs.

d. The graph shows 6 CDs for \$90 and 7 CDs for \$108. Thus with \$100 you could buy 6 CDs.

e. $\text{Average price} = \dfrac{\text{Total price}}{\text{Number bought}}$

For 3 CDs: $\dfrac{\$54}{3\text{ CDs}} = \$18 \text{ per CD}$ For 6 CDs: $\dfrac{\$90}{6\text{ CDs}} = \$15 \text{ per CD}$

17. a. From the graph, the value is approximately \$9000.

b. From the graph, the monthly payment is approximately \$340.

c. From the graph, the payment for a \$15,000 car is about \$320, and the payment for a \$20,000 car is about \$425. The amount of increase is approximately \$425 – \$320 = \$105.

d. The graph would pass through (0, 0) but would lie below the graph in Figure 1.1.8 because the same monthly payment would pay for a smaller loan amount.

19. a. From the graph, it was 3.0%

b. Cost-of-living increase was greatest in 1990 at 5.4%.

c. It was 2.8% in 1994.

d. Benefits increased, but the percentage by which they increased decreased.

21. At birth the baby weighed 7 pounds, so (0, 7) is a point on the weight graph.
After 3 days $\left(\frac{3}{7}\text{ week}\right)$ the baby has lost 7% of its birth weight, thus the weight is 93% of the birth weight: 0.93(7) = 6.51 pounds, and $\left(\frac{3}{7}, 6.51\right)$ is a point on the graph. At 1 week, the weight is again 7 pounds; at 2 weeks, the weight is 7.5 pounds; at 3 weeks, the weight is 8 pounds; and at 4 weeks, the weight is 8.5 pounds. Thus we have the points (1, 7), (2, 7.5), (3, 8), (4, 8.5).

Plotting these points and connecting them with line segments results in the following graph:

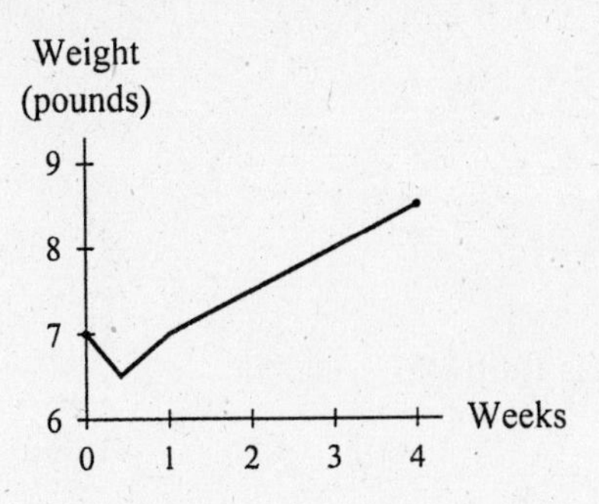

23. a. From the graph, it was about 5 inches deep.

b. It remained the same for approximately 4 days.

c. Snow fell.

d. The snow settled.

e. The snow was deepest (52 inches) around February 21.

f. Warm temperatures probably caused the decline in late February.

g. The most snow fell around February 18 when the graph rises most steeply.

25. a. Because 20% of 500 milligrams is 100 milligrams, the amount decreases by 100 milligrams each day.

x	0	1	2	3	4	5
y	500	400	300	200	100	0

b. $y = 500 - 100x$ mg after x days

c. x is between 0 and 5 days
$(0 \le x \le 5)$
y is between 0 and 500 mg
$(0 \le y \le 500)$

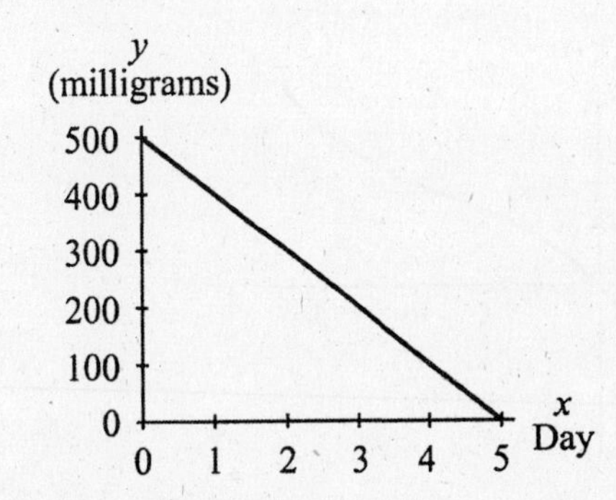

d. The x-intercept is 5 days. The y-intercept is 500 mg.
The y-intercept is the amount of the drug initially in the patient's body (500 mg). The x-intercept is the time when none of the drug remains (after 5 days).

e. y always decreases (for $0 \le x \le 5$).

f. According to the graph, after 3.5 days there will be about 150 mg of the drug remaining. The exact value is $500 - 100(3.5) = 500 - 350 = 150$ mg.

g. According to the graph, the concentration is 60 mg after about 4.5 days. Solve $60 = 500 - 100x$ to find the exact value, $x = 4.4$ days.

27. $s = 5: t = 3.2(5) + 6 = 22$
$s = 10: t = 3.2(10) + 6 = 38$

29. $R(3) = 39.4(1.998^3) \approx 314.255$
$R(0) = 39.4(1.998^0) = 39.4$

31. Solve $R(w) = 78.8$ using technology or algebraically as follows:

$$78.8 = 39.4(1.998^w)$$
$$2 = 1.998^w$$
$$\ln 2 = w \ln 1.998$$
$$w = \frac{\ln 2}{\ln 1.998} \approx 1.001$$

$R(w) = 78.8$ when $w \approx 1.001$.

Solve $R(w) = 394$ using technology or algebraically as follows:

$$394 = 39.4(1.998^w)$$
$$10 = 1.998^w$$
$$\ln 10 = w \ln 1.998$$
$$w = \frac{\ln 10}{\ln 1.998} \approx 3.327$$

$R(w) = 394$ when $w \approx 3.327$.

33. We begin by graphing the function $Q(x) = 0.32x^3 - 7.9x^2 + 100x - 15$ to determine how many solutions there are to the equations $Q(x) = 515$ and $Q(x) = 33.045$.

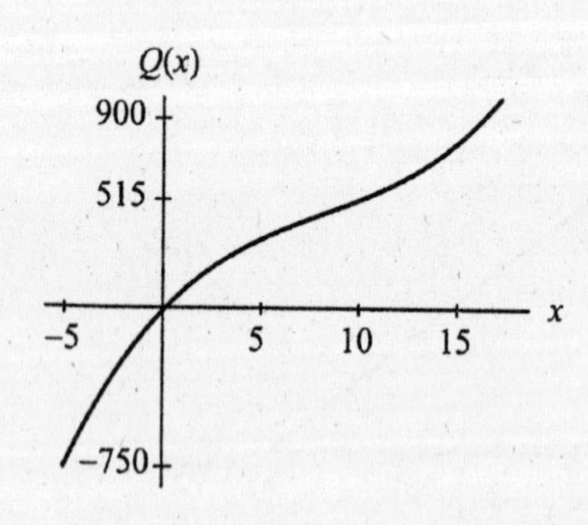

The graph indicates that we seek only one input for each given output. Using technology we find the solutions as
$Q(x) = 515$ when $x = 10$
$Q(x) = 33.045$ when $x \approx 0.5$

35. An input is given.

$$A(15) = 3200e^{0.492(15)}$$
$$\approx 5{,}131{,}487.257$$

The corresponding output is about 5,131,487.257.

37. An output is given. Use technology to solve or find the solution algebraically as

$$2.97 = 39.4(\ln 1.998)(1.998^x)$$

$$\frac{2.97}{39.4\ln 1.998} = 1.998^x$$

$$\ln\left(\frac{2.97}{39.4\ln 1.998}\right) = x\ln 1.998$$

$$x = \left(\frac{1}{\ln 1.998}\right)\ln\left(\frac{2.97}{39.4\ln 1.998}\right)$$

$$x \approx -3.203$$

The corresponding input is ≈ - 3.203.

39. Quill Activity

Section 1.2 Constructed Functions

1. a. $T(x) = K(x) + L(x)$
$= 9088.859 + 1697.717\ln x + 2424.764 + 915.025\ln x = 11{,}513.623 + 2612.742\ln x$ kidney and liver transplants
where x is the number of years since 1990

b. $T(5) = 11{,}513.623 + 2612.742\ln 5 \approx 15{,}719$ transplants

3. $N(t) = M(t) - W(t) = -0.2190t + 45.2325 - \left(0.01685t^2 - 3.48778t + 187.95962\right)$
$= -0.01685t^2 + 3.26878t - 142.72712$ gallons of milk other than whole milk per person per year where t is the number of years since 1900. According to this model, per capita consumption of milk (other than whole milk) is approximately
$N(100) = -0.01685(100)^2 + 3.26878(100) - 142.72712 \approx 15.65$ gallons in 2000.

5. a. Let $n(t)$ be the percent who had heard about, but not planted, hybrid seed corn t years after 1924.

$$n(t) = h(t) - p(t) = \frac{100}{1 + 128.0427e^{-0.7211264t}} - \frac{100}{1 + 913.7241e^{-0.607482t}}$$

b. In 1929, $n(5) = \dfrac{100}{1 + 128.0427e^{-0.7211264(5)}} - \dfrac{100}{1 + 913.7241e^{-0.607482(5)}} \approx 20\%$ of Iowa corn farmers had heard about but not yet planted the hybrid seed corn.

7. Let $c(x)$ be the number of cesarean-section deliveries performed x years after 1980.
$c(x) = n(x)p(x) = (-0.034x^3 + 1.331x^2 + 9.913x + 164.447)(-0.183x^2 + 2.891x + 20.215)$

9. The percentage of debit card transactions that were conducted at the point of sale can be found by dividing the number of point of sale debit transactions by the total number of debit transactions. Thus if $R(y)$ represents the percentage of debit card transactions that were conducted at the point of sale, then, for any year y, $R(y) = \dfrac{P(y)}{D(y)}$ percent per year.

11. The functions can be combined because outputs from C can be used as inputs to P. $(P \circ C)(t) = P(C(t)) =$ profit after t hours of production

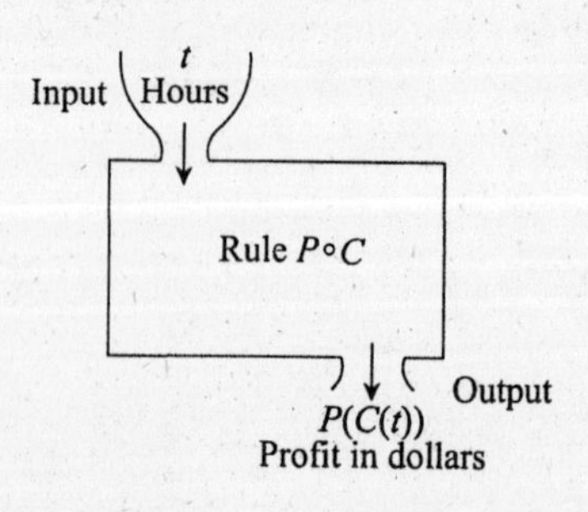

13. The functions can be combined because the outputs from C can be used as inputs to P.
$(P \circ C)(t) = P(C(t))$ = average tips from customers in the restaurant t hours after 4 p.m.

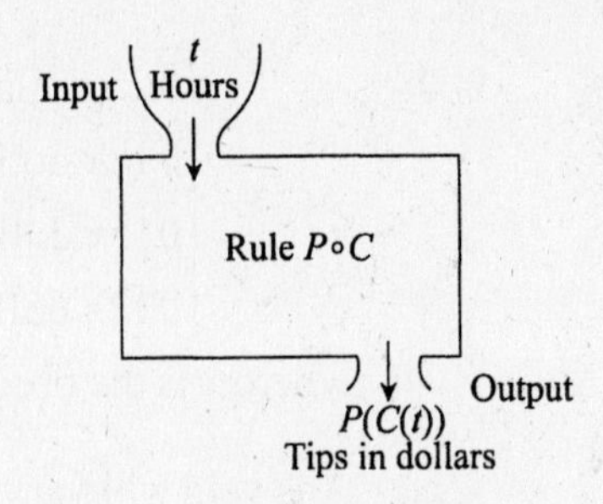

15. $f(t(p)) = f(4p^2) = 3e^{4p^2}$

17. $g(x(w)) = g(4e^w) = \sqrt{7(4e^w)^2 + 5(4e^w) - 2}$

19. $g(t(m)) = g(4m + 17) = 3$

21. a. $S(85) = 12.1(85) - 905.4 = \$123.1$ million
$S(88) = 12.1(88) - 905.4 = \$159.4$ million
$S(89) = -14.8(89) + 1414.9 = \$97.7$ million
$S(92) = -14.8(92) + 1414.9 = \$53.3$ million

b.

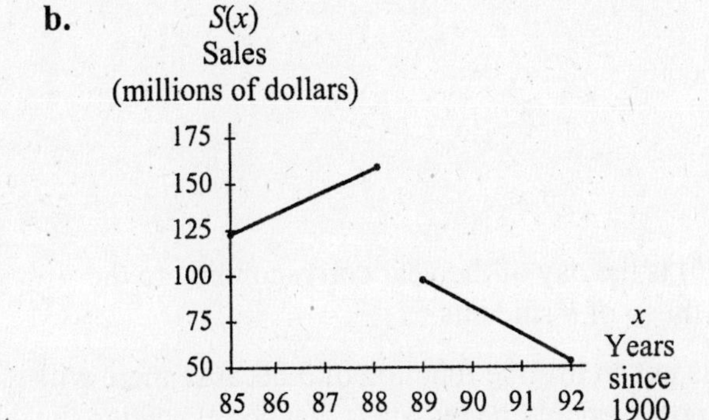

c. S is a function of x because each year corresponds to only one amount of sales.

23. a. The shipping charges encourage larger orders.

b.

Order amount	Shipping
\$17.50	0.20(17.50) = \$3.50
\$37.95	0.18(37.95) ≈ \$6.83
\$75.00	0.15(75.00) = \$11.25
\$75.01	0.12(75.01) ≈ \$9.00

c, d.

$$S(x) = \begin{cases} 0.2x \text{ dollars} & \text{when } 0 \le x \le 20 \\ 0.18x \text{ dollars} & \text{when } 20 < x \le 40 \\ 0.15x \text{ dollars} & \text{when } 40 < x \le 75 \\ 0.12x \text{ dollars} & \text{when } x > 75 \end{cases}$$

where x is the order amount in dollars.

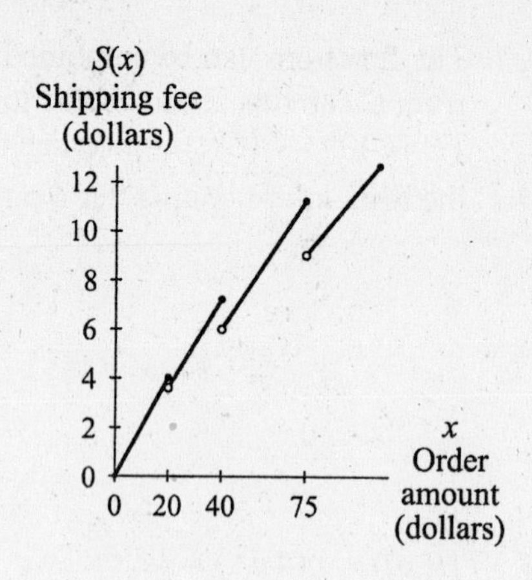

e. Answers will vary.

25.

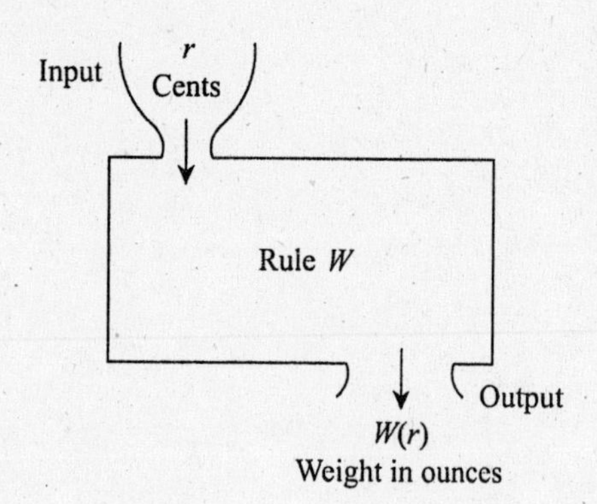

$W(r)$ is the weight in ounces of a first class letter or parcel that costs r cents to mail.

W is not an inverse function of r, because each value of r corresponds to a *range* of values for W, so there are multiple outputs for most inputs.

27.

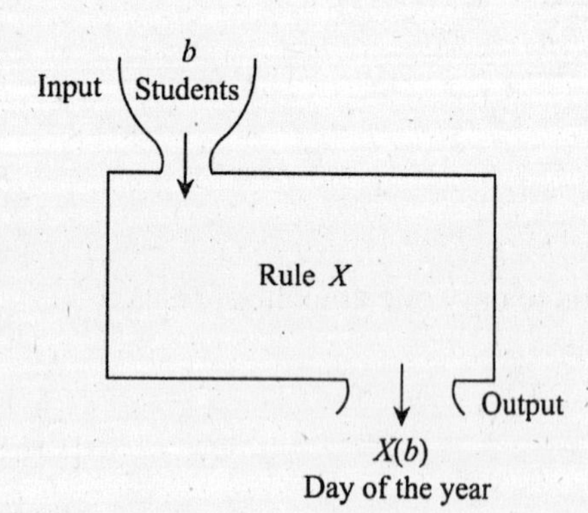

$X(b)$ is the day of the year corresponding to the birthday of b students.

X is not an inverse function of b because there will almost certainly be two days that correspond to the same number of students.

29. Reverse the inputs and outputs.

Percent using computers at work	Age
37.1	18-24
52.5	25-29
53.3	30-39
54.9	40-49
50.7	50-59
32.6	over 59

This is not an inverse function, because each input (percent) corresponds to a range of outputs (ages).

31. Reverse the inputs and outputs.

Person's weight (pounds)	Person's Height
139	5′3″
196	6′1″
115	5′4″
203	6′0″
165	5′10″
127	5′3″
154	5′8″
189	6′0″
143	5′6″

This is a function, but it is not an inverse function because the original table did not represent a function.

33. a. $m(2399) = 5$; In May of 2001, 2399 complaints were received.

b. $C(1) + C(2) + C(3) = 4321$ complaints in the first quarter of the year

c. C graphs as a diagonal line so it satisfies the vertical line test.

35. a.

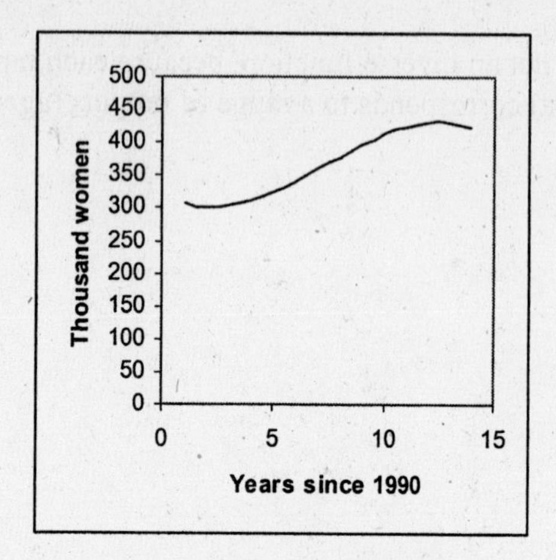

b. No, because the horizontal line at $w = 305$ crosses the graph of the function twice.

c. No, because it is not one-to-one.

37. Quill Activity

Section 1.3 Function Behavior, Limits, and Continuity

1. a. 5

b. 0

c. 2.6

3. a. 3

b. 1

c. Does not exist

d. Does not exist, $\lim_{t\to\infty} m(t) \to -\infty$

e. 6

5. a. $y = -4$ is a horizontal asymptote on the left

b. i. Yes, because the function is defined at $x = -5$, and the limit as x approaches -5 is the same as the function value at $x = -5$.

ii. Yes, because the function is defined at $x = 2$, and the limits as x approaches 2 from the left and right are the same as the function value at $x = 2$.

iii. Yes, because the function is defined at $x = 3$, and the limit as x approaches 3 is the same as the function value at $x = 3$.

iv. No, because the function is not defined at $x = 4$.

7.

$x \to \frac{-1}{3}^-$	$\frac{x^3+6x}{3x+1}$	$x \to \frac{-1}{3}^+$	$\frac{x^3+6x}{3x+1}$
−0.35	42.858	−0.32	−48.819
−0.34	103.965	−0.332	−507.149
−0.334	1020.630	−0.3332	−5090.482
−0.3334	10,187.296	−0.33332	−50,923.814

$\lim_{x\to\frac{-1}{3}^-} \frac{x^3+6x}{3x+1} \to \infty$; $\lim_{x\to\frac{-1}{3}^+} \frac{x^3+6x}{3x+1} \to -\infty$; $\lim_{x\to\frac{-1}{3}} \frac{x^3+6x}{3x+1}$ does not exist

9.

$t \to \infty$	$\frac{14}{1+7e^{0.5t}}$
10	0.01346
20	$9.0799 \cdot 10^{-5}$
30	$6.1180 \cdot 10^{-7}$
40	$4.12231 \cdot 10^{-9}$
50	$2.7776 \cdot 10^{-11}$

$\lim_{t\to\infty} \frac{14}{1+7e^{0.5t}} = 0$

11. Quill Activity

13. a. $A = \left(1+\frac{1}{n}\right)^n$ dollars

b-c.

Compounding	n	Amount
yearly	1	\$2.00
semiannually	2	\$2.25
quarterly	4	\$2.44
monthly	12	\$2.61
weekly	52	\$2.69
daily	365	\$2.71
every hour	8760	\$2.72
every minute	525,600	\$2.72
every second	31,636,000	\$2.72

d. \$2.72

e. $\lim_{n\to\infty} \left(1+\frac{1}{n}\right)^n \approx 2.72$

15. a. The concentration increases rapidly from 0 mg at time 0 to approximately 63 mg one quarter hour after the initial dose. The concentration then decreases again until it is almost non-existent four hours later.

b. $\lim_{t\to\infty} 100\left(e^{-1.3t} - e^{-9.5t}\right) = 100(0-0) = 0$

c. Answers vary. One possible answer: After 4 hours, the amount of Ultram in the blood is about 0.6 milligrams. It is probably safe to take another dose after 4 hours.

17. $P(m)$ can be interpreted continuously assuming that magnitude can be measured as any real number.

19. $Q(t)$ can be used without restriction.

21. a.

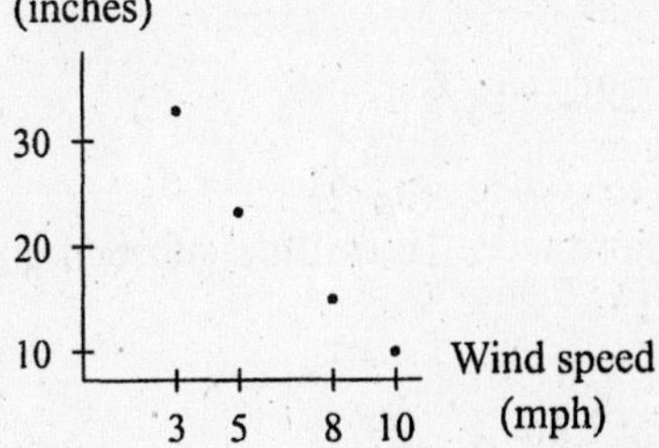

b. Because no vertical line passes through two data points, the table of data represents a function.

c. Because the table does not contain an input of 6, we cannot read this information from the table.

d. Tables of data are always discrete representations.

23. a. $P(25) = 0.043(25^3) - 3.129(25^2)$
$+ 71.133(25) - 315.524 \approx \$179.34$

When 25 calls are made in one day, the average profit is about \$179.

b. Use technology to solve for c in $180 = 0.043c^3 - 3.129c^2 + 71.133c - 315.524$
to obtain $c \approx 13.46$ (There are 3 solutions to this equation, but we consider only the one between 0 and 25.)

Because the number of calls must be a positive integer, this input value does not have a meaningful interpretation in this context. According to the model given, the average daily profit is never exactly \$180. When 13 calls are made, the average profit will be a little less than \$180, and when 14 calls are made, the average profit will be a little more than \$180.

c.

$P(c)$
Profit
(dollars)
Maximum
profit
200
100
0
-300
10
20
30
c
Daily
calls

According to the graph, the maximum occurs when $c \approx 18.2$. Checking the output corresponding to the integer input values on either side of 18.2, we find that
$P(18) \approx \$201.85$ and $P(19) \approx \$201.37$.
The maximum profit of \$201.85 occurs when 18 calls are made.

25. a. Input: r dollars per month is the average basic rate for a cable subscription
Output: S is the number (in millions) of cable subscribers in the United States

b. Input: dollars per month
Output: million subscribers

c. No. This function does not have to be discretely interpreted because input is an average which does not have to be represented in whole cents.

d. $S(15) = -56.115 + 37.201 \ln(15) \approx 44.632$ million subscribers

e. Revenue = (monthly rate)(number of subscribers)
In order to determine the monthly rate for which there were 10 million subscribers, solve $S(r) = -56.115 + 37.201 \ln(r) = 10$ for r: $r \approx 5.913$
Thus,

$$\text{Revenue} \approx (\$5.913)(10 \text{ million}) \approx \$59.128 \text{ million}$$

27. a. Input: the number of hours after the market opened
Output: the price of Microsoft stock

b. Input units: hours
Output units: dollars per share

c. The model can be interpreted without restriction within one trading day because the price changes in an approximately continuous manner throughout a trading day.

d. $P(3) = 0.368(3) + 43.99$
$\approx \$45.09$ per share

e. Solving the equation $47 = 0.368h + 43.99$ for h:

$3.01 = 0.368h$

$\frac{3.01}{0.368} = h$ or $h \approx 8.2$ hours

Because the trading day is less than 8.2 hours long, we conclude that the price was never \$47 per share during this trading day.

29. a. $P(20) = 0.000498(20)^3 - 0.0686(20)^2 + 3.044(20) - 14.952 = 22.472$. Approximately 22.5% of 20-year-old workers have flex schedules.

b. Input units: years (of age)
Output units: percent

c. No, age is a real number and those surveyed were at different ages in their age range at the time of the survey.

31. Quill Activity

Section 1.4 Linear Functions and Models

1. a. $\text{Slope} \approx \frac{-\$2.5 \text{ million}}{5 \text{ years}}$

$= -\$0.5$ million per year

The corporation's profit was declining by approximately a half a million dollars per year during the 5-year period.

b. The rate of change is approximately –\$0.5 million per year.

c. The vertical axis intercept is approximately \$2.5 million. This is the value of the corporation's profit in year zero. The horizontal axis intercept is 5 years. This is the time when the corporation's profit is zero.

3. a. 382.5 donors per year

b.

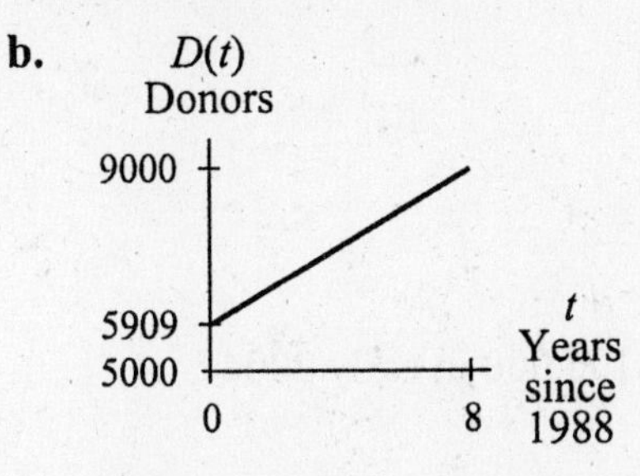

c. The vertical axis intercept is 5909 (when $t = 0$). This is the number of donors in 1988, the starting year.

5. This question cannot be answered because the model does not include a description of the input variable.

7. a. Rate of change of revenue $= \dfrac{\$824.1 - \$744.0 \text{ million}}{1998 - 1997} = \$80.1$ million per year

b. Because $\dfrac{80.1}{4} = 20.025$, the revenue increased by \$20.025 million during each quarter of 1998.

c. Add \$80.1 million to each revenue amount to find the next year's revenue.

Year	Revenue (millions of dollars)
1997	744.0
1998	824.1
1999	904.2
2000	984.3

d. Revenue $= 744.0 + 80.1t$ million dollars t years after 1997

9. a.

$$\text{Rate of change} = \frac{\$97{,}500 - \$73{,}000}{2002 - 1990} = \frac{\$24{,}500}{12 \text{ years}} \approx \$2041.67 \text{ per year}$$

or about \$2042 per year

b. $\$97{,}500 + 3(\$2042) \approx \$103{,}600$

c. Let t be the number of years after the end of 1990. Then the predicted value is given by $V(t) = 2041.667t + 73{,}000$ dollars.

$$V(t) = 75{,}000$$
$$2041.667t + 73{,}000 = 75{,}000$$
$$2041.667t = 2000$$
$$t \approx 0.98$$

$$V(t) = 100{,}000$$
$$2041.667t + 73{,}000 = 100{,}000$$
$$2041.667t = 27{,}000$$
$$t \approx 13.22$$

The value was \$75,000 in late 1991 ($t \approx 0.98$) and \$100,000 in early 2004 ($t \approx 13.22$, shortly after the end of 2003).

d. $V(t) = 2041.667t + 73{,}000$ dollars
t years after the end of 1990
1999: $V(9) = 2041.667(9) + 73{,}000$
$\approx \$91{,}400$
The model assumes the rate of increase of the market value remains constant. This assumption is not necessarily true. (In some markets, home prices fluctuate wildly.)

11. **a.** Rate of increase

$$= \frac{\$1235.8 – \$838.6 \text{ billion}}{1997–1993}$$
$$= \frac{\$397.2 \text{ billion}}{4 \text{ years}} = \$99.3 \text{ billion / year}$$

b. $1235.8 + 3(99.3) = \$1533.7$ billion

c. Answers will vary.

d. If t is the number of years after 1993, then $C(t) = 838.6 + 99.3t$ billion dollars. Because \$2 trillion = \$2000 billion, solve $C(t) = 2000$.

$$838.6 + 99.3t = 2000$$
$$99.3t = 1161.4$$
$$t \approx 11.7$$

The linear model predicts that consumer credit will reach \$2 trillion about 12 years after 1993, in the year 2005.

e. Answers will vary.

13. **a.** 2000 dogs per hour, or 48,000 dogs per day

b. 3500 cats per hour, or 84,000 cats per day

c. If we assume that the existing population cited in the article was determined at the beginning of the year, the equation is $D(t) = 48{,}000t + 54{,}000{,}000$ dogs t days after the beginning of 1993.

d. With the same assumption used in part c, the equation is $C(t) = 84{,}000t + 56{,}000{,}000$ cats t days after the beginning of 1993.

e. Total dogs and cats = $D(365) + C(365) = 71{,}520{,}000 + 86{,}660{,}000 = 158{,}180{,}000$ dogs and cats. This is about 0.16 billion dogs and cats, so it does not agree with the prediction in the article.

15. a. 78 million people per year

b. $P(t) = 6 + 0.078t$ billion people t years after the beginning of 2000

c. Setting the model in part b equal to 12 and solving for t yields $t \approx 76.9$ years after the beginning of 2000, which corresponds to near the end of 2076. The article estimates the world population will be 12 billion in 2050.

d. The prediction in part c assumes that the world will grow at a constant rate of 78 million people per year between now and 2076. In making their prediction, the Census Bureau must have assumed that the growth rate will increase so that the 12 billion population will be reached sooner than our prediction based on the linear model.

17. a. $\dfrac{\$1.37 - \$0.97}{1999 - 1996} = \dfrac{\$0.40}{3} \approx \$0.13/\text{year}$

b. Answers may vary. Three possible models are:

$A1(x) = 0.1333x - 265.1633$ dollars in year x

$A2(x) = 0.1333x + 0.17$ dollars
x years after 1990

$A3(x) = 0.1333x + 0.97$ dollars
x years after 1996

c. The average fee in 1998 can be estimated using any one of the three models in part b:

$A1(1998) \approx \$1.24$
$A2(8) \approx \$1.24$
$A3(2) \approx \$1.24$

The average fee in 2003 can be estimated using any one of the three models in part b:

$A1(2003) \approx \$1.90$
$A2(13) \approx \$1.90$
$A3(7) \approx \$1.90$

d. The 1998 estimate is an interpolation, and the 2003 estimate is an extrapolation.

19. Quill Activity

21. a. Answers may vary. Three possible models are:

$S1(x) = 499.3x - 976{,}088.3$ students in year x

$S2(x) = 499.3x - 27{,}418.3$ students
x years after 1900

$S3(x) = 499.3x + 5036.2$ students
x years after 1965

b. The enrollment in 1970 can be estimated as

$S1(1970) \approx 7533$ students

$S2(70) \approx 7533$ students

$S3(5) \approx 7533$ students

c. Because $7533 - 8038 = -505$, the estimate is 505 students lower than the actual enrollment. Answers vary on whether the error is significant. The error represents about 6% of the actual enrollment. For a school the size of the one in this activity, an error of 500 students could mean a significant increase in student housing and faculty loads.

d. It would not be wise to use these models to extrapolate 31 years beyond the last data point.

23. a. $P(t) = 0.762t + 10.176$ dollars t years after 1981

b. Round to the nearest dollar since the original data were rounded to the nearest dollar.

1984: $P(3) \approx \$12$ interpolation

1992: $P(11) \approx \$19$ extrapolation

1999: $P(18) \approx \$24$ extrapolation

c. The model prediction is $1 less than the actual price. That's fairly accurate.

d, e. Answers will vary.

f. When a linear model is used to extrapolate, the underlying assumption is that the output will continue to increase at a constant rate. Often this may be a valid assumption for short-term extrapolation but not for a long-term extrapolation.

g. Extrapolating from a model must always be done with caution. In order for the extrapolation to be accurate, the model must accurately describe the situation, and the future behavior of the output must match that of the model. Long-term extrapolations are always risky.

25. a. $F(y) = -0.152y + 19.514$ percent
where y is 81 for 1981–82 school year, 82 for the 1982–83 school year, etc.

b. Because -0.152 is multiplied by y in the equation, the rate of change is about -0.152 percentage points per year.

c. $F(93) = -0.152(93) + 19.514 \approx 5.4\%$

d. Solving $F(y) = 5$ gives $y \approx 95.7$, which corresponds to the 1996–97 school year.

27. a. See part *c* for the scatter plot. The scatter plot does reflect the statements about atmospheric release of CFCs.

b. Answers may vary. One possible model is given. Let x be the number of years after 1900.

The linear model for 1974–1980 is $y = -15.37x + 1554.19$ million kg

The linear model for 1980–1988 is $y = 9.165x - 412.5$ million kg

The linear model for 1988–1992 is $y = -34.375x + 3413.283$ million kg

Note that $-15.37(80) + 1554.19 \approx 324.6$ and $9.165(80) - 412.5 \approx 320.7$, so the linear regression model provides the best estimate for 1980 top, and we include the equality $x = 80$ in the top portion of the function.

Similarly, $9.165(88) - 412.5 \approx 391.7$ and $-34.375(88) + 3413.283 \approx 388.3$, so the middle linear model provides the best estimate for 1988, and we include the equality $x = 88$ in the middle portion of the function. Because the middle model provides the closest estimates to the output at the breakpoints, we include both inequalities on the middle portion of the function. The piecewise continuous function is:

$$R(x) = \begin{cases} -15.37x + 1554.19 \text{ million kg} & \text{when } x \le 80 \\ 7.985x - 311.02 \text{ million kg} & \text{when } 80 < x \le 88 \\ -34.375x + 3413.283 \text{ million kg} & \text{when } x > 88 \end{cases}$$

where x is the number of years after 1900.

(Note that in computing the middle model, the 1980 data point was not included, but in computing the bottom model, the 1988 data point was included. This was done to improve the fit of the model. You will have to use your own discretion to determine whether or not to include the points at which the data is divided when finding a piecewise model.)

c.

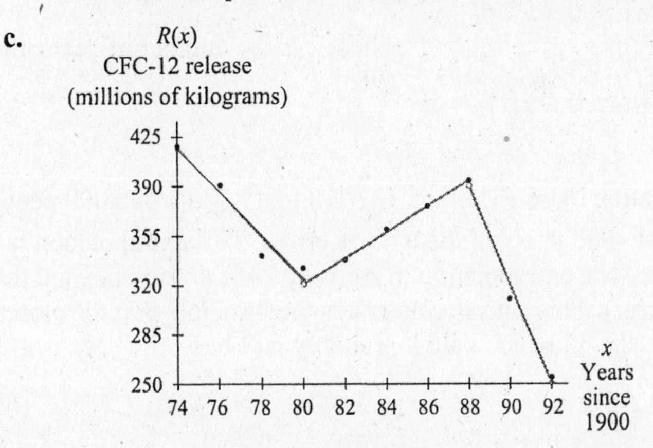

d. i. What was the amount of CFCs released into the atmosphere in 1975? 1995?

$C(75) \approx 401.4$ million kg,
$C(95) \approx 147.7$ million kg

ii. At what rate was the release of CFCs declining between 1974 and 1980?

15.37 million kilograms per year between 1974 and 1980.

iii. On the basis of the data accumulated since 1987, in what year will there no longer be any CFCs released into the atmosphere?

Determining where the far right portion of the function is zero yields $x \approx 99.3$. This answer indicates that according to the model there should have been no release of CFCs by early in 2000.

29. a.

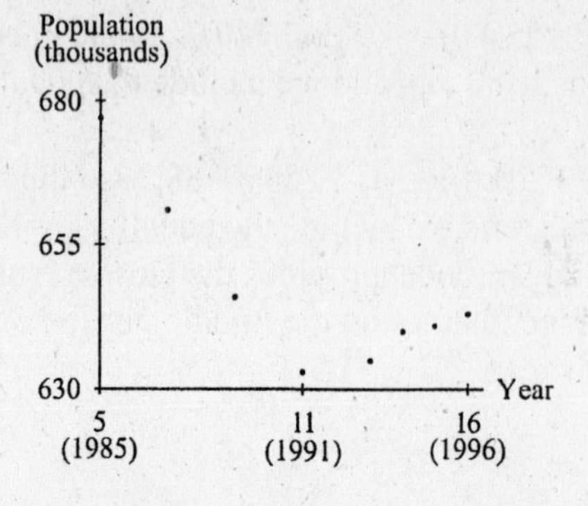

1991 is the last year in which the data decline. In 1993 the data began to rise. We choose 1991 as the dividing point for a piecewise continuous model.

b. Let x be the number of years after 1985.

For the years 1985 to 1991, the linear model is $y = -7.35x + 676.3$ thousand people.

For the years 1991 to 1996, the linear model is $y = 2.122x + 619.730$ thousand people.

The first equation gives a population of 632.2 thousand people in 1991, and the second equation gives a population of 632.5 thousand people in 1991. Because the second equation is closer to the actual population of 633 thousand people, we define the 1991 population using the second equation. The piecewise continuous function is

$$p(x) = \begin{cases} -7.35x + 676.30 \text{ thousand people} & \text{when } 0 \le x < 6 \\ 2.122x + 619.730 \text{ thousand people} & \text{when } 6 \le x \le 11 \end{cases}$$

where x is the number of years after 1985

c. The model estimates the population to be $P(12) = 2.122(12) + 619.730 \approx 645{,}000$ people
This is an overestimate of about 4000 people (about 0.6% error). This extrapolation is only one year beyond the data given. The extrapolation in Activity 24 is 4 years beyond the data given (and results in a 1.7% error). This activity illustrates the principle that the closer an extrapolation is to known data, the more accurate it probably will be.

31. Excel Activity

a. 1954

b. The overall pattern appears reasonably linear.

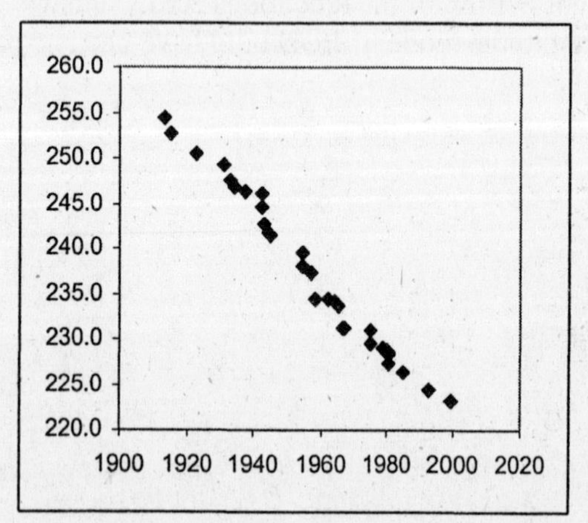

c. 1957 Derek Ibbotson of the UK and 1958 Herb Elliot of Australia

d. $S(x) = -0.390x + 1001.511$ seconds in year x

e. 2029, discussion will vary

33. Quill Activity

Chapter 1 Review Test

1. **a.** Because the graph shows the *gain* in wetlands, and all the outputs values are positive, the number of wetlands increased every year. Thus the number of acres increased between 1991 and 1992 but by a smaller amount than the number increased between 1990 and 1991 or 1992 and 1993.

   b. Between 1989 and 1991, the yearly gain of wetlands was increasing by approximately $\frac{140-75}{2} \approx 33$ thousand acres per year.

2. **a.**

Input: d Day in 2000 → Rule T → Output: $T(d)$ Tickets sold

   b. T is a function of d because for every possible day, there is only one associated number of tickets.

   c. If the inputs and outputs are reversed, the result is not an inverse function because one number of tickets could have more than one day associated with it.

3. **a.** Rate of increase $= \frac{23\% - 18\%}{1995 - 1973} = \frac{5\%}{22 \text{ years}} \approx 0.23$ percentage point per year

   b. Percentage in 2002 $= 23\% + (7 \text{ yrs})\left(\frac{5}{22}\frac{\text{percentage pt}}{\text{per year}}\right) \approx 24.6\%$ in 2002

   This estimate is valid only if the rate of change remains constant through 2002.

4. **a.** **i.** $\lim_{x\to 1^-} f(x) = 2(1)+5 = 7$ and $\lim_{x\to 1^+} f(x) = -2(1)+9 = 7$, thus $\lim_{x\to 1} f(x) = 7$.

   ii. $\lim_{x\to 2^-} f(x) = -2(2)+9 = 5$ and $\lim_{x\to 2^+} f(x) = \lim_{x\to 2^+} \frac{2(x-2)(x+1)}{(x-2)(x+3)} = \lim_{x\to 2^+} \frac{2(x+1)}{x+3} = \frac{6}{5}$

   Thus $\lim_{x\to 2} f(x)$ does not exist.

   iii. $\lim_{x\to\infty} f(x) = \lim_{x\to\infty} \frac{2(x^2-x-2)}{x^2+x-6} = \lim_{x\to\infty} \frac{2x^2}{x^2} = \lim_{x\to\infty} 2 = 2$

   b. The function f is not continuous at $x = 1$ because $f(1)$ does not exist (that is, the function is not defined for $x = 1$). The function f is not continuous at $x = 2$ because the limit of $f(x)$ as x approaches 2 does not exist.

5. **a.** $C(t) = 0.0342t + 11.39$ million square kilometers t years after 1900

   b. C is continuous without restriction.

c. The rate of change of the model is the slope of the equation: 0.0342 million square kilometers of cropland per year. The amount of cropland increased by approximately 0.034 million square kilometers, or 34,000 square kilometers, per year between 1970 and 1990.

d. Answers will vary. One possible answer is that because the data are linear in nature, the rate of increase is "steady" as stated.

e. $C(95) = 0.0342(95) + 11.39 \approx 14.64$ million square kilometers

Chapter 2

Section 2.1 Exponential Functions and Models

1. $f(x) = 2(1.3^x)$ is the black graph. $f(x) = 2(0.7^x)$ is the teal graph.

3. $f(x) = 3(1.2^x)$ is the teal graph. $f(x) = 2(1.4^x)$ is the black graph.

5. Because $1.05 = 1 + 0.05$, f is increasing with a 5% change in output for every unit of input.

7. Because $0.87 = 1 - 0.13$, y is decreasing with a 13% change in output for every unit of input.

9. Because $0.61 = 1 - 0.39$, the number of bacteria declines by 39% each hour.

11. **a.** With starting value $a = 4.81$ quadrillion Btu and the parameter b calculated as $b = 1 + 0.0547$, the model is $P(t) = 4.81(1.0547^t)$ quadrillion Btu in petroleum product imports t years after 2005.

b. Solving the equation $10 = 4.81(1.0547^t)$ for t yields $t \approx 13.7$ years after 2005. Thus petroleum product imports will exceed 10 quadrillion Btu for the first time in September of 2018.

c. $P(t)$ is an increasing exponential function so as the inputs increase without bound the outputs will also increase without bound.

13. **a.** With starting value $a = \$1.50$ and the parameter b calculated as $b = 1 + 0.0746$, the model is $S(y) = 1.5(1.0746^y)$ dollars y years after 1997

b. $S(13) \approx \$3.82$

15. **a.** Using starting value $a = 3.3$ and calculating b as $b = 1 - 0.0146 = 0.9854$ produces the model $W(t) = 3.3(0.9854^t)$ workers per beneficiary t years after 1996.

b. $W(34) \approx 2.00$ workers per beneficiary. Fewer workers per beneficiary will mean that Social Security will have to find other means of supplementing payments rather than relying solely on Social Security withholdings from workers' wages.

17. a. $P(t) = 1.269(1.015646^t)$ billion people t years after 1900.

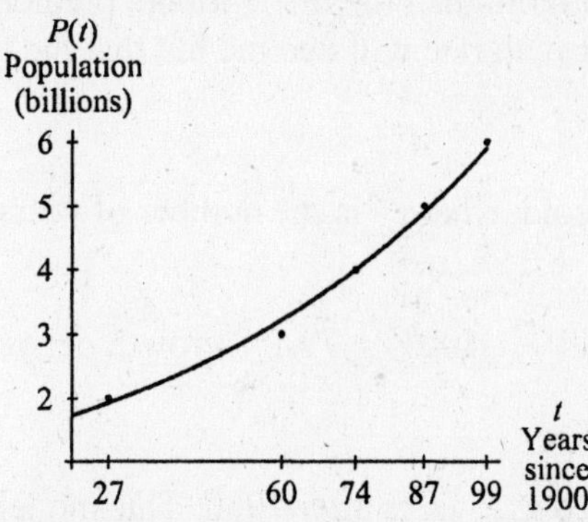

The function is a good fit for the data. The 1960 data point is the one farthest from the function. The data points for 1974, 1987, and 1999 appear close to the graph.

b. The rate of change given in the article is 78 million people per year (0.078 billion people per year). Using a starting population of 6 billion people, the linear model is $W(x) = 6 + 0.078x$ billion people x years after 1999.

c. Using the relationship: $x = t - 99$, we have $W(x) = 6 + 0.078x$ rewritten as $W(t) = 6 + 0.078(t - 99)$ billion people t years after 1900.

d.

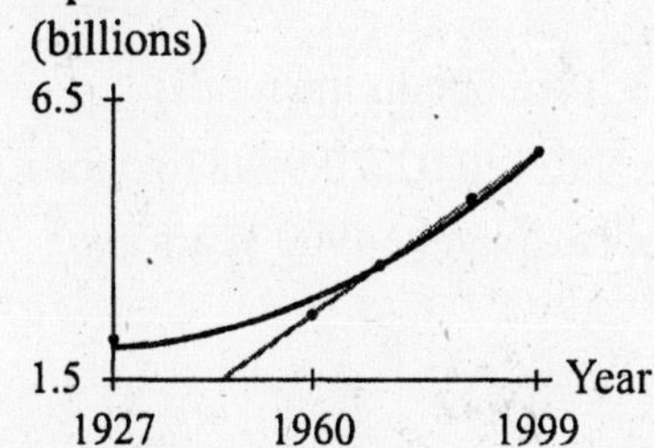

The linear model appears to be a better fit for the years 1960–1999; the exponential model is a better overall fit.

e.

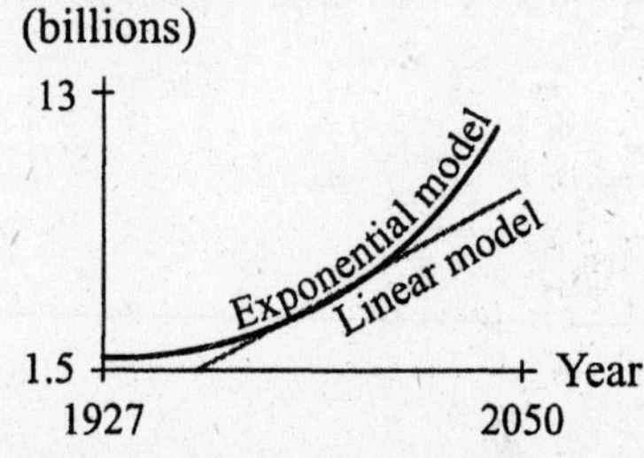

For years beyond 1999, the exponential model rises dramatically compared to the linear model. Both models should give reasonably close answers for populations between 1960 and 2000, but the exponential model should be used from 1900 through 1960. Which model will be more appropriate for years beyond 2000 depends on what happens to the world population growth rate in the future. See part *f* for a more detailed answer.

f. The exponential model gives the population as approximately 6.0 billion in 2000 and 13.0 billion in 2050. The linear model gives 6.1 billion people in 2000 and 10.0 billion people in 2050. The linear model assumes that the growth rate (people per year) remains constant at the 1999 level. The exponential model assumes that the percentage growth rate remains constant at the level obtained by modeling the 1927 through 1999 data.

g. The Census Bureau predicts a population of 12 billion in 2050. This is 1 billion less than the exponential model prediction and 2 billion more than the linear model prediction. The Census Bureau assumed that the percentage growth rate will decline but that the growth rate will increase.

19. a. $C(y) = 6.673(1.791044^y)$ million CD singles sold, where y is the number of years since 1993

b. The model reflects an approximately $(1.791044 - 1)100\% \approx 79.1\%$ growth per year in the sales of CD singles.

21. a. $F(x) = 345.957(0.942378^x)$ farms with milk cows x years after 1980. This model indicates a $(1 - 0.942378)100\% \approx 5.8\%$ decline per year in the number of farms with milk cows.

b. $\lim_{x\to\infty} 345.957(0.942378^x) = 0$; It is not reasonable that the number of farms with milk cows will approach zero in the future.

c. Answers vary. Possible reasons: The decline in farms with milk cows is probably indicative of an overall decline in farms because of the nature of the economy over the past decades. Specialization has led to fewer nondairy farms owning milk cows. The emphasis on decreasing fat and cholesterol in food may have produced a decrease in demand for dairy products.

23. Using the two data points (0, 0.027) and (5580, 0.0135) we obtain the model $C(t) = 0.027(0.999876^t)$ grams after t years. Solving $0.05 = 0.027(0.999876^t)$ for t, we obtain $t \approx -4960$ years. The sample contained 0.05 gram approximately 4960 years ago.

25. a. Using the points (0, 250) and (0.5, 125), we obtain the model $P(h) = 250(0.25^h)$ mg of penicillin left after h hours.

b. Solving $1 = 250(0.25)^h$ gives $h \approx 3.98$ hours. The second dose of penicillin should be taken approximately 4 hours after the first dose.

27. a.

$$2P = P\left(1 + \frac{r}{n}\right)^{nt}$$

$$2 = \left(1 + \frac{0.063}{12}\right)^{12t}$$

$$\ln 2 = 12t \ln\left(1 + \frac{0.063}{12}\right)$$

$$t = \frac{\ln 2}{12 \ln\left(1 + \frac{0.063}{12}\right)} \approx 11.03$$

It will take just over 11 years (that is, 11 years 1 month).

b. $2P = Pe^{rt}$

$$2 = e^{0.08t}$$

$$\ln 2 = 0.08t$$

$$t = \frac{\ln 2}{0.08} \approx 8.66$$

It will take about 8.66 years (8 years 8 months).

c. $2P = P\left(1 + \frac{r}{n}\right)^{nt}$

$$2 = \left(1 + \frac{0.0685}{4}\right)^{4t}$$

$$\ln 2 = 4t \ln\left(1 + \frac{0.0685}{4}\right)$$

$$t = \frac{\ln 2}{4 \ln\left(1 + \frac{0.0685}{4}\right)} \approx 10.21$$

It will take 10 years 3 months.

29. Quill Activity

Section 2.2 Logarithmic Functions and Models

1. $f(x) = 2\ln x$ is the teal graph. $f(x) = -2\ln x$ is the black graph.

3. $f(x) = 2\ln x$ is the teal graph. $f(x) = 4\ln x$ is the black graph.

5. a. $L(d) = 158.574 - 42.877\ln d$ ppm for soil that is d meter from the road

b. $L(12) = 158.574 - 42.877\ln 12$

≈ 52 ppm

c. $E(d) = 123.238\left(0.932250^{d}\right)$ ppm for soil that is d meters from the road. The log model appears to be a better fit for small distances than the exponential model. The two models are similar for distances between 5 and 20 meters. The log model will eventually be negative, while the exponential model will approach zero. The exponential end behavior seems to better describe the lead concentration as a function of distance from a road.

7. a. $B(t) = 8.435 - 0.639\ln x$ percent for a maturity time of t years

b. The model estimates 15-year bond rates as $B(15) \approx 6.70\%$, which is 0.3 percentage point less than the fund manager's estimate.

9. a. $S(x) = -38{,}217.374 + 23{,}245.372\ln x$ dollars x years after 1970

b. In 1993 the salary is estimated as $S(23) \approx \$34{,}668$ and in 2000, the salary is estimated as $S(30) \approx \$40{,}845$. The 1993 estimate is probably the more accurate one because it is an interpolation, whereas the 2000 estimate is an extrapolation.

c. Solving $S(x) = 40{,}000$ yields

$x \approx 28.9$ years after 1970. Salaries are expected to have reached $40,000 in 1999.

11. a. A scatter plot of the data is increasing but concave down. Thus it can be neither linear nor exponential.

b. Peach consumption is $P(i) = 5.005 + 2.001911\ln i$ pounds per person per year where I is the yearly income of the consumer's family in tens of thousands of dollars.

c. $P(3.5) = 5.005 + 2.001911\ln(3.5) \approx 7.5$ pounds per person per year

13. a. $S(x) = -350{,}193.616 + 78{,}843.360 \ln x$ million dollars x years after 1990

The graph of this equation appears linear when graphed on the scatter plot.

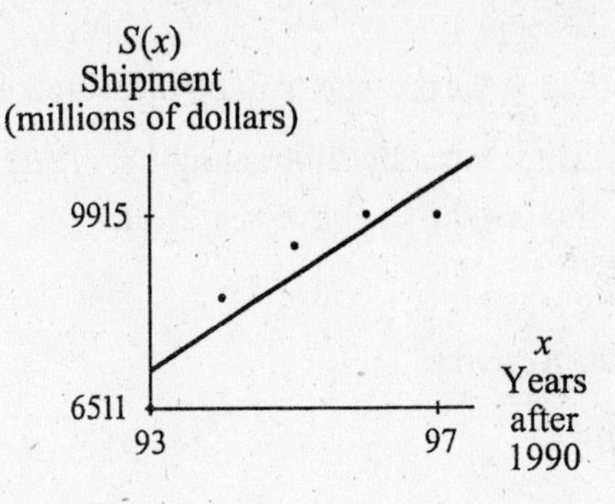

b. $C(t) = 6721.530 + 2213.132 \ln t$ million dollars t years after 1992

The graph of this equation is a better fit for the data than the equation in part *a*; however, it still is not as concave down as the scatter plot.

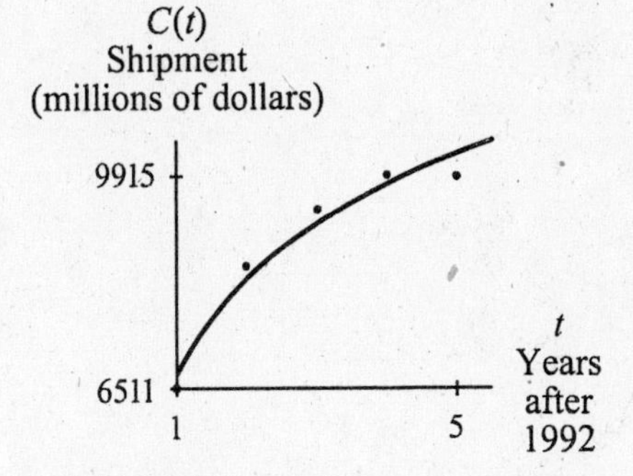

c. $A(y) = 7734.238 + 1633.996 \ln y$ million dollars where y is the number of years since 1993 plus 0.5.

This equation fits the data on the left better than the one in part *b*.

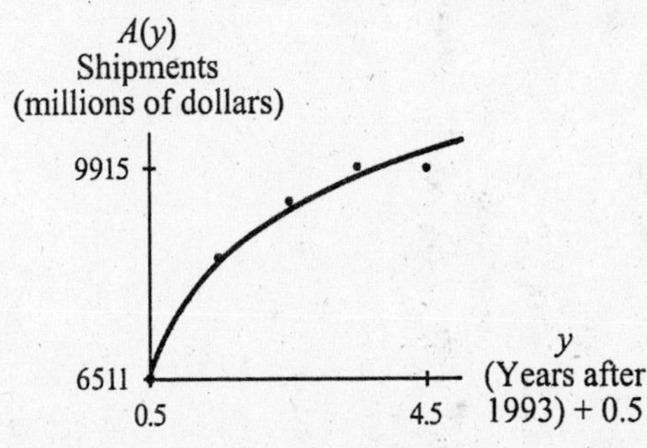

d. All three models will increase infinitely as the input increases. The data show a decline from 1996 to 1997. A quadratic model may be a better model for these data.

e. A log model must have inputs greater than zero.

15. a. $pH(x) = -9.792 \cdot 10^{-5} - 0.434 \ln x$
where x is the H_3O^+ concentration in moles per liter

b. $pH(1.585 \cdot 10^{-3}) =$
$-9.792 \cdot 10^{-5} - 0.434 \ln(1.585 \cdot 10^{-3})$
≈ 2.8

c. Solving the equation $5.0 = -9.792 \cdot 10^{-5} - 0.434 \ln x$ for x gives $x \approx 1.0 \cdot 10^{-5}$ moles per liter.

d. The pH of beer is approximately $pH(3.162 \cdot 10^{-5}) \approx 4.5$, which means it is acidic.

17. a. $S(15) = -56.1105 + 37.201 \ln(15) \approx 44.6$ million subscribers
$S(30) = -56.1105 + 37.201 \ln(30) \approx 70.4$ million subscribers
$S(45) = -56.1105 + 37.201 \ln(45) \approx 85.5$ million subscribers

b. $R(s) = 4.527(1.027226^s)$ dollars per month when there are s million subscribers.

19. a. Yes. Air pressure can be measured at different altitudes.

b. Align input by subtracting 18. $P(a) = 16.443 - 3.713 \ln a$ inches of mercury at $a+18$ thousand feet above sea level.

21. Quill Activity

Section 2.3 Logistic Functions and Models

1. The concave-up, decreasing shape could be either exponential or logarithmic.

3. The increasing, concave-down shape is that of a logarithmic function.

5. The scatter plot is none of these types. It exhibits change in concavity, so it cannot be linear, logarithmic, or exponential. It does not level off, so it is not logistic. It is possible to model this data with a piecewise continuous model using three linear pieces.

7. Note that $f(x) = \dfrac{L}{1 + Ae^{-Bx}}$ where
$L = 100$, $A = 9.8$, and $B = 0.98$. Because L, A, and B are positive, f is increasing to a limiting value of $L = 100$. The horizontal asymptotes are $y = 0$ and
$y = 100$.

9. Note that $h(g) = \dfrac{L}{1 + Ae^{-Bg}}$ where
$L = 39.2$, $A = 0.8$, and $B = -0.325$. Because L and A are positive and B is negative, h is decreasing from a limiting value $L = 39.2$. The horizontal asymptotes are $y = 0$ and $y = 39.2$.

11. a. $C(t) = \dfrac{37.195}{1 + 21.374^{-0.182968t}}$ countries t years after 1840. The model is a good fit.

b.

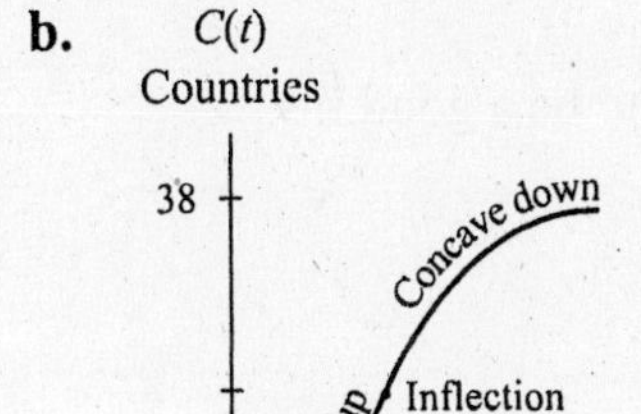
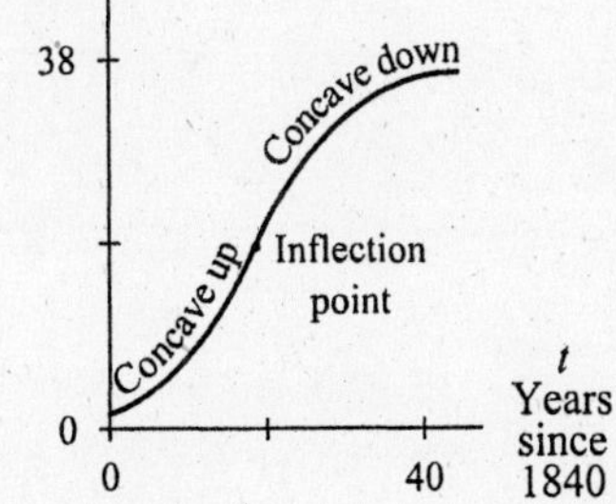

13. a. A plow sulky is a horse-drawn plow with a seat so that the person plowing can ride instead of walk. It was a precursor to the tractor.

b. $P(t) = \dfrac{2591.299}{1 + 16.848e^{-0.194833t}}$ patents t years after 1871

c. Patents begin with one innovative idea and grow almost exponentially as more and more improvements and variations are patented. Eventually, however, the patent market becomes saturated and the number of new patents dwindles to none as newer ideas and products are invented.

15. a.

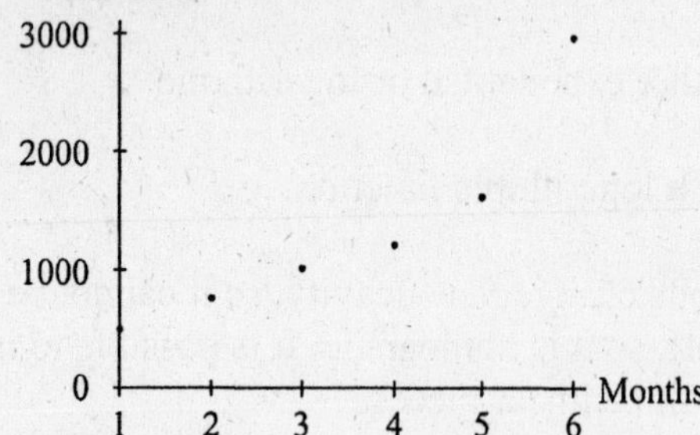

The data are concave down from January through April and concave up from April through June. This is not the concavity exhibited by a logistic model

b.

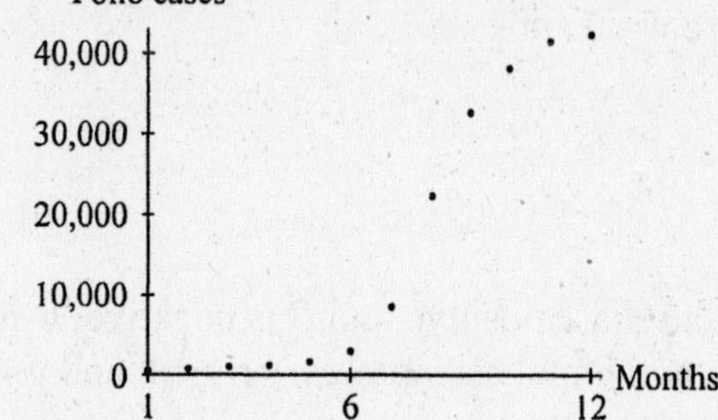

The entire data set does appear to be logistic.

c. $P(t) = \dfrac{42{,}183.911}{1 + 21{,}484.253e^{-1.248911t}}$

polio cases t months after Dec. 1948. The model appears to be a good fit.

d.

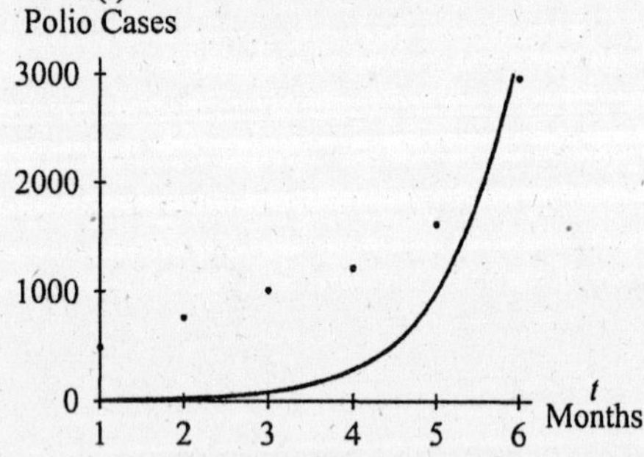

The model is a poor fit for the January through June data.

17. a. The limiting value is approximately 2U/100 μg. The inflection point occurs at approximately 9 minutes. (Answers may vary.)

b. $A(m) = \dfrac{1.937}{1 + 29.064e^{-0.421110m}}$ U/100 μg after m minutes.

The limiting value is about 1.94 U/100 μg.

c. $A(11) - A(7) \approx 1.51 - 0.77$

≈ 0.74 U/100 μg

19. a. $P(x)=\dfrac{11.742}{1+154.546e^{-0.025538x}}$ billion people x years after 1800. The equation is a good fit for the later (1960–2071) data but a poor fit for the early (1800–1960) data.

b. According to the model, the world population will level off at 11.7 billion. This is probably not an accurate prediction of future world population.

c. The model will probably be a poor estimate of the 1850 population because it is a poor fit for the years between 1800 and 1960. It is a good fit around 1990, however, so it will probably give a good estimate of population in that year.

21. Excel Activity

a. The inflection point may be near 1983.

b. $L(y)=\dfrac{536.848}{1+0.257026e^{0.179043y}}$ thousand beneficiaries over 190 thousand in the year 1975+y

c. $B(y)=\dfrac{536.848}{1+0.257026e^{0.179043y}}+190$ thousand beneficiaries in the year $1975+y$

$B(27)=\dfrac{536.848}{1+0.257026e^{0.179043(27)}}+190\approx 206$ thousand beneficiaries

23. a. The data appear to be concave up from 1990 through 1994 and concave down from 1995 through 2000. The concavity changes between 1994 and 1995. Either a logistic or a cubic function could be used to model these data.

b. $D(t)=\dfrac{552.278}{1+2.773e^{-0.306931t}}$ billion dollars of debt t years after 1990. This equation is not a good fit for the data.

c. $C(t)=\dfrac{311.126}{1+51.770e^{-0.866231t}}$ billion dollars over \$165 billion t years after 1990. This equation is an excellent fit for the data.

d. $F(t)=\dfrac{311.126}{1+51.770e^{-0.866231t}}+165$ billion dollars t years after 1990

e. $y=165$ and $y\approx 311.126+165=476.126$. The function in part *b* has asymptotes $y=0$ and $y\approx 552.278$.

25.a,b

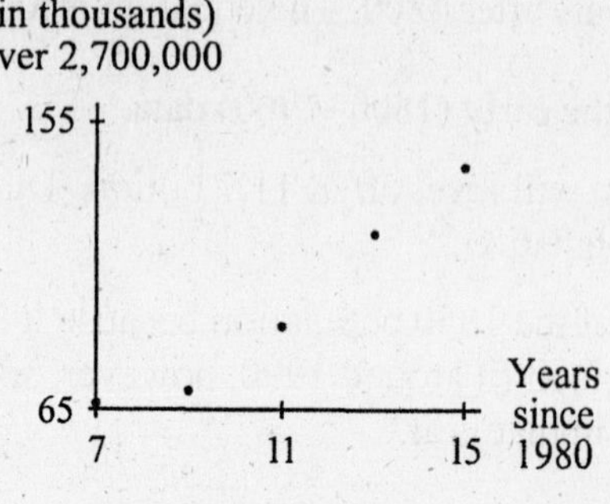

A scatter plot shows an inflection point with leveling-off behavior at both ends.

$P(t) = \dfrac{235.641}{1 + 9.637e^{0.172050t}}$ thousand people above 2,700,000 t years after 1980. This equation is not a good fit for the data. It does not exhibit the same inflection point or the same horizontal asymptotes as the data.

c. $I(t) = \dfrac{97.500}{1 + 1405.904e^{0.591268t}}$ thousand people above 2,760,000 t years after 1980. This equation is an excellent fit for the data.

d. $N(t) = \dfrac{97.500}{1 + 1405.904e^{0.591268t}} + 2760$ thousand people t years after 1980. The horizontal asymptotes are $y = 2760$ and $y = 2857.5$.

27. a. $p(x) = 15.501(0.972249^x)$ percent of all babies born to mothers who had gained x pounds had low birth weight.

b. $B(x) = 78.196(0.863408^x)$ percent above 5 percent when mothers gained x pounds

c. $B(x) = 78.196(0.863408^x) + 5$ percent of all babies born to mothers who had gained x pounds had low birth weight. $B(14) \approx 15.0\%$

Section 2.4 Polynomial Functions and Models

1. Concave up, decreasing from $x = 0.75$ to $x = 3$, increasing from $x = 3$ to $x = 4$

3. Concave up, decreasing from $x = 13.5$ to $x = 18$, increasing $x = 18$ to $x = 22.5$

5. Concave down, always decreasing

7. a. Calculate the first and second differences for the height data:

128 140 144 140 128 108 80

12 4 – 4 – 12 – 20 – 28

– 8 – 8 – 8 – 8 – 8

Second differences are constant, so the data are quadratic.

b. Work from bottom to top to continue the pattern above.

108 80 44 0

– 28 – 36 – 44

– 8 – 8

After 3.5 seconds the height is 44 feet. After 4 seconds the height is 0 feet.

c. $H(s) = -16s^2 + 32s + 128$ feet after s seconds

d. Solving $-16s^2 + 32s + 128 = 0$ yields $s = -2$ and $s = 4$. Because negative times values do not make sense in this context, we conclude that the missile hits the water after 4 seconds.

9. a. $J(12) = 30.571(12^2) - 67.029(12) + 155.6 \approx 3729$ jobs

b. Answers will vary. Possible answer: No, it does not seem reasonable that the roofing company would have 12 jobs in January and 374 jobs in December.

c. Answers will vary.

11. a. The first differences in the ages are $20.8 - 20.3 = 0.5$, $22 - 20.8 = 1.2$, and $23.9 - 22 = 1.9$, so the second differences are $1.2 - 0.5 = 0.7$ and $1.9 - 1.2 = 0.7$. Because the input data are evenly spaced and the second differences are constant, the data are perfectly quadratic.

b. The next first difference should be $1.9 + 0.7 = 2.6$, so the median age for 2000 should be $23.9 + 2.6 = 26.5$ years.

c. $A(x) = 0.0035x^2 - 0.405x + 32$ years of age x years after 1900

d. $A(100) = 0.0035(100^2) - 0.405(100) + 32 = 26.5$ years of age.

e, f. Answers will vary.

13. a.

Cost (dollars)
18
13
8
3
1000 3000 4500 7000
Ball bearings

The data do not appear to be concave up or concave down, so they should be modeled with a linear model: $B(x) = 0.0023525x + 1.880$ dollars to make x ball bearings

b. The overhead is $B(0) \approx \$1.88$.

c. $B(5000) = 0.0023525(5000) + 1.880 \approx \$13.64$; The cost is \$13.64.

d. $B(5100) - B(5000) \approx 13.88 - 13.64 \approx \$0.24$
The additional cost is \$0.24. This answer could also be found by multiplying the slope, \$0.0023525 per ball bearing, by 100 ball bearings. This value is called *marginal cost.*

e. $C(u) = 0.0023525(500u) + 1.880 = 1.176u + 1.88$ dollars to make u cases of ball bearings

15. a. $V(t) = 0.092t^2 + 0.720t + 149.554$ pounds per person t years after 1980

b. The model appears to be a good fit.

c. $V(21) = 0.092(21)^2 + 0.720(21) + 149.554 \approx 205$ pounds per person; discussion may vary.

d. Solving $V(t) = 225$ for t yields $t \approx 24.95$ or the year 2005.

e. Answers will vary.

17. a.

Lead usage (thousand of tons)
70
0
0 10 20 30 40
Years since 1940

Quadratic model:

$Q(x) = 0.057x^2 - 3.986x + 69.429$

thousands of tons of lead x years after 1940

Exponential model:

$E(x) = 211.196(0.821575^x)$ thousand tons of lead x years after 1940.

The quadratic equation is a good fit. The exponential equation is a poor fit.

b. The data suggest that lead usage is approaching zero as time increases. The exponential function approaches zero as time increases.

c. The quadratic model is the more appropriate one to use to interpolate because the fit is so much better than the exponential function. $Q(15) \approx 22$ thousand tons of lead

19.

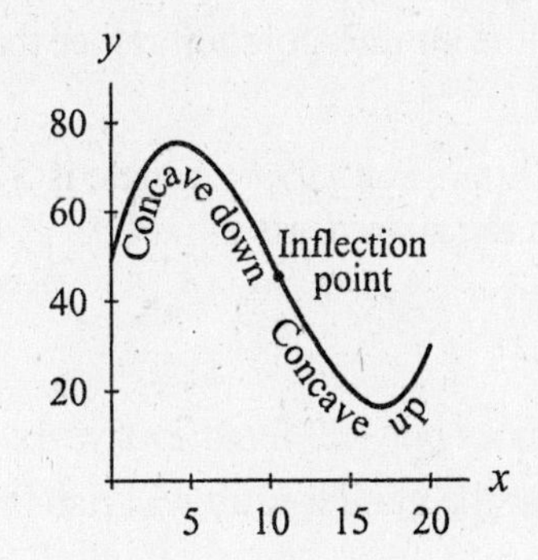

21.

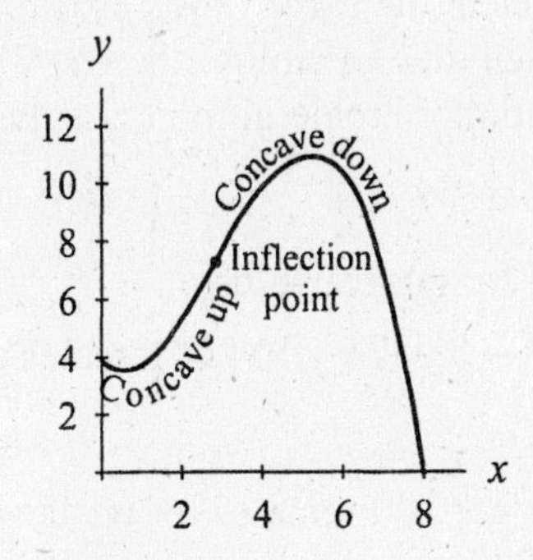

23.

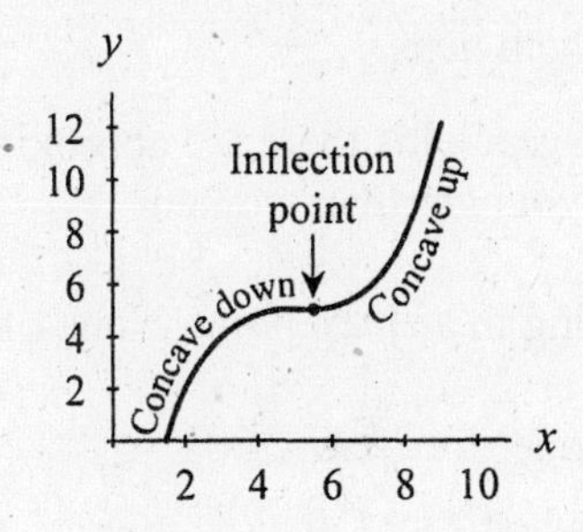

25.

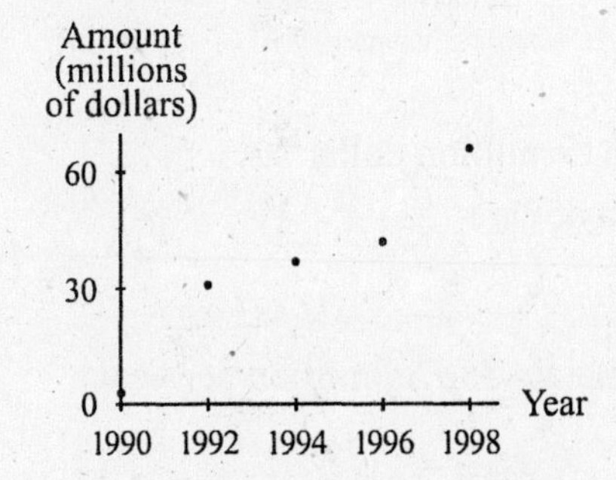

The scatter plot indicates an inflection point.

b. $A(t) = 0.427t^3 - 5.286t^2 + 22.827t + 3.014$ million dollars t years after 1990

c. $A(3) \approx \$35$ million $A(9) \approx \$92$ million

The 1993 estimate is more likely to be accurate because it is an interpolation rather than an extrapolation.

d. The 1993 estimate exceeded the actual amount by \$1 million. The 1999 estimate is \$7 million short of the actual amount. These figures confirm the statements in part *c*.

27. a. The scatter plot suggests an inflection point.

b. $G(x) = (5.051 \cdot 10^{-5})x^3 - 0.007x^2 + 0.085x + 105.027$ males per 100 females x years after 1900. The graph of G rises beyond 1990. It is unlikely that the gender ratio will rise in this manner.

c. Answers will vary. Possible answers include: Health issues in the early 1900s probably contributed to a greater death rate among women who often died in childbirth. The western expansion and gold and silver rush days probably attracted more male immigrants than female ones.

29. a. The number of females and males is approximately equal for 40-year-olds.

b. Cubic model: $C(a) = (-9.590 \cdot 10^{-5})a^3 + (8.421 \cdot 10^{-4})a^2 - 0.037a + 104.601$ males per 100 females for an age of a years

Logistic model: $L(a) = \dfrac{104.30}{1 + (9.817 \cdot 10^{-4})e^{0.081848a}}$ males per 100 females for an age of a years

The cubic equation fits the data well but the logistic fits them better.

c. Solving $L(a) = 50$ gives $a \approx 85.6$ years of age. Among 87-year-olds there are approximately twice as many women as men. This implies that men die younger than women.

31. a. The behavior of the data changes abruptly in 1999, resulting in a sharp point. None of the five models we have studied will reflect that behavior.

b. Use a quadratic model for 1992–1999, and use a linear model for 1999–2001.

$$V(t) = \begin{cases} 22.929t^2 - 133.310t + 429.488 \text{ million dollars} & \text{when } 2 \le t \le 9 \\ -567.5t + 6340.33 \text{ million dollars} & \text{when } 9 < t \le 11 \end{cases}$$

where t is the number of years since 1990

c. 1991: $V(1) = 22.929(1)^2 - 133.310(1) + 429.488 \approx 319$ million dollars

2002: $V(12) = -567.5(12) + 6340.33 \approx -470$ million dollars

d. Answers will vary.

e. The model indicates a constant decline of approximately \$567.5 million per year.

f. Answers will vary.

33. Quill Activity

Section 2.5 Choosing a Function to Fit Data

1. Because the data are concave up and display a minimum, a quadratic model is the only appropriate one.

3. The data appear to be essentially linear. Any concavity is probably not obvious enough to warrant the use of a more complex model.

5. Because of the indicated inflection point, a cubic or logistic model would be an appropriate choice. The choice would depend on the desired behavior of the model outside the range of the data.

7. a. Linear data have constant first differences and lie in a line.

b. Quadratic data have constant second differences and are either concave up or concave down.

c. Exponential data have constant percentage differences and are concave up, approaching the horizontal axis.

9. a. Discussion will vary.

b. Quadratic model: $Q(t) = 0.005t^2 - 1.148t + 118.971$ seconds t years after 1900

Log model: $L(t) = 172.311 - 25.273 \ln t$ seconds t years after 1900

The quadratic model fits the data better.

c. $A(t) = 92.134 - 8.838 \ln t$ seconds t years after 1946

This log model fits the data better than the log model in part *b*.

d. Discussion will vary. The log model would be a better choice if end behavior is a concern.

11. a. A scatter plot of the data is slightly concave down. The best model choices are quadratic and log.

Quadratic model: $Q(t) = -31.286t^2 + 1292.171t + 3321.286$ million dollars worth of CDs shipped t years after 1990

Log model: $L(t) = -29{,}492.053 + 14{,}221.876 \ln t$ million dollars worth of CDs shipped t years after 1990

b. Quadratic model: $Q(11) = -31.286(11)^2 + 1292.171(11) + 3321.286 \approx 13{,}750$ million dollars

Log model: $L(11) = -29{,}492.053 + 14{,}221.876 \ln(11) \approx 13{,}807$ million dollars

Quadratic model: $Q(15) = -31.286(15)^2 + 1292.171(15) + 3321.286 \approx 15{,}665$ million dollars

Log model: $L(15) = -29{,}492.053 + 14{,}221.876 \ln(15) \approx 16{,}286$ million dollars

13. a. The scatter plot is decreasing. It is concave up to the left of approximately 50 thousand miles and concave up to the right. The presence of an inflection point and the end behavior (leveling off) indicate that a logistic equation may be a good fit. There appears to be a lower limiting value that is higher than 0. We align the data by subtracting 12,000 from the outputs in order to obtain a better fitting logistic model.

$R(m) = \frac{8147.558}{1 + 0.084e^{0.051302m}} + 12{,}000$ dollars when the Jeep has accumulated m thousand miles.

b. $R(52) = \frac{8147.558}{1 + 0.084e^{0.051302(52)}} + 12{,}000 \approx 15{,}676$ dollars

15. a. $E(y) = -0.009y^3 + 0.203y^2 + 0.126y + 21.618$ billion dollars y years after 1990

b. The data appear to have an inflection point near 1998 and the context does not suggest that the expenditure will level off past 2001.

17. a. The scatter plot is concave down from 1980 through 1987. From 1987 through 1997, the scatter plot appear to be logistic.

b. The data exhibit two changes in concavity: one abrupt change around 1987 and a smoother change between 1991 and 1993. This concavity description does not match that of a cubic model that has one concavity change.

c. A cubic model is an extremely poor fit to the data.

d. Answers will vary. However, one possible model is

$$P(t) = \begin{cases} -2.1t^2 - 6.3t + 2914 \text{ thousand people} & \text{when } 0 \le t \le 7 \\ \frac{97.5}{1 + 1405.903e^{-0.591268t}} + 2760 \text{ thousand people} & \text{when } 7 < t \le 17 \end{cases}$$

where t is the number of years since 1980. Note the constant term on the end of the logistic portion of the model. In order to obtain a good fitting logistic equation, we aligned the outputs of the data by subtracting 2760 before fitting an equation.

e. Answers will vary.

19. Quill Activity

21. Quill Activity

23. Excel Activity

a. Both scatter plots are concave up and increasing and appear to be either quadratic or exponential.

b. $T(x) = 44.641x^2 - 175{,}902.443x + 173{,}280{,}348.731$ million dollars in year x; This equation seems to fit the data well after 1980 but does not curve as much as the data before 1980.

c. $B(x) = -0.769x^3 + 4624.357x^2 - 9{,}267{,}369.976x + 6{,}189{,}586{,}383.933$ million dollars in year x; This equation fits the data well. However, a quadratic is also a reasonable fit.

d. $T(x) - B(x) = 44.641x^2 - 175{,}902.443x + 173{,}280{,}348.731 -$
$$(-0.769x^3 + 4624.357x^2 - 9{,}267{,}369.976x + 6{,}189{,}586{,}383.933)$$
$$= -0.769x^3 - 4579.716x^2 + 9{,}091{,}467.533x - 6{,}016{,}306{,}035.203$$

million dollars in year x.

$$T(2003) - B(2003) = -0.769(2003)^3 - 4579.716(2003)^2 + 9{,}091{,}467.533(2003) - 6{,}016{,}306{,}035.203$$
$$\approx 8413 \text{ million dollars}$$

Chapter 2 Review Test

1. a. $C(x) = 2.2(1.021431^x)$ million children x years after 1970

   b. Approximately 2.1% per year

   c. Solving $C(x) = 5$ gives $x \approx 38.7$, which corresponds to the fall of 2009. The equation can be solved using technology or algebraically as follows:

$$5 = 2.2(1.021431^x)$$

$$\frac{5}{2.2} = 1.021431^x$$

$$\ln\left(\frac{5}{22}\right) = \ln(1.021431^x)$$

$$\ln\left(\frac{5}{22}\right) = x\ln 1.021431$$

$$x = \frac{\ln\left(\frac{5}{22}\right)}{\ln 1.021431} \approx 38.7$$

   d. Answers will vary.

2. a. The data are increasing and concave up, indicating that either an exponential or a quadratic model is appropriate. The data suggest a limiting value on the left, so an exponential model may be the better choice.

   b. $P(t) = 0.975(1.540440^t)$ dollars t years after 1989

   c. Answers will vary.

3. a.

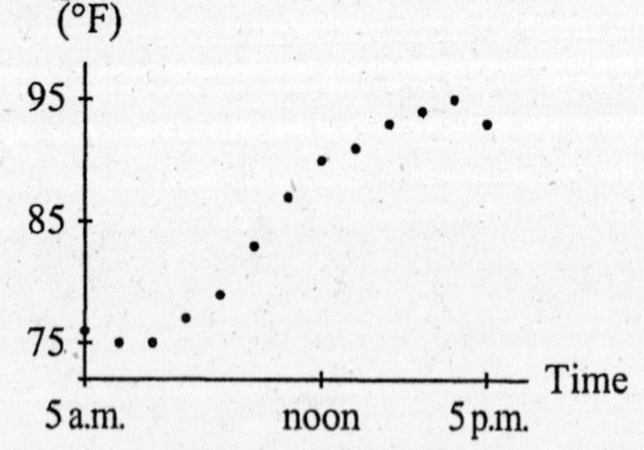

The scatter plot is essentially concave up and then concave down. A cubic model appears to be appropriate.

   b. $F(t) = -0.049t^3 + 1.560t^2 - 13.485x + 110.175$ °F t hours after midnight August 27

   c. $F(17.5) \approx 91$°F

   d. Solving $F(t) = 90$ gives three solutions: $t \approx 1.88$, 12.37, and 17.82. The first solution lies outside the time frame of the data. The other two solutions correspond to 12:22 p.m. and 5:49 p.m.

4. a. The statement is false because the input data are not evenly spaced.

b. Population (billions)

10

6

1999   2071   Year

The scatter plot suggests a concave-down shape. This shape could be modeled by a quadratic or log function. It is also possible to use that the right side of a logistic function to model the data.

c. $Q(t) = (-4.988 \cdot 10^{-4})t^2 + 0.0191t - 7.995$ billion people t years after 1900. The graph of this equation is an excellent fit for the data.

$L(t) = \dfrac{10.764}{1 + 20.429e^{-0.032771t}}$ billion people t years after 1900. The graph of this equation is also an excellent fit for the data.

$G(t) = -27.890 + 7.408 \ln t$ billion people t years after 1900. The graph of this equation is a reasonably good fit for the data, although not so good a fit as the other two models. Shifting the input data to the left might produce a better fit.

d. $\lim_{t\to\infty} Q(t) \to -\infty$; $\lim_{t\to\infty} L(t) \approx 10.8$ billion people; $\lim_{t\to\infty} G(t) \to \infty$

Because the United Nations study suggested that the population will stabilize, the logistic model is the most appropriate one even though the data do not suggest an inflection point.

Chapter 3

Section 3.1 Change, Percentage Change, and Average Rates of Change

1. $\dfrac{\$2.30}{5\text{ days}} = \$0.46$ per day

The stock price rose an average of 46 cents per day during the 5-day period.

3. $\dfrac{\$25{,}000}{3\text{ months}} \approx \$8333.33$ per month

The company lost an average of $8333.33 per month during the past three months.

5. $\dfrac{4\text{ percentage points}}{3\text{ years}} \approx 1.3$ percentage points per year

Unemployment has risen an average of about 1.3 percentage points per year in the past three years. *Note*: Whenever you are writing a change for a function whose output is a percentage, the correct label (unit of measure) is *percentage points*. The same is true for a rate of change. The phrase *percentage points per year* indicates that the effect is additive, whereas the phrase percent per year indicates a
multiplicative effect.

7. Change: $20.7 - 20.3 = 0.3$

The ACT composite average for females increased by 0.3 point between 1990 and 2002.

Percentage change: $\left(\dfrac{0.3\text{ point}}{20.3\text{ point}}\right)100\% \approx 1.5\%$

The ACT composite average for females increased by about 1.5% between 1990 and 2002.

Average rate of change: $\dfrac{0.3\text{ point}}{2002-1990} = \dfrac{0.3\text{ point}}{12\text{ years}} \approx 0.025$ point per year

The average female score increased by an average of 0.025 point per year between 1990 and 2002.

9. Change: $12.0 - 0.9 = 11.1$ million users

The number if Internet users in China grew by 11.1 million between 1997 and 2000.

Percentage change: $\left(\dfrac{11.1\text{ million users}}{0.9\text{ million users}}\right)100\% \approx 1233\%$

The number of Internet users grew by about 1233% between 1997 and 2000.

Average rate of change: $\dfrac{11.1\text{ million users}}{2000-1997} = \dfrac{11.1\text{ million users}}{3\text{ years}} \approx 3.7$ million users per year

The number of Internet users increased at an average rate of 3.7 million users per year between 1997 and 2000.

11. a. Slope of secant line:

$$\frac{6229.09-6228.98}{1996-1982}=\frac{0.11 \text{ foot}}{14 \text{ years}} \approx 0.008 \text{ foot / year}$$

b. In the 14-year period from 1982 through 1996, the lake level rose an average of 0.008 foot per year.

c. The lake level dropped below the natural rim because of drought conditions in the early 1990s but rose again to normal elevation by 1996. The average rate of change tells us that the level of the lake in 1996 was close to the 1982 level. Although the average rate of change is nearly zero, the graph shows that the lake level changed dramatically during the 14-year period.

13. a. The balance increased by
$1908.80 – $1489.55 = $419.25.

b. Average rate of change:

$$\frac{\$419.25}{4 \text{ years}} \approx \$104.81 \text{ per year}$$

Between the end of year 1 and the end of year 5, the balance increased at an average rate of $104.81/ year.

c. You could estimate the amount in the middle of the fourth year, but doing so wouldn't be as accurate as using a model to find the amount.

d. $A(t)=1400(1.063962^t)$ dollars after t years
Amount in middle of year 4: $A(3.5)=\$1739.28$
Amount at end of year 4: $A(4)=\$1794.04$
Average rate of change =

$$\frac{\$1794.04-\$1739.28}{\frac{1}{2}\text{ year}}=\$109.52 \text{ per year}$$

15. a. Average rate of change: $\frac{11.8 \text{ years}-68.3 \text{ years}}{70} \approx -0.81$ year per year

b. Average rate of change between ages 10 and 20: $\frac{50 \text{ years}-59.6 \text{ years}}{10} \approx -0.96$ year per year

Average rate of change between ages 20 and 30: $\frac{41.5 \text{ years}-50 \text{ years}}{10} \approx -0.89$ year per year

The magnitude of the average rate of change between ages 10 and 20 is greater than the magnitude of the average rate of change between ages 20 and 30.

17. a. $p(55)-p(40) \approx 31.70-21.45$ million people ≈ 10.3 million people

$$\text{Percentage change: } \left[\frac{p(55)-p(40)}{p(40)}\right]100\% \approx \left(\frac{10.3 \text{ million people}}{21.45 \text{ million people}}\right)100\% \approx 48\%$$

b. $\dfrac{p(85)-p(83)}{85-83} \approx \dfrac{69.259 - 65.744 \text{ million people}}{2 \text{ years}} \approx 1.8 \text{ million people per year}$

19. a.

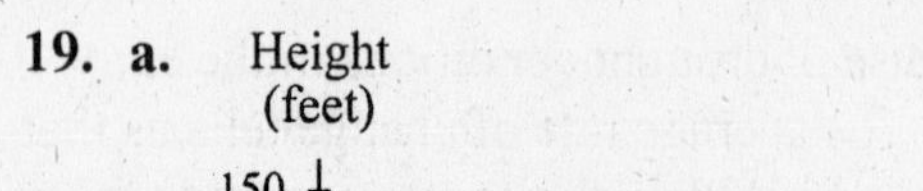

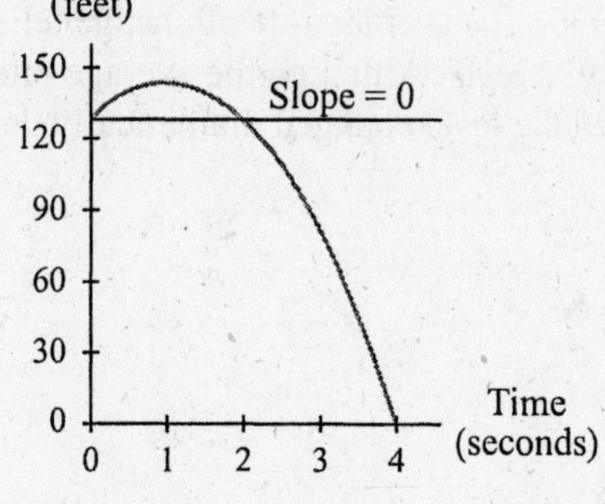

b.

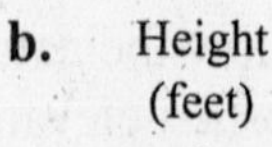

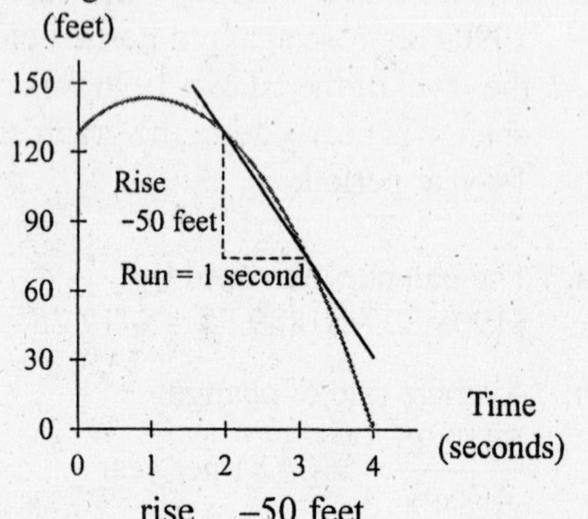

$\text{Slope} = \dfrac{\text{rise}}{\text{run}} = \dfrac{0 \text{ feet}}{2 \text{ seconds}} = 0 \text{ feet/second}$

$\text{Slope} = \dfrac{\text{rise}}{\text{run}} \approx \dfrac{-50 \text{ feet}}{1 \text{ second}} = -50 \text{ feet/second}$

c. $\dfrac{-50 \text{ feet}}{1 \text{ second}} \cdot \dfrac{3600 \text{ seconds}}{1 \text{ hour}} \cdot \dfrac{1 \text{ mile}}{5280 \text{ feet}} \approx -34 \text{ miles per hour}$

21. a. Average rate of change $\approx \dfrac{1.6 \text{ billion dollars}}{3.8 \text{ years}} = \$0.42$ billion

Kelly's interest income decreased by approximately \$0.42 billion per year on average between 1996 and 2001.

b. From the graph we estimate the interest income in 1996 as \$2.9 billion and in 2001 as \$4.5 billion. We calculate the percentage change as $\left(\dfrac{\$1.6 \text{ billion}}{\$3.8 \text{ billion}}\right)100\% \approx 42\%$.

23. 1996: $s(6) \approx 16.9\%$ 1999: $s(11) \approx 95.6\%$

Percentage change $= \left(\dfrac{95.6 - 16.9}{15.9}\right)100\% \approx 501\%$

25. a. i. $\dfrac{y(3)-y(1)}{3-1} = \dfrac{13-7}{2} = \dfrac{6}{2} = 3$

ii. $\dfrac{y(6)-y(3)}{6-3} = \dfrac{22-13}{3} = \dfrac{9}{3} = 3$

iii. $\dfrac{y(10)-y(6)}{10-6} = \dfrac{34-22}{4} = \dfrac{12}{4} = 3$

b. i. $\left[\dfrac{y(3)-y(1)}{y(1)}\right]100\% = \left(\dfrac{6}{7}\right)100\% \approx 85.7\%$

ii. $\left[\dfrac{y(5)-y(3)}{y(3)}\right]100\% = \left(\dfrac{6}{13}\right)100\% \approx 46.2\%$

iii. $\left[\dfrac{y(7)-y(5)}{y(5)}\right]100\% = \left(\dfrac{6}{19}\right)100\% \approx 31.6\%$

c. The average rate of change of any linear function over any interval will be constant because the slope of a line (and, therefore, of any secant line) is constant. The percentage change is not constant.

27. a,b.

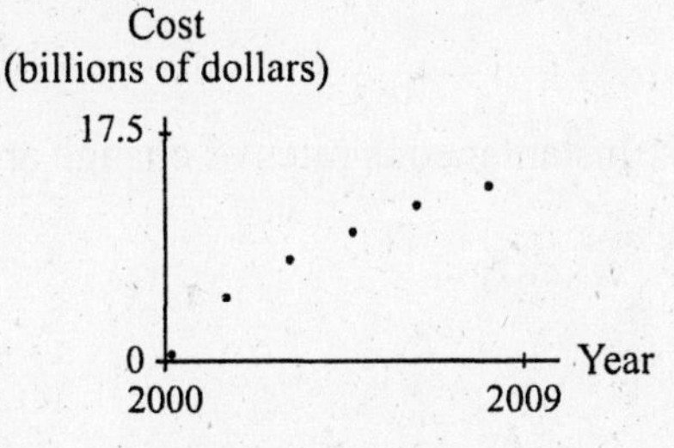

The data suggest a log model: concave down with no sign of leveling off.

$C(t) = 0.689 + 7.258\ln t$ billion dollars
t years after 1999

c. $C(6) - C(1) \approx \$13$ billion

d. $\dfrac{C(6)-C(1)}{6-1} \approx \$2.6$ billion per year

e. $\dfrac{C(6)-C(1)}{C(1)} \cdot 100\% \approx 1888\%$

29. a. Solve $\dfrac{152.6-x}{x} = 0.197$ for x to obtain $x \approx 127.5$ births per 100,000 in 1995.

Solve $\dfrac{152.6-x}{x} = 3.124$ for x to obtain $x \approx 37.0$ births per 100,000 in 1980.

b. $M(y) = \dfrac{176.984}{1+968.131e^{-0.291929y}} + 29$ births per 100,000 y years after 1970

c. $M(25) \approx 135.9$ births per 100,000 in 1995
$M(10) \approx 32.3$ births per 100,000 in 1980

d. Both estimates are interpolations.
The increased use of fertility drugs is the primary factor in the rise of the multiple-birth rate. Women having children later in life also contributes to this rise. (Answers will vary.)

31. Quill Activity

Section 3.2 Instantaneous Rates of Change

1. **a.** A continuous graph or model is defined for all possible input values on an interval. A continuous model with discrete interpretation has meaning for only certain input values on an interval. A continuous graph can be drawn without lifting the pencil from the paper. A discrete graph is a scatter plot. A continuous model or graph can be used to find average or instantaneous rates of change. Discrete data or a scatter plot can be used to find average rates of change.

   b. An average rate of change is a slope between two points. An instantaneous rate of change is the slope at a single point on a graph.

   c. A secant line connects two points on a graph. A tangent line touches the graph at a point and is tilted the same way the graph is tilted at that point.

3. Average rates of change are slopes of secant lines. Instantaneous rates of change are slopes of tangent lines.

5. Average speed =
$$\frac{19-0\text{ miles}}{17\text{ minutes}}\cdot\frac{60\text{ minutes}}{\text{hour}}\approx 67.1\text{ mph}$$

7. **a.** The slope is positive at A, negative at B and E, and zero at C and D.

   b. The graph is slightly steeper at point B than at point A.

9. **a.**

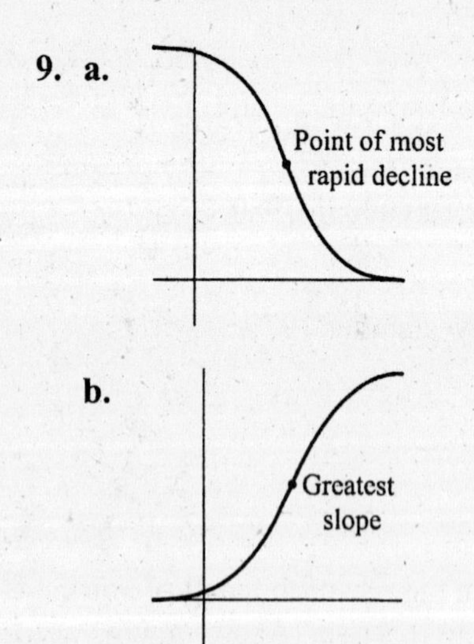

   b.

11. The lines at A and C are not tangent lines.

13.

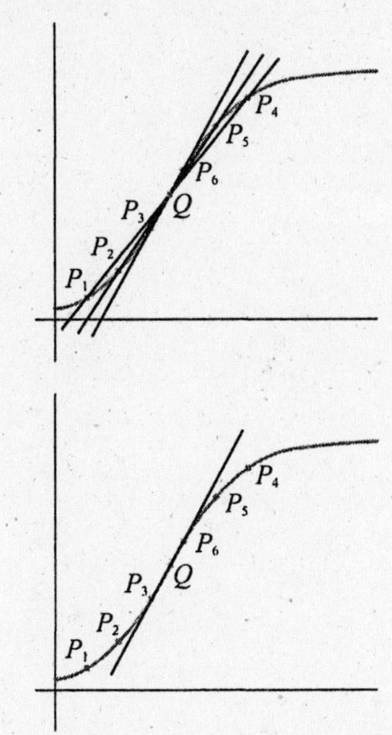

15. **a,b.** A: concave down, tangent line lies above the curve

B: inflection point, tangent line lies below the curve on the left, above the curve on the right

C: inflection point, tangent line lies above the curve on the left, below the curve on the right.

D: concave up, tangent line lies below the curve.

c.

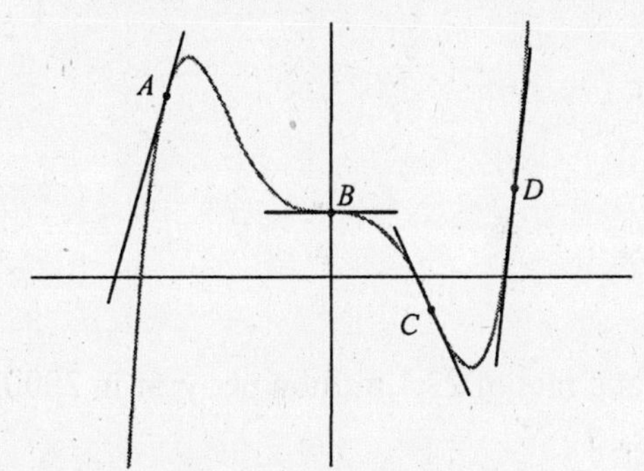

d. A, D: Positive slope

C: Negative slope (inflection point)

B: Zero slope (inflection point)

17.

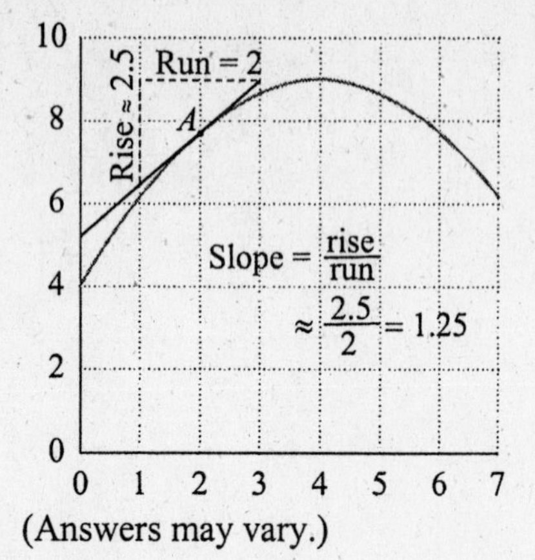

(Answers may vary.)

19.

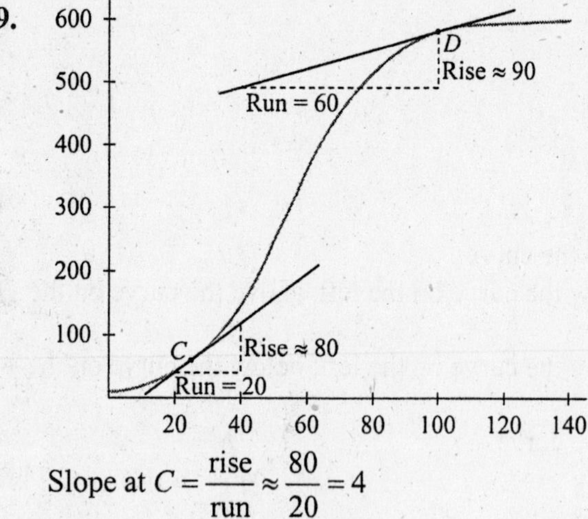

$$\text{Slope at } C = \frac{\text{rise}}{\text{run}} \approx \frac{80}{20} = 4$$

$$\text{Slope at } D = \frac{\text{rise}}{\text{run}} \approx \frac{90}{60} = 1.5$$

(Answers may vary.)

21. a. Million subscribers per year

b. The number of subscribers was growing at a rate of 23.1 million per year in 2000.

c. 23.1 million subscribers per year

d. 23.1 million subscribers per year

23. a, b. *A*: 1.3 mm per day per °C

B: 5.9 mm per day per °C

C: –4.2 mm per day per °C

c. The growth rate is increasing by 5.9 mm per day per °C.

d. The slope of the tangent line at 32°C is –4.2 mm per day per °C.

e. At 17°C, the instantaneous rate of change is 1.3 mm per day per °C.

25. a. Declination of sun

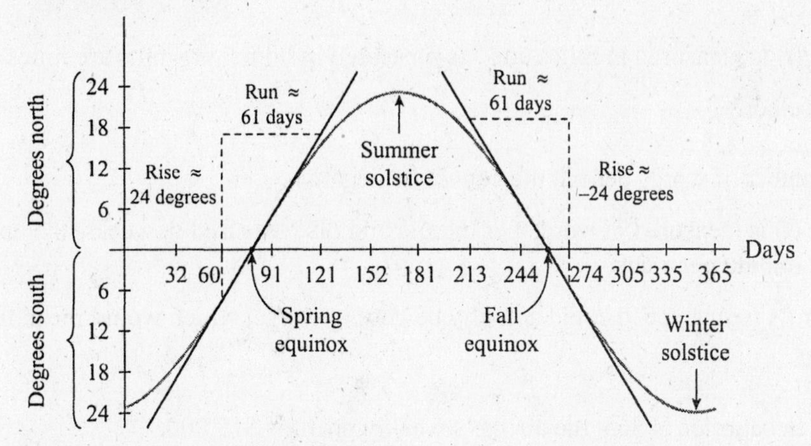

The slope at the solstices is zero.

b. The steepest points on the graph are those where the graph crosses the horizontal axis. The slopes are estimated as

$$\frac{24 \text{ degrees}}{61 \text{ days}} \approx 0.4 \text{ degree per day and } \frac{-24 \text{ degrees}}{61 \text{ days}} \approx -0.4 \text{ degree per day}$$

A negative slope indicates that the sun is moving from north to south.

c. The points with greatest slope identified in part *b* correspond to the spring and fall equinoxes.

27. a. Because the model is linear, the line to be sketched is the same as the model itself. From the equation, its slope is approximately 2370 thousand people per year.

b. Any line tangent to a graph of p is the graph of p.

c. Any line tangent to this graph has a slope of about 2370 thousand people per year.

d. The slope of the graph at every point will be 2370 thousand people per year.

e. The instantaneous rate of change is 2370 thousand people per year.

29. a. Slope in 1994 $\approx \frac{450 \text{ employees}}{2 \text{ years}} = 225$ employees per year

b. The slope of the graph at 1995 does not exist because the graph has a sharp point at 1995. There is not enough information given to estimate a rate of change of the underlying situation in 1995.

c. Slope in 1998 $\approx \frac{850 \text{ employees}}{2 \text{ years}} = 425$ employees per year

31. a. $\frac{31-47 \text{ million subscribers}}{2000-1998} =$ -8 million subscribers per year

b. Slope $\approx \frac{-14 \text{ million subscribers}}{2 \text{ years}} =$ -7 million subscribers per year

33. Quill Activity

Section 3.3 Derivatives

1. a. Because $P(t)$ is measured in miles and t is measured in hours, the units are miles per hour.

b. Speed or velocity

3. a. No, the number of words per minute cannot be negative.

b. Because $w(t)$ is measured in words per minute and t is measured in weeks, the units are words per minute per week.

c. The student's typing speed could actually be getting worse, which would mean that $W'(t)$ is negative.

5. a. When the ticket price is \$65, the airline's weekly profit is \$15,000.

b. When the ticket price is \$65, the airline's weekly profit is increasing by \$1500 per dollar of ticket price. Raising the ticket price a little will increase profit.

c. When the ticket price is \$90, the profit is declining by \$2000 per dollar of ticket price. An increase in price will decrease profit.

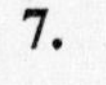

7.

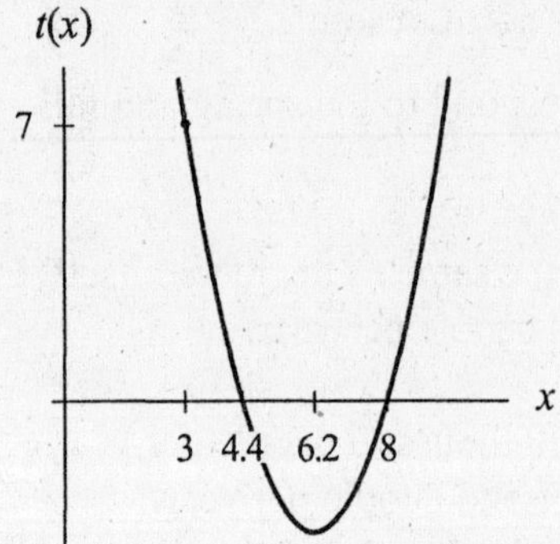

9. a. At the beginning of the diet you weigh 167 pounds.

b. After 12 weeks of dieting your weight is 142 pounds.

c. After 1 week of dieting your weight is decreasing by 2 pounds per week.

d. After 9 weeks of dieting your weight is decreasing by 1 pound per week.

e. After 12 weeks of dieting your weight is neither increasing nor decreasing.

f. After 15 weeks of dieting you are gaining weight at a rate of a fourth of a pound per week.

g.

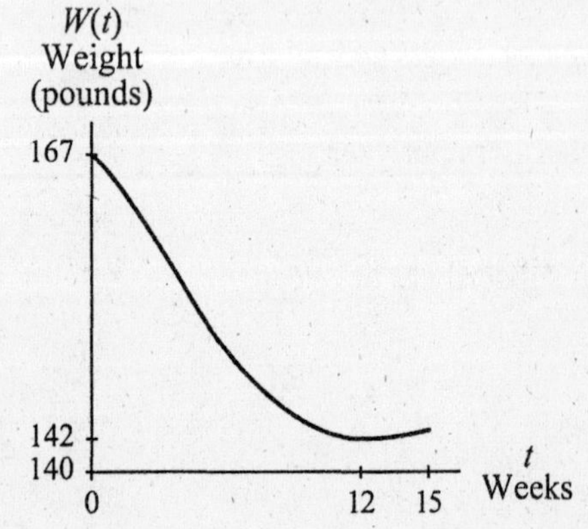

11. Because 21 + 12 = 32 and
12 + 10(0.6) = 18, we know that the following points are on the graph:
(1940, 4) (1970, 12)
(2000, 33) (1980, 18)
We also know the graph is concave up between 1940 and 1990 and concave down between 1990 and 1995

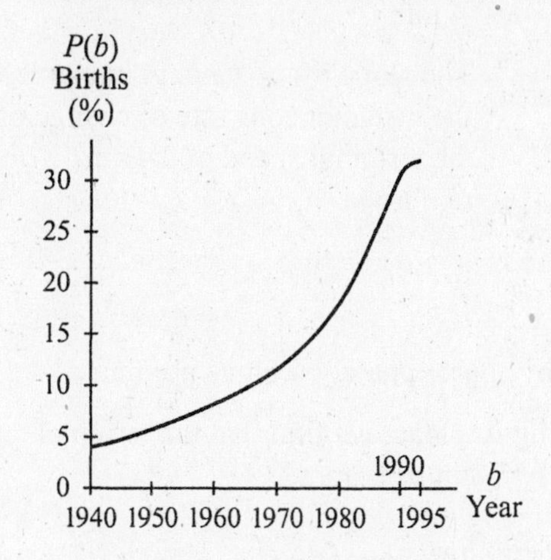

13. **a.** It is possible for profit to be negative if costs are more than revenue.

b. It is possible for the derivative to be negative if profit declines as more shirts are sold (because the price is so low, the revenue is less than the cost associated with the shirt.)

c. If $P'(200) = -1.5$, the fraternity's profit is declining. In other words, selling more shirts would result in less profit. Profit may still be positive (which means the fraternity is making money), but the negative rate of change indicates they are not making the most profit possible (they could make more money by selling fewer shirts).

15. **a.** Because $D(r)$ is measured in years and r is measured in percentage points, the units on $\frac{dD}{dr}$ are years per percentage point.

b. As the rate of return increases, the time it takes the investment to double decreases.

c. **i.** When the interest rate is 9%, it takes 7.7 years for the investment to double.

ii. When the interest rate is 5%, the doubling time is decreasing by 2.77 years per percentage point. A one-percentage-point increase in the interest rate will decrease the doubling time by approximately 2.8 years.

iii. When the interest rate is 12%, the doubling time is decreasing by 0.48 year per percentage point. A one-percentage-point increase in the interest rate will decrease the doubling time by approximately half of a year.

17. a,b.

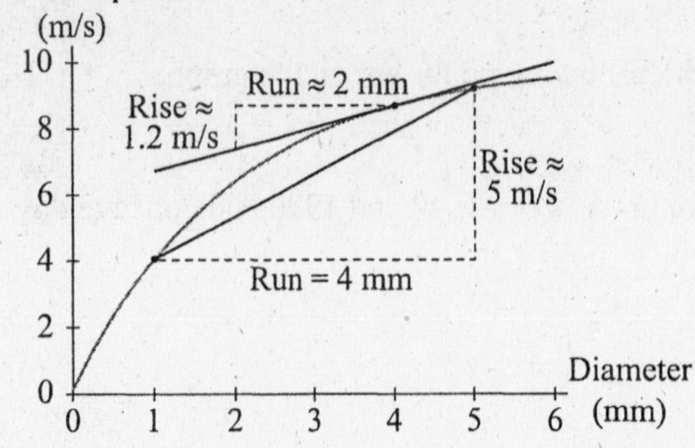

The slope of the secant line gives the average rate of change. Between 1 mm and 5 mm, the terminal speed of a raindrop increases an average of about

$$\frac{5 \text{ m/s}}{4 \text{ mm}} = 1.2 \text{ m/s per mm.}$$

The slope of the tangent line gives the instantaneous rate of change of the terminal speed of a 4 mm raindrop.

c. $\text{Slope} = \frac{\text{rise}}{\text{run}} \approx \frac{1.2 \text{ m/s}}{2 \text{ mm}} = 0.6 \text{ m/s per mm}$

A 4 mm raindrop's terminal speed is increasing by approximately 0.6 m/s per mm.

d. By sketching a tangent line at 2 mm and estimating its slope, we find that the terminal speed of a raindrop is increasing by approximately 1.8 m/s per mm.

e. $\text{Percentage rate of change} = \frac{1.8 \text{ m/s per mm}}{6.4 \text{ m/s}} \cdot 100\% \approx 28\% \text{ per mm}$

The terminal speed of a 2 mm raindrop is increasing by about 28% per mm as the diameter increases.

19. a,b.

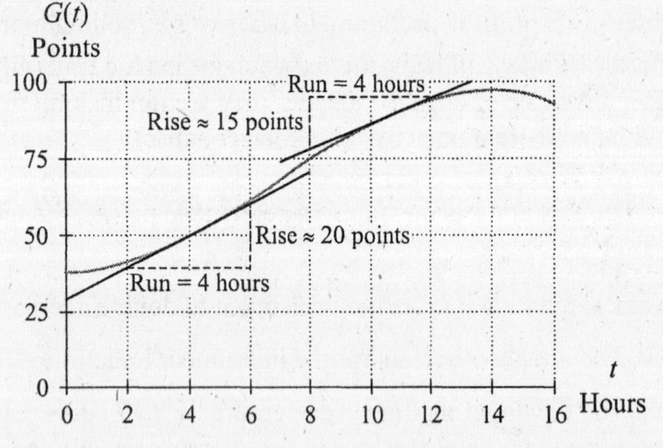

$$\text{Slope at 4 hours} \approx \frac{20 \text{ points}}{4 \text{ hours}} = 5 \text{ points per hour}$$

$$\text{Slope at 11 hours} \approx \frac{15 \text{ points}}{4 \text{ hours}} = 3.75 \text{ points per hour}$$

(Answers will vary.)

c. Estimate the two points on the graph: (4, 50) and (10, 86).

$$\text{Average rate of change} \approx \frac{86 \text{ points} - 50 \text{ points}}{10 \text{ hours} - 4 \text{ hours}} = \frac{36 \text{ points}}{6 \text{ hours}} = 6 \text{ points per hour}$$

d. $\text{Percentage rate of change} \approx \frac{5 \text{ points per hour}}{50 \text{ points}} \cdot 100\% = 10\% \text{ per hour}$

When you have studied for 4 hours, the number of points you will make on the test is improving by about 10% per hour.

e. $50 + 0.6(5) = 53$ points

21. a.

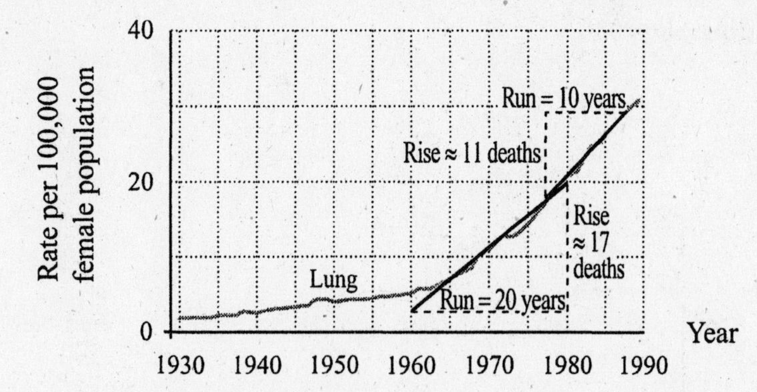

1970: $\text{Slope} = \dfrac{\text{rise}}{\text{run}} \approx \dfrac{17 \text{ deaths}}{20 \text{ years}} = 0.85$ death per 100,000 women per year

$$\text{Percentage rate of change} \approx \frac{0.85 \text{ death per 100,000 women per year}}{11 \text{ deaths per 100,000 women}} \cdot 100\%$$
$$\approx 7.7\ \% \text{ per year}$$

1980: $\text{Slope} = \dfrac{\text{rise}}{\text{run}} \approx \dfrac{11 \text{ deaths}}{10 \text{ years}} = 1.1$ deaths per 100,000 women per year

$$\text{Percentage rate of change} \approx \frac{1.1 \text{ deaths per 100,000 women per year}}{21 \text{ deaths per 100,000 women}} \cdot 100\%$$
$$\approx 5.2\ \% \text{ per year}$$

b. Answers will vary.

23. Rate of change in 1995 $\approx \dfrac{158 - 78.2 \text{ thousand deaths}}{4 \text{ years}} \approx 20$ thousand deahts per year

In 1995 the number of alcohol-related deaths was increasing by about approximately 20 thousand deaths per year.

25. Quill Activity

27. Quill Activity

Section 3.4 Numerically Finding Slopes

1. a.

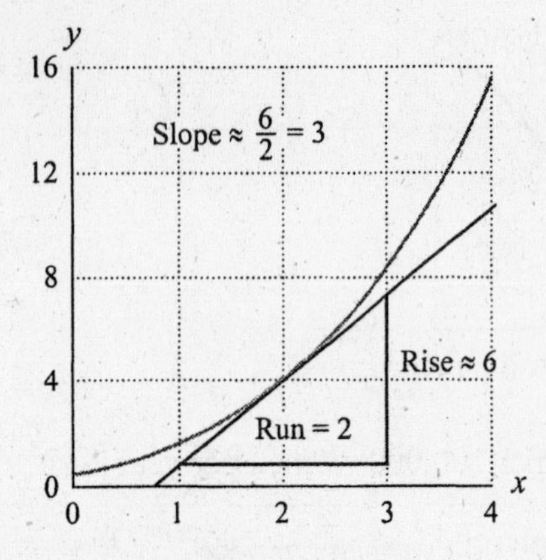

b. Sample calculation:

Point at $x = 2$: (2, 4)

Point at $x = 1.9$: $(1.9,\ 2^{1.9}) \approx (1.9, 3.73213)$

Secant line slope $\approx \dfrac{3.732132-4}{1.9-2} = \dfrac{-0.267868}{-0.1} = 2.67868$

Input of close point on left	Slope	Input of close point on right	Slope
1.9	2.67868	2.1	2.87094
1.99	2.76300	2.01	2.78222
1.999	2.77163	2.001	2.77355
1.9999	2.77249	2.0001	2.77268
1.99999	2.77258	2.00001	2.77260
Limit ≈ 2.77		**Limit ≈ 2.77**	

The slope of the line tangent to $y = 2^x$ at $x = 2$ is approximately 2.77.

3. a, b. Sample calculation:

For $g(x) = x^3 + 2x^2 + 1$

Point at $x = 2$: $(2,\ 2^3 + 2(2)^2 + 1) = (2, 17)$

Secant lines to the left:

Point at $x = 1.9$ using $g(x) = x^3 + 2x^2 + 1$:

$(1.9,\ 1.9^3 + 2(1.9)^2 + 1) = (1.9, 15.079)$

$$\text{Secant line slope} = \frac{15.079-17}{1.9-2} = \frac{-1.021}{-0.1} = 19.21$$

For $h(x) = x^2 + 16x$

Point at $x = 2$: $(2,\ 2^2 + 16(2)) = (2, 36)$

Secant lines to the right:

Point at $x = 2.1$ using $h(x) = x^2 + 16x$

$(2.1,\ 2.1^2 + 16(2.1)) = (2.1, 38.01)$

$$\text{Secant line slope} = \frac{38.01-36}{2.1-2} = \frac{2.01}{0.1} = 20.1$$

Input of close point on left	Slope	Input of close point on right	Slope
1.9	19.21	2.1	20.1
1.99	19.9201	2.01	20.01
1.999	19.99200	2.001	20.001
1.9999	19.99920	2.0001	20.0001
1.99999	19.99992	2.00001	20.00001
Limit = 20		Limit = 20	

c. Although the limit of the secant line slopes of g using points to the left of $x = 2$ is the same as the limit of secant lines slopes of h using point to the right of $x = 2$, the graph is not continuous at $x = 2$ because the pieces do not join at $x = 2$ (that is, the limit of f as x approaches 2 does not exist because the left and right limits are not the same value: $\lim_{x \to 2^-} f(x) = 17$ and $\lim_{x \to 2^+} f(x) = 36$). The graph of f is not smooth at $x = 2$ because it is not continuous at $x = 2$.

d. Because the graph is not continuous at $x = 2$, it is not possible to sketch a tangent line at $x = 2$.

5. a. See figure in Answer Key page A-19 in back of Text.

b.

Input of close point on left	Slope	Input of close point on right	Slope
3.9	–6.2828	4.1	–5.8463
3.99	–6.0849	4.01	–6.0412
3.999	–6.0652	4.001	–6.0609
3.9999	–6.0633	4.0001	–6.0628
3.99999	–6.0631	4.00001	–6.0630
Limit from either direction ≈ –6.06 thousand cases/year			

c. In 1998 the number of AIDS cases diagnosed was decreasing by approximately 6.06 thousand cases per year.

7. a. See figure in Answer Key (page A-19) in back of Text.

b.

Input of close point on left	Slope	Input of close point on right	Slope
6.9	10.15555	7.1	10.18681
6.99	10.17156	7.01	10.17468
6.999	10.17298	7.001	10.17330
6.9999	10.17312	7.0001	10.17315
6.99999	10.17314	7.00001	10.17314
Limit from either direction ≈ 10.17 percentage points per year			

c. In 1987 the percentage of households with VCRs was increasing by approximately 10.2 percentage points per year.

d. The tangent line approximation in part *a* is fast but can be inaccurate. The numerical approximation in part *b* is somewhat tedious but usually accurate.

9. a.

Input of close point on left	Slope	Input of close point on right	Slope
9.9	7.37148	10.1	7.50528
9.99	7.43131	10.01	7.44469
9.999	7.43733	10.001	7.43867
9.9999	7.43793	10.0001	7.43806
9.99999	7.43799	10.00001	7.43800
Limit from either direction ≈ $7.44 billion per year			

b. In 2000 the annual total U.S. factory sales of electronics was increasing by approximately $7.44 billion per year.

11. a. $\frac{32.9-43.1}{2000-1998} = \frac{-10.2 \text{ violent crimes per 1000 persons}}{2 \text{ years}}$ = -5.1 violent crimes per thousand persons per year

b. $V(t) = -4.57t + 55.386$ violent crimes per thousand persons per year t years after 1995

Slope = -4.57 violent crimes per thousand persons per year. The symmetric difference quotient is the average rate of change between 1998 and 2000. The slope of the linear equation can be thought of as an average of the rates of change for all years between 1997 and 2001. The answer to part *a* is the more accurate answer.

13. a. $\frac{318.7-294.1}{1999-1997} = \frac{24.6 \text{ points}}{2 \text{ years}} = 12.3$ index points per year

b. $L(x) = 65.411 + 144.812 \ln x$ points x years after 1985

$Q(x) = -0.498x^2 + 25.586x + 58.985$ points x years after 1985

c. Log model estimate: 11.1 index points per year

Quadratic model estimate: 12.6 index points per year

d. The symmetric difference quotient is faster than the numerical estimation and gives an answer comparable to that obtained from the quadratic model.

15. a. Sample calculation:

Point at $m = 5$: $(5,\ 0.3172(5)^3 - 2.0820(5)^2 - 1.7895(5) + 98.6398) \approx (5, 77.2923)$

Point at $m = 4.9$: $(4.9,\ 0.3172(4.9)^3 - 2.0820(4.9)^2 - 1.7895(4.9) + 98.6398) \approx$

$(4.9, 77.2007)$

Secant line slope $\approx \frac{77.2007 - 77.2923}{4.9 - 5} \approx$ \$0.916 per 100 pounds per month

Input of close	Slope

point on left	
4.9	0.916072
4.99	1.153772
4.999	1.177824
4.9999	1.180232
4.99999	1.180473
4.999999	1.180499
Limit ≈ $1.18 per 100 pounds per month	

b. The slope of the line is \$1.67 per 100 pounds per month.

c. Because the slopes of the portions of the graph to the left and right of $m = 5$ are not the same, the derivative of p does not exist at $m = 5$. Note that you could also come to this conclusion by observing that the function is not continuous at $m = 5$.

d. Because we do not have data available, we can use the model to estimate cattle prices in August and October of 1998 and find the symmetric difference quotient using those two points:

$$\frac{p(6)-p(4)}{6-4} = \frac{78.918-78.4706}{6-4} = \frac{\$0.4474}{2 \text{ months}} \approx \$0.22 \text{ per 100 pounds per month}$$

There are other valid methods of estimating this rate of change.

17. a. $C(P(x)) = \frac{P(x)}{1.5786} = \frac{1.02^x}{1.5786}$ U.S. dollars when x mountain bikes are sold

b. $P(400) = 1.02^{400} \approx 2755$ Canadian dollars

$$C(P(400)) \approx \frac{2755}{1.5786} \approx 1745 \text{ U.S. dollars}$$

c. We estimate the rate of change using the numerical method:

Input of close point on left	Slope	Input of close point on right	Slope
399.9	34.5214	400.1	34.5899
399.99	34.5522	400.01	34.5590
399.999	34.5553	400.001	34.5560
399.9999	34.55564	400.0001	34.55571
399.99999	34.55568	400.00001	34.55568
Limit from either direction ≈ $34.56 per mountain bike			

When 400 mountain bikes are sold, the profit (in U.S. dollars) is increasing at a rate of approximately \$34.56 per mountain bike sold.

19. Excel Activity

a. $$W(x) = \begin{cases} -2.999x^2 + 128.496x + 3864.377 & \text{when } 0 \le x < 21 \\ -6.130x^3 + 441.07x^2 - 10612.866x + 90334.048 & \text{when } 21 \le x \le 26 \end{cases}$$

thousand beneficiaries x years after 1975.

b. The number of widow-widower beneficiaries was decreasing by approximately 57.49 thousand beneficiaries per year in 1998.

c. Limit from left ≈ 2.55, Limit from right ≈ -197.53. Thus the slope does not exist at $x = 21$.

d. The derivative of W does not exist at $x = 21$.

e. Using the symmetric difference from 1995 and 1997, we obtain the following estimate of the rate of change of beneficiaries in 1996: -87.73. The number of widow-widower beneficiaries was decreasing by approximately 87.73 thousand beneficiaries per year in 1996.

20. Quill Activity

Section 3.5 Algebraically Finding Slopes

1. a. $T(13) = 67.946$ seconds

b. $T(13+h) = 0.181(13+h)^2 - 8.463(13+h) + 147.376 = 0.181h^2 - 3.757h + 67.946$

c. $\dfrac{T(13+h) - T(13)}{13+h-13} = \dfrac{0.181h^2 - 3.757h + 67.946 - 67.946}{h} = \dfrac{0.181h^2 - 3.757h}{h}$

d. $\lim\limits_{h\to 0} \dfrac{0.181h^2 - 3.757h}{h} = \lim\limits_{h\to 0}(0.181h - 3.757) = 0.181(0) - 3.757$

$= -3.757$ seconds per year of age

The swim time for a 13-year-old is decreasing by 3.757 seconds per year of age. This tells us that as a 13-year-old athlete gets older, the athlete's swim time improves.

3. a. $c(8) = -0.498(8)^2 + 20.603(8) + 174.458 = 307.41$

b. $c(8+h) = -0.498(8+h)^2 + 20.603(8+h) + 174.458$

$= -0.498(64 + 16h + h^2) + 20.603(8+h) + 174.458$

$= 307.41 + 12.635h - 0.498h^2$

c. $\dfrac{c(8+h) - c(8)}{8+h-8} = \dfrac{12.635h - 0.498h^2}{h}$

d. $\lim\limits_{h\to 0} \dfrac{12.635h - 0.498h^2}{h} = 12.635$ index points per year

e. The consumer price index for college tuition was increasing by 12.635 index points per year in 1998.

f. Because we are dealing with a model that is not an exact fit to the data, none of these estimates is exact.
Assuming that you both have the data and can find a good model for the data, then

i. Because the algebraic method uses a rounded model, the numerical method with an unrounded model will probably produce the most accurate answer as long as the model is an excellent fit to the data.

ii. Use the data and a symmetric difference quotient to obtain a quick, rough estimate.

iii. Because both the numerical method and the algebraic method are somewhat time-consuming, use the data and a symmetric difference quotient to obtain a fairly good estimate without taking much time.

5. i. $g(4) = -89$

ii. $g(4+h) = -6(4+h)^2 + 7 = -6h^2 - 48h - 89$

iii. $\dfrac{c(4+h) - c(4)}{4+h-4} = \dfrac{-6h^2 - 48h}{h}$

iv. $\lim_{h\to 0} \dfrac{-6h^2-48h}{h} = \lim_{h\to 0}(-6h-48) = -48$

$\dfrac{dg}{dt} = -48$ when $t = 4$

7. a. i. $H(t) = -16t^2 + 100$

ii. $H(t+h) = -16(t+h)^2 + 100 = -16(t^2 + 2th + h^2) + 100 = -16h^2 - 32th - 16t^2 + 100$

iii. $\dfrac{H(t+h) - H(t)}{t+h-t} = \dfrac{-16h^2 - 32th - 16t^2 + 100 - (-16t^2 + 100)}{h} = \dfrac{-16h^2 - 32th}{h}$

iv. $\lim_{h\to 0} \dfrac{-16h^2 - 32th}{h} = \lim_{h\to 0}(-16h - 32t) = -16(0) - 32t = -32t$

$\dfrac{dH}{dt} = -32t$ feet per second t seconds after the object is dropped

b. After 1 second, the velocity of the object was $-32(1) = -32$ feet per second; that is, the object was falling at a rate of 32 feet per second.

9. a. $D(a) = -0.045a^2 + 1.774a - 16.064$ million drivers age a years

b. $D(a+h) = -0.045(a+h)^2 + 1.774(a+h) - 16.064$

$= -0.045a^2 - 0.09ah - 0.045h^2 + 1.774a + 1.774h - 16.064$

$\dfrac{D(a+h) - D(a)}{a+h-a} = \dfrac{-0.09ah - 0.045h^2 + 1.774h}{h}$

$\lim_{h\to 0} \dfrac{-0.09ah - 0.045h^2 + 1.774h}{h} = \lim_{h\to 0}(-0.09a - 0.045h + 1.774) = -0.09a + 1.774$

Thus $D'(a) = -0.09a + 1.774$ million drivers per year of age where a is the age in years

c. $D'(20) = -0.026$ million drivers per year of age. For 20-year-olds, the number of licensed drivers is decreasing by 26,000 driver per year of age. This tells us that there are fewer 21-year-olds who are licensed drivers than there are 20-year-olds.

11. i. $f(x) = 3x - 2$ **ii.** $f(x+h) = 3(x+h) - 2 = 3x + 3h - 2$

iii. $\dfrac{f(x+h) - f(x)}{x+h-x} = \dfrac{3h}{h}$

iv. $\lim_{h\to 0} \dfrac{3h}{h} = \lim_{h\to 0} 3 = 3$. Therefore, $\dfrac{df}{dx} = 3$.

13. i. $f(x) = 3x^2$ **ii.** $f(x+h) = 3(x+h)^2 = 3x^2 + 6xh + 3h^2$

iii. $\dfrac{f(x+h) - f(x)}{x+h-x} = \dfrac{6xh + 3h^2}{h}$

iv. $\lim_{h\to 0} \frac{6xh+3h^2}{h} = \lim_{h\to 0}(6x+3h) = 6x$. Therefore, $f'(x) = 6x$.

15. i. $f(x) = x^3$ **ii.** $f(x+h) = (x+h)^3 = x^3 + 3x^2h + 3xh^2 + h^3$

iii. $\frac{f(x+h)-f(x)}{x+h-x} = \frac{3x^2h+3xh^2+h^3}{h}$

iv. $\lim_{h\to 0} \frac{3x^2h+3xh^2+h^3}{h} = \lim_{h\to 0}\left(3x^2+3xh+h^2\right) = 3x^2$. Therefore, $f'(x) = 3x^2$.

17. Quill Activity

Chapter 3 Review Test

1. **a.** **i.** A, B, C **ii.** E **iii.** D

   b. The graph is steeper at B than it is at A, C, or D.

   c. Below: C, D, E
   Above: A
   At B: above to the left of B, below to the right of B

   d.

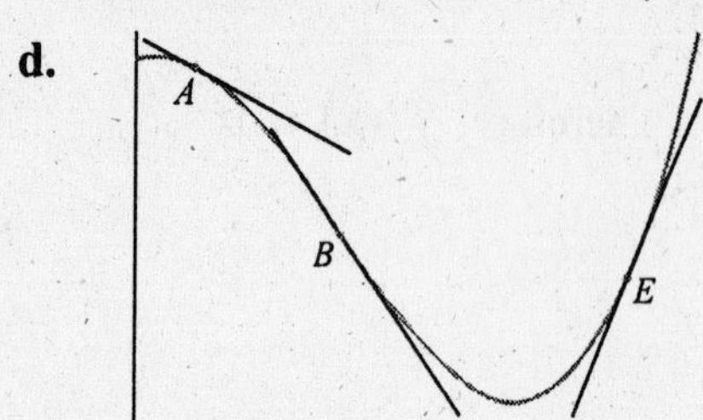

   e. **i.** Feet per second per second. This is acceleration.

   ii, iii. The roller coaster was slowing down until D, when it began speeding up.

   iv. The roller coaster's speed was slowest at D.

   v. The roller coaster was slowing down most rapidly at B.

   f.

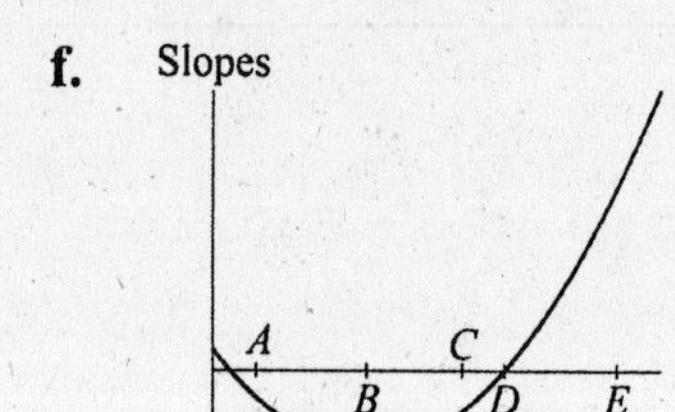

3. **a, b.** The slope of the secant line gives the average rate of change between 1993 and 1997. The slope of the tangent line gives the instantaneous rate of change in 1998.

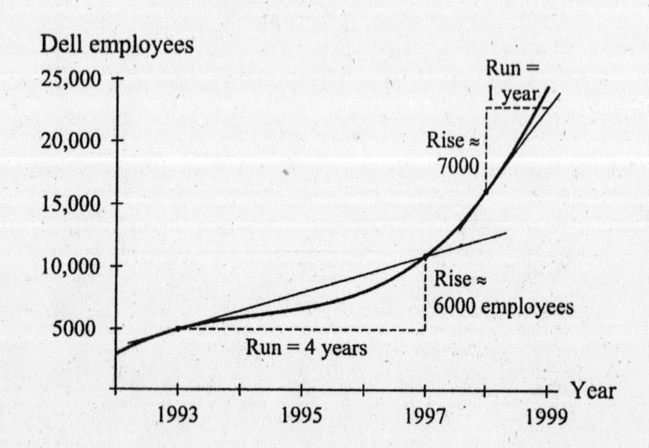

c. In 1998 the number of employees was increasing by approximately 7000 employees per year. If we estimate the number of employees in 1998 as 16,000, then this rate represents an increase of approximately 44% per year.

d. Between 1993 and 1997, the number of Dell employees increased at an average rate of 1500 employees per year.

5. **a.** $M(t) = 0.982(1.531417^t)$ dollars per share t years after 1989

b.

Close point on left	Slope	Close point on right	Slope
7.9	12.3996	8.1	12.9395
7.99	12.6388	8.01	12.6927
7.999	12.6630	8.001	12.6684
7.9999	12.6655	8.0001	12.6660
7.99999	12.6657	8.00001	12.6657
Limit ≈ 12.67		**Limit ≈ 12.67**	

c. In 1997 the yearly high stock price for Microsoft was increasing by approximately $12.67 per share per year.

d. $M'(t) = 0.982(\ln 1.531417)(1.531417^t)$ dollars per share per year t years after 1989
$M'(8) \approx \$12.67$ per share per year

Chapter 4

Section 4.1 Drawing Rate-of-Change Graphs

1. The slopes are negative to the left of A and positive to the right of A. The slope is zero at A.

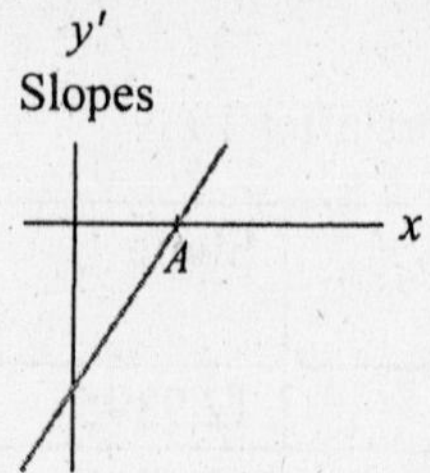

3. The slopes are positive everywhere, near zero to the left of zero, and increasingly positive to the right of zero.

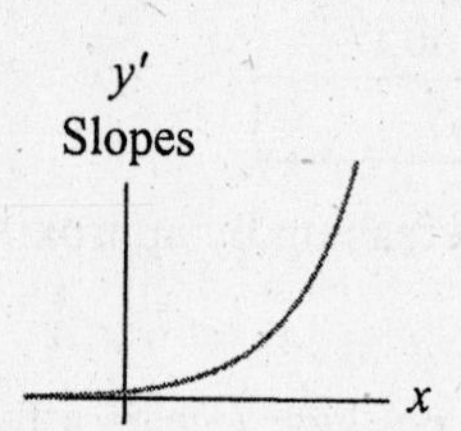

5. The slope is zero everywhere.

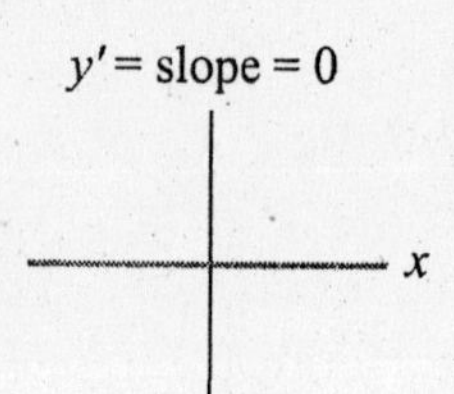

7. The slopes are negative everywhere. The magnitude is large close to $x = 0$ and is near zero to the far right.

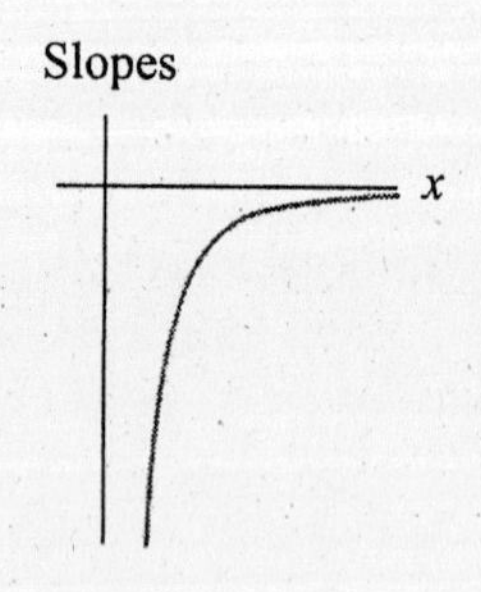

9. The slopes are negative to the left and right of A. The slope appears to be zero at A.

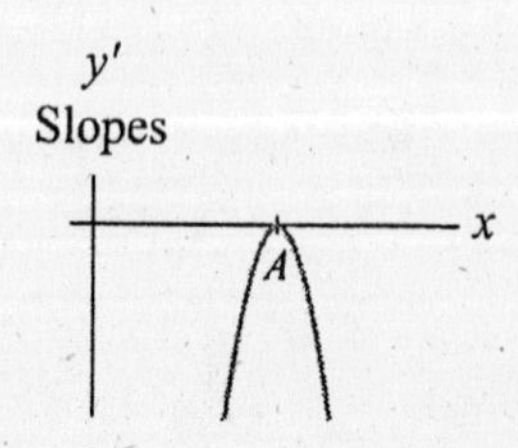

11. a.

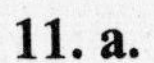

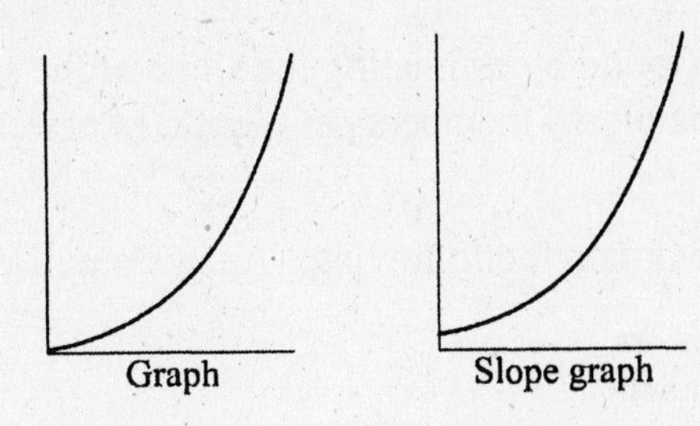

b.

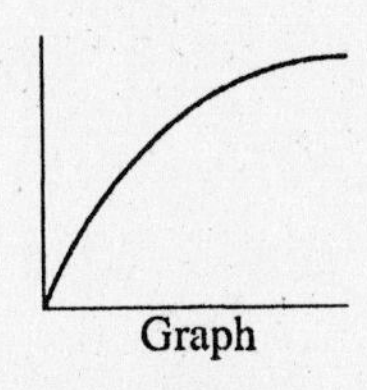

Slope graph

13. a, b.

Year	Slope
1991	–6.6
1993	–6.2
1997	–2.5
1999	1.0
2000	5.4

(Table values may vary.)

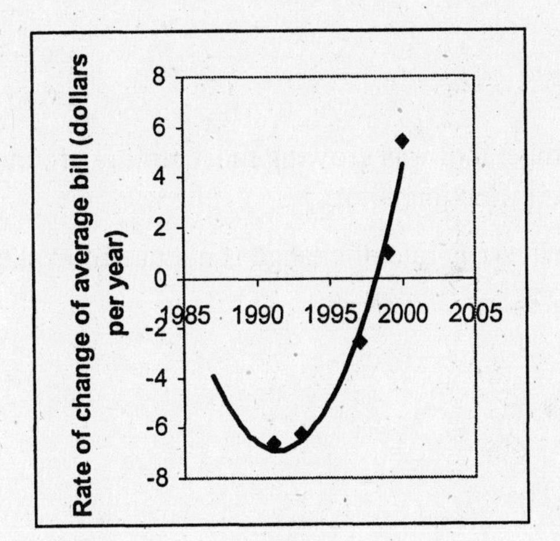

15. a. (Table values may vary.)

Year	Slope
1985	7.8
1990	46.8
1995	79.2
1997	55.2
2000	21.4

b. Graph may vary, but its basic shape should be concave down with a maximum between 1993 and 1995.

17. **a.** The average rate of change during the year (found by estimating the slope of the secant line drawn from September to May) is approximately 14 members per month. (Answers will vary.)

b, c. By estimating the slopes of tangent lines we obtain the following. (Answers will vary.)

Month	Slope (members per month)
Sept	98
Nov	–9
Feb	30
Apr	11

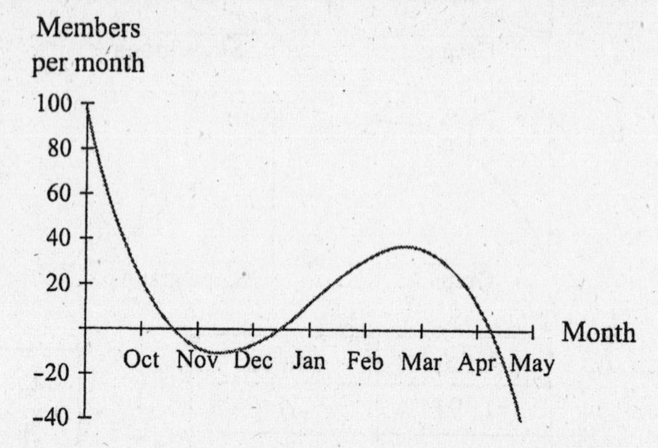

d. Membership was growing most rapidly around March. This point on the membership graph is an inflection point.

e. The average rate of change is not useful in sketching an instantaneous-rate-of-change graph.

19.

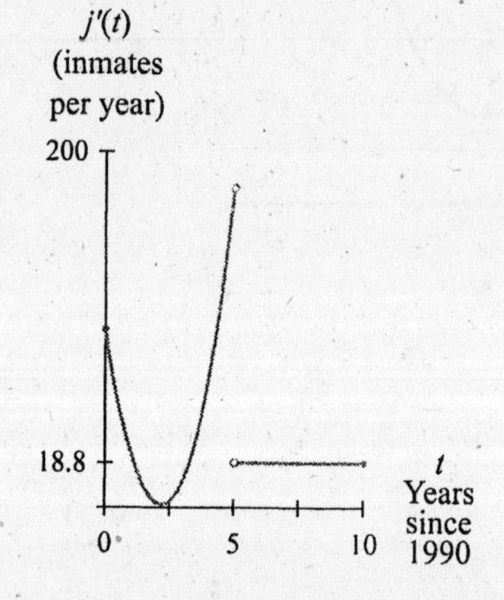

21. **a.** Slope $\approx \dfrac{\$18{,}000}{30 \text{ cars}} = \$600$ per car

Profit is increasing on average by approximately \$600 per car.

b. By sketching tangent lines we obtain the following estimates. (Answers will vary.)

Number of cars	Slope (dollars per car)
20	0
40	160
60	750
80	10
100	–1200

c. Average monthly profit (dollars per car)

1200, 900, 600, 300, -300, -600, -900, -1200

10 20 40 50 60 70 80 90 100 110 120 Cars sold

d. The average monthly profit is increasing most rapidly for about 60 cars sold and is decreasing most rapidly when about 100 cars are sold. The corresponding points on the graph are inflection points.

e. Average rates of change are not useful when graphing an instantaneous-rate-of-change graph.

23. The derivative does not exist at $x = 0$, $x = 3$, and $x = 4$ because the graph is not continuous at those input values.

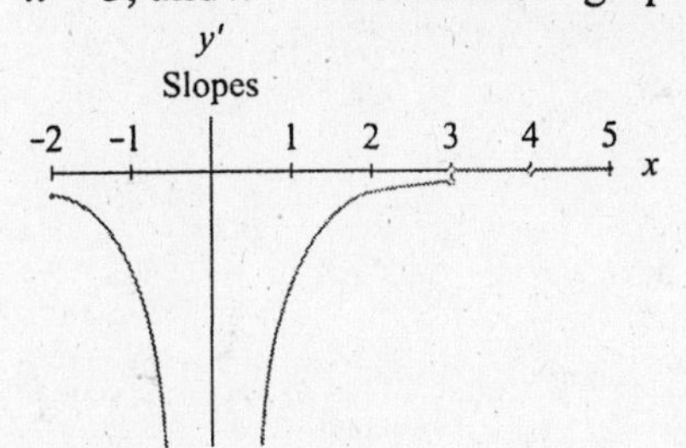

25. The derivative does not exist at $x = 2$ and $x = 3$ because the slopes from the left and right are different at those inputs.

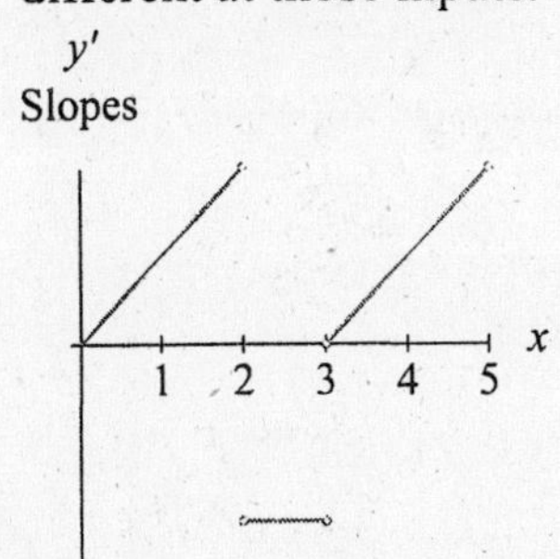

31. a. $p'(m)$

m

b. i. $p(m) = m + \sqrt{m}$.

ii. $p(m+h) = m + h + \sqrt{m+h}$

iii. $$\frac{p(m+h)-p(m)}{m+h-m} = \frac{h+\sqrt{m+h}-\sqrt{m}}{h} = \left(\frac{h+\sqrt{m+h}-\sqrt{m}}{h}\right)\frac{\left(\sqrt{m+h}+\sqrt{m}\right)}{\left(\sqrt{m+h}+\sqrt{m}\right)}$$

$$= \frac{h\left(\sqrt{m+h}+\sqrt{m}\right)+(m+h)-m}{h\left(\sqrt{m+h}+\sqrt{m}\right)} = \frac{h\left(\sqrt{m+h}+\sqrt{m}\right)+h}{h\left(\sqrt{m+h}+\sqrt{m}\right)}$$

iv. $$\lim_{h\to 0}\frac{h\left(\sqrt{m+h}+\sqrt{m}\right)+h}{h\left(\sqrt{m+h}+\sqrt{m}\right)} = \lim_{h\to 0}\frac{\left(\sqrt{m+h}+\sqrt{m}\right)+1}{\sqrt{m+h}+\sqrt{m}} = \frac{2\sqrt{m}+1}{2\sqrt{m}} = 1+\frac{1}{2\sqrt{m}}$$

$\frac{dp}{dm} = 1 + \frac{1}{2\sqrt{m}}$. The graph of $\frac{dp}{dm}$ is the same as the one in part *a*.

33. Quill Activity

Section 4.2 Simple Rate-of-Change Formulas

1.

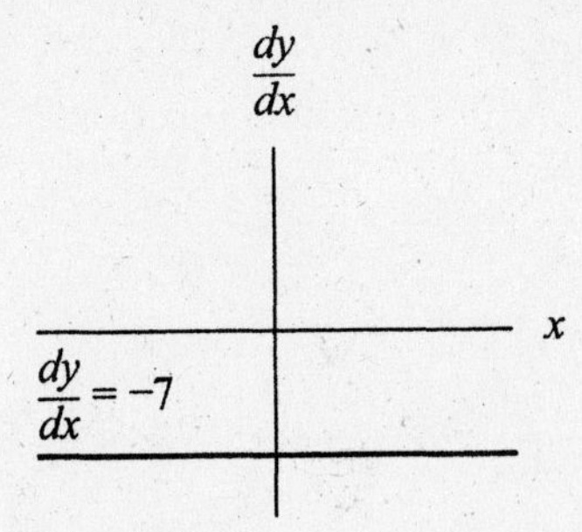

The function $y = 2 - 7x$ is a line with slope −7. Thus $\frac{dy}{dx} = -7$.

3.

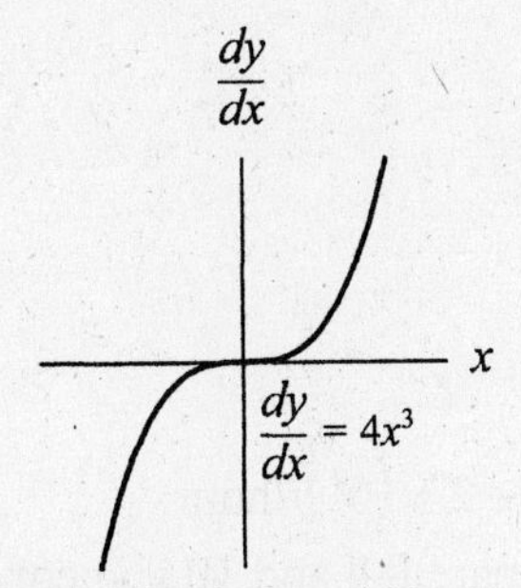

The slope formula is

$\frac{dy}{dx} = 4x^{4-1} = 4x^3$.

5.

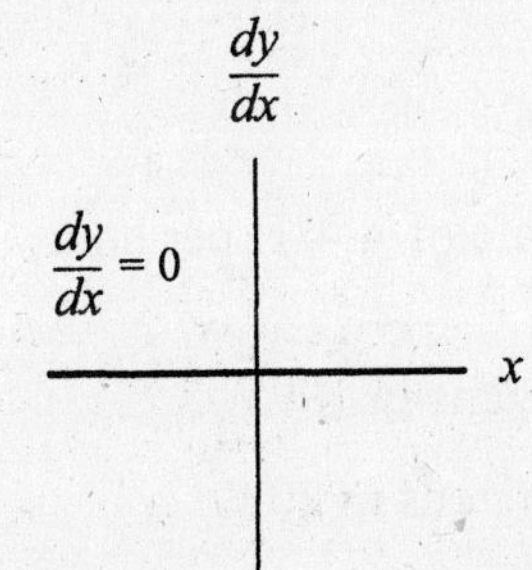

The slope of any horizontal line is 0.

7. $\frac{dy}{dx} = 7(2)x^{2-1} - 12x + 0 = 14x - 12$

9. $\frac{dy}{dx} = 5(3)x^{3-1} + 3(2)x^{2-1} - 2$

$= 15x^2 + 6x - 2$

11. $\frac{dy}{dx} = -9(-2)x^{-2-1} = 18x^{-3} = \frac{18}{x^3}$

13. $h'(x) = 0 + -0.049(\frac{1}{2}x^{1-\frac{1}{2}})$

$= -0.0245x^{-\frac{1}{2}} = \frac{-0.0245}{\sqrt{x}}$

15. a. $A'(t) = 0.1333$ dollar per year t years after 1990

b. $A(10) = 0.1333(10) + 0.17 \approx \$1.50$

c. $A'(9) = \$0.1333$ per year
The transaction fee in 1999 was increasing by about \$0.13 per year.

17. a. $P'(t) = 15.48$ thousand people/year t years after 1950

b. $P(20) = 15.48(2) + 485.4$
$= 795$ thousand people

c. $P'(40) = 15.48$ thousand people/year

19. a. $T'(x) = -16t + 2$ °F per hour t hours after noon

$T'(1.5) = -1.6(1.5) + 2 = -0.4°$F per hour ; $T'(-5) = -1.6(-5) + 2 = 10°$ F/hour
Steepness does not consider the sign of the slope, so we compare 0.4 and 10 and conclude that the graph is steeper at 7 a.m. than it is at 1:30 p.m.

b. $T'(-5) = -1.6(-5) + 2 = 10°$F per hour

c. $T'(0) = -1.6(0) + 2 = 2°$F per hour

d. $T'(4) = -1.6(4) + 2 = -4.4°$F per hour

Because this value is negative, the temperature is falling at a rate of 4.4°F per hour.

21. a. $N(-5) = 0.03(5)^2 + 0.315(-5) + 34.23 \approx 33.4$ million senior citizens in 1995
$N(30) = 0.03(30)^2 + 0.315(30) + 34.23 \approx 70.7$ million senior citizens in 2030

b. $N'(x) = 0.03(2x) + 0.315 + 0 = 0.06x + 0.315$ million senior citizens per year x years after 2000
$N'(-5) = 0.06(-5) + 0.315 = 0.015$ million senior citizens per year in 1995
$N'(30) = 0.06(30) + 0.315 \approx 2.1$ million senior citizens per year in 2030

c. $\frac{N'(30)}{N(30)} \cdot 100\% = \frac{2.115}{70.68} \cdot 100\% \approx 3\%$ per year in 2030

d. Solve $70.68 = 0.201P$ for P to obtain $P = \frac{70.68}{0.201} \approx 351.6$ million senior citizens in 2030.

23. a,b. $B'(x) = 0.2685(3x^2) - 15.6(2x) + 94.684 + 0 = 0.8055x^2 - 31.2x + 94.684$ births per year where x is the number of years since 1950.

$B'(20) = 0.8055(20)^2 - 31.2(20) + 94.684 = -207.12$ births per year in 1970

$B'(45) = 0.8055(45)^2 - 31.2(45) + 94.684 = 321.82$ births per year in 1995

The number of live births in 1970 was falling by about 207 births per year. The number of live births was rising in 1995 by about 322 births per year.

25. a. $R(x) = -3.68x^3 + 47.958x^2 - 80.759x + 166.98$ billion dollars of revenue when \$$x$ billion is spent on advertising.

b. $R'(x) = -3.68(3x^2) + 47.958(2x) - 80.759(1) + 0 = -11.039x^2 + 95.916x - 80.759$ billion dollars per billion dollars (billion dollars of revenue per billion dollars of advertising) when \$$x$ billion is spent on advertising

c. $R'(5) = -11.039(5)^2 + 95.916(5) - 80.759 \approx \$122.8$ billion per billion dollars

d. $R(5) = -3.68(5)^3 + 47.958(5)^2 - 80.759(5) + 166.98 \approx \$502.2$ billion

e. $\frac{R'(4.6)}{R(4.6)} 100\% \approx 25.3\%$ per billion advertising dollars

27. a. $P(x) = 175 - (0.0146x^2 - 0.7823x + 46.9125 + \frac{49.6032}{x})$ dollars when x windows are produced each hour

b. $P'(x) = 0 - \left[0.0146(2)x - 0.7823 + 0 + 49.6032(-1x^{-2})\right]$

$= -0.0292x + 0.7823 + \frac{49.6032}{x^2}$ dollars per window

when x windows are produced each hour

c. $P(80) = 175 - \left[0.0146(80)^2 - 0.7823(80) + 46.9125 + \frac{49.6032}{80}\right] \approx \$96.61$ profit

d. $P'(80) == -0.0292(80) + 0.7823 + \frac{49.6032}{(80)^2} \approx -\$1.55$ per window produced

When 80 units are produced each hour, the profit from the sale of one window is decreasing by \$1.55 per unit produced. In other words, if production is increased from 80 to 81 windows per hour, then the average profit per window will decrease by about \$1.55. This indicates that if the company wishes to maximize the profit per window, they should not increase production above 80 units per hour.

29. a. $f'(x) = \begin{cases} 44.408x - 108.431 \text{ million pounds per year} & \text{when } 0 \le x < 10 \\ 171.244x - 2253.951 \text{ million pounds per year} & \text{when } 10 < x \le 20 \end{cases}$

x years after 1970

b. In 1970: $f'(0) = -108.431$ million pounds per year

In 1980: $f'(10)$ does not exist. The rate of change of the amount of fish produced cannot be calculated directly from this model. A graph of f indicates that production rose until 1980 and then began to decline. It is reasonable to conclude that in 1980, the rate of change of fish production was approximately 0 million pounds per year. (Answers will vary.)

In 1990: $f'(20) = 1170.93$ million pounds per year

c. $\dfrac{f'(20)}{f(20)} \cdot 100\% \approx 16.9\%$ per year

31. Quill Activity

Section 4.3 More Simple Rate-of-Change Formulas

1.

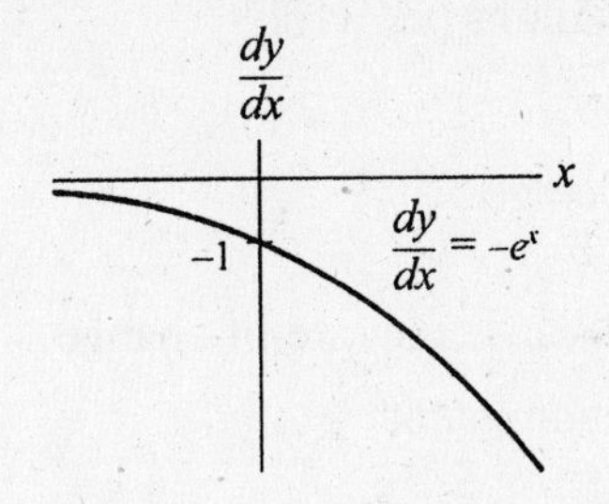

3.

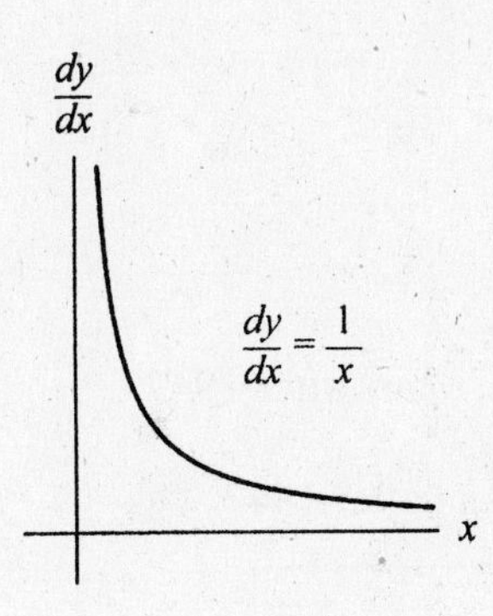

5.

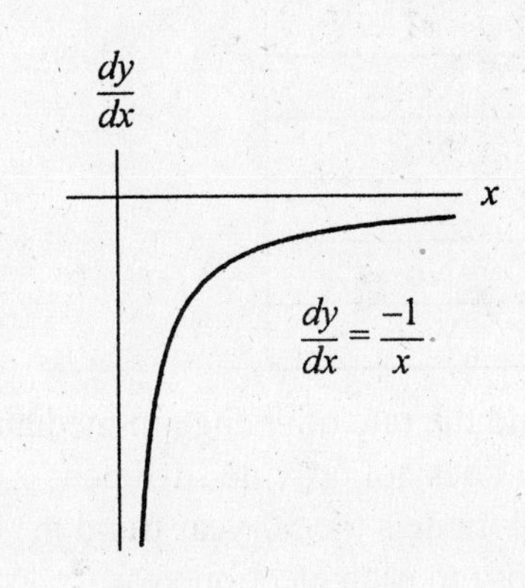

7. $h'(x) = 14x + \dfrac{13}{x}$

9. $\dfrac{dg}{dx} = 17(\ln 4.962)(4.962^x)$

11. $f'(x) =$

$100{,}000 \ln\left(1+\frac{0.05}{12}\right)^{12}\left(1+\frac{0.05}{12}\right)^{12x}$

$\approx 4989.61218\left(1+\frac{0.05}{12}\right)^{12x}$

$\approx 4989.61218(1.05116^x)$

13. $\dfrac{dy}{dx} = \dfrac{-4.2}{x} + 3.3(\ln 2.9)(2.9^x)$

15. a. Solving $2.5 = e^h$ for h yields $h = \ln 2.5 \approx 0.916$ hours

b. $\dfrac{dV}{dh} = e^h$ quarts per hour

c. $24 \text{ min} = 24 \text{ min} \cdot \dfrac{1 \text{ hr}}{60 \text{ min}} = 0.4 \text{ hr}$

$V(0.4) \approx 1.49$ quarts per hour

Convert to quarts per minute by multiplying by $\dfrac{60 \text{ min}}{1 \text{ hr}}$:

$V(0.4) \approx 0.025$ quarts per minute

$42 \text{ min} = 42 \text{ min} \cdot \dfrac{1 \text{ hr}}{60 \text{ min}} = 0.7 \text{ hr}$

$V(0.7) \approx 2.01$ quarts per hour

≈ 0.034 quart per minute

$55 \text{ min} = 55 \text{ min} \cdot \dfrac{1 \text{ hr}}{60 \text{ min}} = \dfrac{55}{60} \text{ hr}$

$V\left(\frac{55}{60}\right) \approx 2.5$ quarts per hour

≈ 0.042 quart per minute

17. a. $\dfrac{dy}{dx} = (\ln e)e^x = (1)e^x = e^x$

b. $\dfrac{dy}{dx} = \left(\ln e^k\right)e^{kx} = k(\ln e)e^{kx}$

$= k(1)e^{kx} = ke^{kx}$

19. a. $w(7) = 11.3 + 7.37 \ln 7 \approx 25.64$ grams; $w'(7) = \dfrac{7.37}{7} \approx 1.05$ grams per week

b. $\dfrac{w(9) - w(4)}{9-4} \approx \dfrac{5.977 \text{ grams}}{5 \text{ weeks}} \approx 1.195$ grams per week on average

c. The older the mouse is, the more slowly it gains weight because the rate-of-change formula, $w'(x) = \frac{7.37}{x}$, has the age of the mouse in the denominator.

21. a. $CPI(x) = -351.521 + 227.777 \ln x$, where x is the number of years since 1980

b. $\dfrac{318.7 - 294.1}{1999 - 1997} = 12.3$ index points per year

c. $CPI'(15) = \dfrac{227.777}{18} \approx 12.7$ index points per year

23. a. $M(t) = -0.216t^3 + 7.383t^2 + 661.969t + 19{,}839.889$ dollars t years after 1947

b. $M'(t) = -0.649t^3 + 14.765t^2 + 661.969$ dollars per year t years after 1947

c.

Year	t	$M'(t)$	$\frac{M'(t)}{M(t)} \cdot 100\%$	Reelected
1972	25	626	1.7% per year	Yes
1980	33	443	1.1% per year	No
1984	37	320	0.7% per year	Yes
1992	45	12.9	0.3% per year	No
1996	49	−172	−0.4% per year	Yes

d. There seems to be no relationship between re-election and the rate of change in median family income. This may be an indication that the model does not provide sufficient information to answer this question accurately because it models the 50-year trend in median family income rather than the median income close to each election year.

25. a. $\dfrac{dA}{dr} = 1000 \ln(e^{10})e^{10r} = 1000(10)e^{10r} = 10{,}000e^{10r}$ dollars per one unit of r

$= 10{,}000e^{10r}$ dollars per 100 percentage points when the interest rate is $100r\%$

b. $A'(0.07) = 10{,}000e^{10(0.07)} \approx \$20{,}137.53$ per 100 percentage points. The rate of change is large because it approximates by how much the value will increase when r increases by 1. Because the interest rate is input in decimals, an increase of 1 in r corresponds to a change in the interest rate of 100 percentage points.

c. $\dfrac{dA}{dr} = 1000 \ln(e^{0.1})e^{0.1r} = 1000(0.1)e^{0.1r} = 100e^{0.1r}$ dollars per percentage points when the interest rate is r%. The constant multiplier and exponent multiplier are both $\frac{1}{100}$ of what they were in the function in part *a*.

d. $A'(7) = 100e^{0.1(7)} \approx \$201.38$ per percentage point. This answer is $\frac{1}{100}$ of the answer to part *b*.

27. a. $p'(t) = \begin{cases} -23.73t^2 + 241.92t + 193.92 \text{ people per year} & \text{when } 0.7 \le t < 13 \\ 45{,}544(\ln 0.8474)(0.8474^t) \text{people per year} & \text{when } 13 < t \le 55 \end{cases}$

where t is the number of years since the beginning of 1860

b. 1870: $p'(10) \approx 240$ people per year

1873: $p'(13)$ does not exist

1900: $p'(40) \approx -10$ people per year

The rate of change of the population in 1873 could be estimated by

i. Calculating the model estimate for the populations in 1872 and 1874: 5954 and 4484

ii. Finding the average rate of change (symmetric difference quotient) for those years:

$$\frac{4484 - 5954}{1874 - 1872} \approx -735 \text{ people per year}$$

There are other valid methods of estimating this rate of change.

29. Quill Activity

Section 4.4 The Chain Rule

1. a. $f(x(2)) = f(6) = 140$

b. When $x = 6$, $\frac{df}{dx} = f'(6) = -27$

c. When $t = 2$, $\frac{dx}{dt} = x'(2) = 1.3$

d. When $t = 2$, $x = x(2) = 6$, so $\frac{df}{dt} = f'(6)x'(2) = (-27)(1.3) = -35.1$

3. Let t denote the number of days from today, w denote the weight in ounces, and v denote the value of the gold in dollars. We know that $\frac{dv}{dw} = \$395.70$ per troy ounce and $\frac{dw}{dt} = 0.2$ troy ounce per day. We seek the value of $\frac{dv}{dt}$:

$\frac{dv}{dt} = \frac{dv}{dw}\frac{dw}{dt} = $ (\$323.10 per troy ounce)(0.2 troy ounce per day)

$= \$64.62$ per day

The value of the investor's gold is increasing at a rate of \$64.62 per day.

5. a. $R(476) = 10{,}000$ deutsche marks (dm)
On January 5, 2001, sales were 476 units, producing a revenue of 10,000 dm.

b. $D(10{,}000) = \$20{,}412$
On January 5, 2001, 10,000 dm were worth \$20,412.

c. $\frac{dR}{dx} = 2.6$ dm per unit; Revenue was increasing by 2.6 dm per unit sold.

d. $\frac{dD}{dr} = \$2.0412$ per dm; The exchange rate was \$2.0412 per dm.

e. $\frac{dD}{dx} = \frac{dD}{dr}\frac{dR}{dx}$

$=$ (\$2.0412 per dm)(2.6 dm per unit)

$= \$5.31$ per unit

On January 5, 2001, revenue was increasing at a rate of \$5.31 per unit sold.

7. a. In 2010, $t = 10$ and $p(10) = \frac{130}{1+12e^{-0.02(10)}} \approx 12.009$ thousand people

In 2010 the city had a population of approximately 12,000 people.

b. $g(p(10)) \approx g(12.009) \approx 2(12.009) - 0.001(12.009^3) \approx 22$ garbage trucks

In 2010 the city owned 22 garbage trucks.

c. Let $u = 1 + 12e^{-0.02t}$ and $v = -0.02t$. Then $p = 130u^{-1}$ and $u = 1 + 12e^{v}$.

$$p'(t) = \frac{dp}{dt} = \frac{dp}{du}\frac{du}{dt} = \frac{dp}{du}\frac{du}{dv}\frac{dv}{dt}$$

$$= \frac{d}{du}(130u^{-1})\frac{d}{dv}(1+12e^{v})\frac{d}{dt}(-0.02t) = (-130u^{-2})(12e^{v})(-0.02)$$

$$= \frac{31.2e^{v}}{u^2} = \frac{31.2e^{-0.02t}}{(1+12e^{-0.02t})^2} \text{ thousand people per year}$$

$$p'(10) = \frac{31.2e^{-0.02(10)}}{(1+12e^{-0.02(10)})^2} \approx 0.22 \text{ thousand people per year}$$

In 2010 the population was increasing at a rate of approximately 220 people per year.

d. $g'(p) = 2 - 0.001(3p^2)$

$= 2 - 0.003p^2$ trucks per thousand people

$g'(p(10)) \approx g'(12.009)$

$= 2 - 0.003(12.009^2) \approx 1.6$ trucks per thousand people

In 2010 when the population was about 12,000, the number of garbage trucks needed by the city was increasing by 1.6 trucks per thousand people.

e. $$\frac{dg}{dt} = \frac{dg}{dp}\frac{dp}{dt} = g'(p(10))p'(10)$$

$\approx$ (1.6 trucks per thousand people)(0.22 thousand people per year)

$\approx$ 0.34 trucks per year

In 2010 the number of trucks needed by the city was increasing at a rate of approximately 0.34 truck per year, or 1 truck every 3 years.

f. See interpretations in parts *a* through *e*.

9. $c(x(t)) = 3(4-6t)^2 - 2$

$$\frac{dc}{dt} = \frac{dc}{dx}\frac{dx}{dt} = (6x)(-6) = 6(4-6t)(-6) = -144 + 216t$$

11. $$h(p(t)) = \frac{4}{1+3e^{-0.5t}}$$

Let $u = 0.5t$. Then $p = 1 + 3e^{u}$.

$$\frac{dh}{dt} = \frac{dh}{dp}\frac{dp}{dt} = \frac{dh}{dp}\frac{dp}{du}\frac{du}{dt} = \frac{d}{dp}(4p^{-1})\frac{d}{du}(1+3u^{4})\frac{d}{dt}(-0.5t)$$

$$= (-4p^{-2})(3e^{u})(-0.5) = \frac{6e^{u}}{p^2} = \frac{6e^{-0.5t}}{(1+3e^{-0.5t})^2}$$

13. $k(t(x)) = 4.3(\ln x)^3 - 2(\ln x)^2 + 4\ln x - 12$

$$\frac{dk}{dx} = 4.3\left[3(\ln x)^2\right]\frac{1}{x} - 2(2\ln x)\frac{1}{x} + 4\frac{1}{x} - 0$$

$$= \frac{12.9(\ln x)^2}{x} - \frac{4\ln x}{x} + \frac{4}{x} = \frac{12.9(\ln x)^2 - 4\ln x + 4}{x}$$

15. $p(t(k)) = 7.9(1.046)^{14k^3 - 12k^2}$

$$\frac{dp}{dk} = 7.9(\ln 1.046)(1.046)^{14k^3 - 12k^2}(42k^2 - 24k)$$

17. Inside function: $u = 3.2x + 5.7$ Outside function: $f = u^5$

$$\frac{df}{dx} = \frac{df}{du}\frac{du}{dx} = \frac{d}{du}(u^5)\frac{d}{dx}(3.2x + 5.7)$$

$$= (5u^4)(3.2)$$

$$= 5(3.2x + 5.7)^4(3.2) = 16(3.2x + 5.7)^4$$

19. Inside function: $u = x^2 - 3x$ Outside function: $f = \sqrt{u} = u^{\frac{1}{2}}$

$$\frac{df}{dx} = \frac{df}{du}\frac{du}{dx} = \frac{d}{du}(u^{\frac{1}{2}})\frac{d}{dx}(x^2 - 3x)$$

$$= \left(\frac{1}{2}u^{-\frac{1}{2}}\right)(2x - 3)$$

$$= \frac{2x - 3}{2\sqrt{u}} = \frac{2x - 3}{2\sqrt{x^2 - 3x}}$$

21. Inside function: $u = 35x$ Outside function: $f = \ln u$

$$\frac{df}{dx} = \frac{df}{du}\frac{du}{dx}$$

$$= \frac{d}{du}(\ln u)\frac{d}{dx}(35x)$$

$$= \left(\frac{1}{u}\right)(35) = \frac{1}{35x}(35) = \frac{1}{x}$$

23. Inside function: $u = 16x^2 + 37x$ Outside function: $f = \ln u$

$$\frac{df}{dx} = \frac{df}{du}\frac{du}{dx} = \frac{d}{du}(\ln u)\frac{d}{dx}(16x^2 + 37x)$$
$$= \left(\frac{1}{u}\right)(32x + 37)$$
$$= \frac{1}{16x^2 + 37x}(32x + 37)$$
$$= \frac{32x + 37}{16x^2 + 37x}$$

25. Inside function: $u = 0.695x$ Outside function: $f = 72.378e^u$

$$\frac{df}{dx} = \frac{df}{du}\frac{du}{dx}$$
$$= \frac{d}{du}(72.378e^u)\frac{d}{dx}(0.695x)$$
$$= (72.378e^u)(0.695)$$
$$= (72.378e^{0.695x})(0.695)$$
$$= 50.30271e^{0.695x}$$

27. Inside function: $u = 0.0856x$ Outside function: $f = 1 + 58.32e^u$

$$\frac{df}{dx} = \frac{df}{du}\frac{du}{dx}$$
$$= \frac{d}{du}(1 + 58.32e^u)\frac{d}{dx}(0.0856x)$$
$$= (58.32e^u)(0.0856)$$
$$= (58.32e^{0.0856x})(0.0856) = 4.992192e^{0.0856x}$$

29. Inside function: $u = 4x + 7$ Outside function: $f = \frac{350}{u} = 350u^{-1}$

$$\frac{df}{dx} = \frac{df}{du}\frac{du}{dx}$$
$$= \frac{d}{du}(350u^{-1})\frac{d}{dx}(4x + 7)$$
$$= (-350u^{-2})(4)$$
$$= -350(4x + 7)^{-2}(4)$$
$$= \frac{-1400}{(4x + 7)^2}$$

31. Inside function: $u = 1 + 8.976e^{-1.243x} \begin{cases} \text{inside: } w = -1.243x \\ \text{outside: } u = 1 + 8.976e^{w} \end{cases}$

Outside function: $f = \dfrac{3706.5}{u} + 89{,}070 = 3706.5u^{-1} + 89{,}070$

$$\frac{df}{dx} = \frac{df}{du}\frac{du}{dx} = \frac{df}{du}\frac{du}{dw}\frac{dw}{dx}$$
$$= \frac{d}{du}(3706.5u^{-1} + 89070)\frac{d}{dw}(1 + 8.976e^{w})\frac{d}{dx}(-1.243x)$$
$$= (-3706.5u^{-2})(8.976e^{w})(-1.243)$$
$$= -3706.5(1 + 8.976e^{-1.243x})^{-2}(8.976e^{-1.243x})(-1.243)$$
$$= \frac{41{,}354.04319e^{-1.243x}}{(1 + 8.976e^{-1.243x})^{2}}$$

33. Inside function: $u = \sqrt{x} - 3x$ Outside function: $f = 3^{u}$

$$\frac{df}{dx} = \frac{df}{du}\frac{du}{dx}$$
$$= \frac{d}{du}(3^{u})\frac{d}{dx}\left(x^{\frac{1}{2}} - 3x\right) = 3^{u}(\ln 3)\left(\tfrac{1}{2}x^{\frac{-1}{2}} - 3\right)$$
$$= 3^{x^{\frac{1}{2}} - 3x}(\ln 3)\left(\tfrac{1}{2}x^{\frac{-1}{2}} - 3\right)$$
$$= 3^{\sqrt{x} - 3x}(\ln 3)\left(\frac{1}{2\sqrt{x}} - 3\right)$$

35. Inside function: $u = 2^{x}$ Outside function: $f = \ln u$

$$\frac{df}{dx} = \frac{df}{du}\frac{du}{dx}$$
$$= \frac{d}{du}(\ln u)\frac{d}{dx}(2^{x})$$
$$= \frac{1}{u}(\ln 2)2^{x} = \frac{1}{2^{x}}(\ln 2)2^{x}$$
$$= \ln 2$$

37. Inside function: $u = 1 + Ae^{-Bx} \begin{cases} \text{inside: } w = -Bx \\ \text{outside: } u = 1 + Ae^{w} \end{cases}$

Outside function: $f = \dfrac{L}{u} = Lu^{-1}$

$$\frac{df}{dx} = \frac{df}{du}\frac{du}{dx} = \frac{df}{du}\frac{du}{dw}\frac{dw}{dx}$$
$$= \frac{d}{du}(Lu^{-1})\frac{d}{dw}(1 + Ae^{w})\frac{d}{dx}(-Bx)$$

$$= (-Lu^{-2})(Ae^{w})(-B)$$
$$= -L(1+Ae^{-Bx})^{-2}(Ae^{-Bx})(-B)$$
$$= \frac{LABe^{-Bx}}{(1+Ae^{-Bx})^2}$$

39. a. $A(t) = 1500e^{0.04t}$ dollars after t years

b. Let $u = 0.04t$. Then $A = 1500e^{u}$.

$$\frac{dA}{dt} = \frac{dA}{du}\frac{du}{dt} = \frac{d}{du}(1500e^{u})\frac{d}{dt}(0.04t) = (1500e^{u})(0.04) = 60e^{u}$$
$$= 60e^{0.04t} \text{ dollars per year after } t \text{ years}$$

c. $A'(1) = 60e^{0.04(1)} \approx$ \$62.45 per year; $A'(2) = 60e^{0.04(2)} \approx$ \$65.00 per year

d. The rates of change in part c tell you approximately how much interest your account will earn during the second and third years. The actual amounts are $A(2) - A(1) =$ \$63.71 during the second year and $A(3) - A(2) =$ \$66.31 during the third year.

41. a. $R'(q) = 0.0314(0.62285)e^{0.62285q}$ million dollars/quarter q quarters after the start of 1998

b.

Quarter Ending	June 1998	June 1999	June 2000
$R(q)$ million dollars	3.0	4.2	18.8
$R'(q)$ million dollars per quarter	0.07	0.82	9.92
$\frac{R'(q)}{R(q)} \cdot 100$ % per quarter	2.3	19.5	52.7

43. a. A logistic model is probably a better model because of the leveling-off behavior, although neither model should be used to extrapolate beyond one 24-hour day.

b. $C(t) = \dfrac{1342.077}{1+36.797e^{-0.258856t}}$ calls t hours after 5 a.m.

c. We use the formula developed in Activity 37: $C'(t) = \dfrac{LABe^{-Bt}}{(1+Ae^{-Bt})^2}$

with $L = 1342.077$, $A = 36.797$, and $B = 0.258856$.

$$C'(t) = \frac{LABe^{-Bt}}{(1+Ae^{-Bt})^2} = \frac{(1342.077)(36.797)(0.258856)e^{-0.258856t}}{(1+36.797e^{-0.258856t})^2}$$
$$\approx \frac{12{,}783.365e^{-0.258856t}}{(1+36.797e^{-0.258856t})^2} \text{ calls per hour } t \text{ hours after 5 a.m.}$$

d. Noon: $C'(7) \approx 42$ calls per hour
10 p.m.: $C'(17) \approx 74$ calls per hour

Midnight: $C'(19) \approx 58$ calls per hour
4 a.m.: $C'(23) \approx 28$ calls per hour

e. The rates of change give approximate hourly calls. This information could be used to determine how many dispatchers would be needed each hour.

45. a. We use the formula developed in Activity 37: $c'(t) = \dfrac{LABe^{-Bt}}{(1+Ae^{-Bt})^2}$ with $L = 93{,}700$, $A = 5095.9634$, and $B = 1.097175$.

$$c'(t) = \frac{LABe^{-Bt}}{(1+Ae^{-Bt})^2} = \frac{(93{,}700)(5095.9634)(1.097175)e^{-1.097175t}}{(1+5095.9634e^{-1.097175t})^2}$$

$$\approx \frac{523{,}892{,}033e^{-1.097175t}}{(1+5095.9634e^{-1.097175t})^2} \text{ deaths per week } t \text{ weeks after August 31, 1918}$$

$$c'(4) = \frac{523{,}892{,}033e^{-1.097175(4)}}{(1+5095.9634e^{-1.097175(4)})^2} \approx 1575 \text{ deaths per week}$$

b. $\dfrac{c'(4)}{c(4)} \cdot 100\% \approx \dfrac{1575 \text{ deaths per week}}{1458 \text{ deaths}} \cdot 100\% \approx 108\%$ increase per week

c. $c'(8) \approx 25{,}331$ deaths per week

$\dfrac{c'(8)}{c(8)} \cdot 100\% \approx \dfrac{25{,}331 \text{ deaths per week}}{52{,}473 \text{ deaths}} \cdot 100\% \approx 48\%$ increase per week

d. Although the rate of change is larger, it represents a smaller proportion the total number of deaths that had occurred at that time.

47. a. The data are essentially concave up, which indicates that a quadratic or exponential model may be appropriate. But the percentage differences vary from about –3% (from 5915 units per week in 1990 to 5750 units in 1991) to about 20% (from 10,245 units in 1996 to 12,210 units in 1997), so an exponential model may not fit the data very well. On the other hand, the second differences are all between 350 and 360. Because the second differences are almost constant, a quadratic model is the better choice.

b. $u(x) = 177.356x^2 - 342.240x + 5914.964$ units per week x years after 1990

c. $C(u(x)) = 3250.23 + 74.95\ln(177.356x^2 - 342.240x + 5914.964)$ dollars per week x years after 1990

d. $\dfrac{dC}{dx} = \dfrac{dC}{du}\dfrac{du}{dx}$

$$= \frac{d}{du}(3250.23 + 74.95\ln u)\frac{d}{dx}(177.356x^2 - 342.240x + 5914.964)$$

$$= \left(\frac{74.95}{u}\right)(354.712x - 342.240 + 0)$$

$$= \left(\frac{74.95}{177.356x^2 - 342.240x + 5914.964}\right)(354.712x - 342.240) \text{ dollars per week per year}$$

x years after 1990

e. Use the expressions from parts *c* and *d*.

Year	**2002**	**2003**	**2004**	**2005**
x	12	13	14	15
$C(u(x))$ ($/week)	4015.95	4026.40	4036.31	4045.72
$\frac{dC}{dx}$ **($/week/year)**	10.73	10.18	9.66	9.17

f, g.

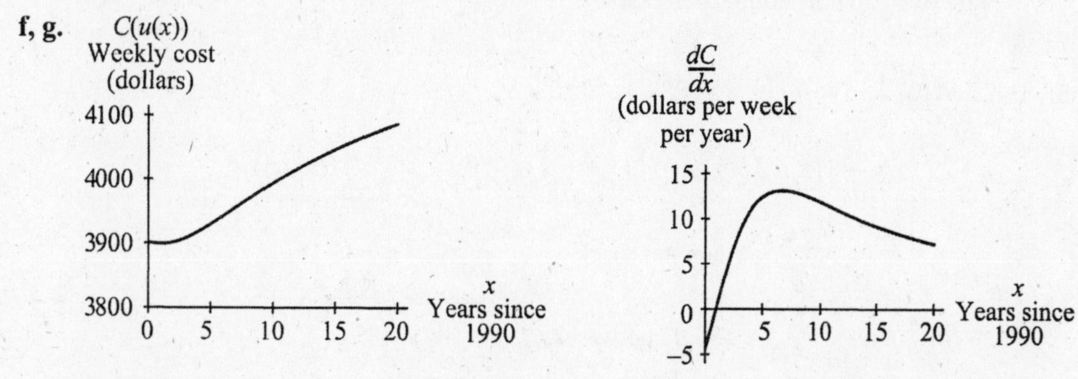

Although a graph of C may not appear ever to decrease, a graph of $\frac{dC}{dx}$ is negative between $x = 0$ and $x \approx 0.965$, indicating that cost was decreasing from the end of 1990 to (almost) the end of 1991. A close-up view of the graph of C between $x = 0$ and $x = 2$ confirms this.

49. Quill Activity

Section 4.5 The Product Rule

1. $h'(2) = f'(2)g(2) + f(2)g'(2) = -1.5(4) + 6(3) = 12$

3. a. i. In 2007 there were 75,000 households in the city.

ii. In 2007 the number of households was declining at a rate of 1200 per year.

iii. In 2007, 52% of households owned a computer.

iv. In 2007 the percentage of households with a computer was increasing by 5 percentage points per year.

b. Input: the number of years since 1995
Output: the number of households with computers

c. $N(2) = h(2)c(2) = (75{,}000)(0.52) = 39{,}000$ households with computers.
$N'(2) = h'(2)c(2) + h(2)c'(2) = (-1200)(0.52) + (75{,}000)(0.05)$
$= 3126$ households per year
In 2007 there were 39,000 households with computers, and that number was increasing at a rate of 3126 households per year.

5. a. i. $S(10) = 15 + \dfrac{2.6}{10+1} \approx \$15.24$

$S'(x) = \dfrac{d}{dx}[15 + 2.6(x+1)^{-1}] = 0 + 2.6(-1)(x+1)^{-2}\dfrac{d}{dx}(x+1)$

$= \dfrac{-2.6}{(x+1)^2}$ dollars per week

$S'(10) = \dfrac{-2.6}{(10+1)^2} \approx -\$0.02$ per week

After 10 weeks, 1 share is worth \$15.24, and the value is declining by \$0.02 per week.

ii. $N(10) = 100 + 0.25(10^2) = 125$ shares

$N'(x) = \dfrac{d}{dx}(100 + 0.25x^2) = 0.5x$ shares per week after x weeks

$N'(10) = 0.5(10) = 5$ shares per week

After 10 weeks, the investor owns 125 shares and is buying 5 shares per week.

iii. $V(10) = S(10)N(10) \approx (15.24)(125) \approx \$1904.55$

$V'(10) \approx S'(10)N(10) + S(10)N'(10)$

$\approx (-0.02)(125) + (15.24)(5) \approx \$73.50$ per week

After 10 weeks, the investor's stock is worth approximately \$1905, and the value is increasing at a rate of \$73.50 per week.

b. $V'(x) = S'(x)N(x) + S(x)N'(x)$

$$= \left(\frac{-2.6}{(x+1)^2}\right)(100 + 0.25x^2) + \left(15 + \frac{2.6}{x+1}\right)(0.5x)$$

$$= -\frac{0.65x^2 + 260}{(x+1)^2} + \frac{1.3x}{x+1} + 7.5x \text{ dollars per week after } x \text{ weeks}$$

7. Let $A(t)$ be the number of acres of corn and let $B(t)$ be the number of bushels of corn per acre, where t is the number of years from the current year. The total number of bushels of corn produced is given by $C(t) = A(t)B(t)$, where $A(0) = 500$ acres, $A'(0) = 50$ acres per year, $B(0) = 130$ bushels per acre, and $B'(0) = 5$ bushels per acre per year. The rate of change is:

$C'(0) = A'(0)B(0) + A(0)B'(0)$

$= (50 \text{ acres/year})(130 \text{ bushels/acre}) + (500 \text{ acres})(5 \text{ bushels/acre/year})$

$= 9000$ bushels per year

9. a. $(17{,}000)(0.48) = 8160$ voters

b. $(8160)(0.57) \approx 4651$ votes for candidate A

c. Let $v(t)$ be the proportion of registered voters who plan to vote, and let $a(t)$ be the proportion who support candidate A, where t is the number of weeks from today and both quantities are expressed as decimals. Then the number of votes for candidate A is given by

$N(t) = 17{,}000v(t)a(t)$, where $v(0) = 0.48$, $a(0) = 0.57$, $v'(0) = 0.07$ and $a'(0) = -0.03$.

The rate of change of N is:

$$N'(t) = \frac{d}{dt}[17{,}000v(t)a(t)] = 17{,}000\frac{d}{dt}[v(t)a(t)] = 17{,}000[v'(t)a(t) + v(t)a'(t)]$$

When $t = 0$,

$N'(0) = 17{,}000[(0.07)(0.57) + (0.48)(-0.03)] \approx 434$ votes for candidate A per week

11. $$f'(x) = \left[\frac{d}{dx}(3x^2 + 15x + 7)\right](32x^3 + 49) + (3x^2 + 15x + 7)\frac{d}{dx}(32x^3 + 49)$$

$$= (6x + 15)(32x^3 + 49) + (3x^2 + 15x + 7)(96x^2)$$

$$= 480x^4 + 1920x^3 + 672x^2 + 294x + 735$$

13. $$f'(x) = \left[\frac{d}{dx}(12.8893x^2 + 3.7885x + 1.2548)\right][29.685(1.7584^x)]$$

$$+ (12.8893x^2 + 3.7885x + 1.2548)\frac{d}{dx}[29.685(1.7584^x)]$$

$$= (25.7786x + 3.7885)[29.685(1.7584^x)]$$

$$+ (12.8893x^2 + 3.7885x + 1.2548)[29.685(\ln 1.7584)(1.7584^x)]$$

15. Note that $f(x) = g(x)h(x)$, where $g(x) = (5.7x^2 + 3.5x + 2.9)^3$ and $h(x) = (3.8x^2 + 5.2x + 7)^{-2}$.

$$g'(x) = 3(5.7x^2 + 3.5x + 2.9)^2 \frac{d}{dx}(5.7x^2 + 3.5x + 2.9)$$
$$= 3(5.7x^2 + 3.5x + 2.9)^2(11.4x + 3.5)$$

$$h'(x) = -2(3.8x^2 + 5.2x + 7)^{-3} \frac{d}{dx}(3.8x^2 + 5.2x + 7)$$
$$= -2(3.8x^2 + 5.2x + 7)^{-3}(7.6x + 5.2)$$

$$f'(x) = g'(x)h(x) + g(x)h'(x)$$
$$= [3(5.7x^2 + 3.5x + 2.9)^2(11.4x + 3.5)](3.8x^2 + 5.2x + 7)^{-2}$$
$$+ (5.7x^2 + 3.5x + 2.9)^3[-2(3.8x^2 + 5.2x + 7)^{-3}(7.6x + 5.2)]$$

17. $f'(x) = \frac{d}{dx}[12.624(14.831^x)(x^{-2})]$

$$= 12.624 \frac{d}{dx}[(14.831^x)(x^{-2})]$$
$$= 12.624\left[\left(\frac{d}{dx}(14.831^x)\right)x^{-2} + (14.831^x)\frac{d}{dx}x^{-2}\right]$$
$$= 12.624\left[(\ln 14.831)(14.831^x)(x^{-2}) + (14.831^x)(-2x^{-3})\right]$$
$$= 12.624(14.831^x)\left(\frac{\ln 14.831}{x^2} - \frac{2}{x^3}\right) = \frac{12.624(14.831^x)}{x^3}[x(\ln 14.831) - 2]$$

19. Note that $f(x) = g(x)h(x)$, where $g(x) = 79.32x$,

$h(x) = 1984.32(1 + 7.68e^{-0.859347x})^{-1} + 1568$, and $g'(x) = 79.32$

Using the formula for the derivative of a logistic function, we know that

$$h'(x) = \frac{LABe^{-Bx}}{(1 + Ae^{-Bx})^2} \quad \text{where } L = 1984.32,\ A = 7.68, \text{ and } B = -0.859347$$

Thus $h'(x) = \dfrac{1984.32(7.68)(0.859347)e^{-0.859347x}}{(1 + 7.68e^{-0.859347x})^2}$.

$$f'(x) = g'(x)h(x) + g(x)h'(x)$$
$$= 79.32\left(\frac{1984.32}{1 + 7.68e^{-0.859347x}} + 1568\right) + (79.32x)\left(\frac{1984.32(7.68)(0.859347)e^{-0.859347x}}{(1 + 7.68e^{-0.859347x})^2}\right)$$
$$\approx \frac{157{,}396}{1 + 7.68e^{-0.859347x}} + 124{,}374 + \frac{1{,}038{,}781xe^{-0.859347x}}{(1 + 7.68e^{-0.859347x})^2}$$

21. Note that $f(x) = g(x)h(x)$, where $g(x) = 430(0.62^x)$ and $h(x) = \left[6.42 + 3.3(1.46^x)\right]^{-1}$.

$$g'(x) = 430(\ln 0.62)(0.62^x)$$

$$h'(x) = -\left[6.42 + 3.3(1.46^x)\right]^{-2} 3.3(\ln 1.46)(1.46^x) = \frac{-3.3(\ln 1.46)(1.46^x)}{\left[6.42 + 3.3(1.46^x)\right]^2}$$

$$f'(x) = g'(x)h(x) + g(x)h'(x)$$

$$= 430(\ln 0.62)(0.62^x)\frac{1}{6.42+3.3(1.46^x)} + 430(0.62^x)\frac{-3.3(\ln 1.46)(1.46^x)}{\left[6.42+3.3(1.46^x)\right]^2}$$

23. $f'(x) = 4\sqrt{3x+2} + 4x\frac{d}{dx}(3x+2)^{½}$

$$= 4\sqrt{3x+2} + 4x\left[\frac{1}{2}(3x+2)^{-½}(3)\right]$$

$$= 4\sqrt{3x+2} + \frac{6x}{\sqrt{3x+2}}$$

25. Note that $f(x) = g(x)h(x)$ where $g(x) = x$ and h is a logistic function with L = 14,000, A = 12.6, and B = 0.73 and derivative of the form $h'(x) = \frac{LABe^{-Bx}}{(1+Ae^{-Bx})^2}$:

$$h'(x) = \frac{128,772e^{-0.73x}}{\left(1+12.6e^{-0.73x}\right)^2}$$

$$f'(x) = g'(x)h(x) + g(x)h'(x) = (1)\frac{14,000}{1+12.6e^{-0.73x}} + (x)\frac{128,772e^{-0.73x}}{\left(1+12.6e^{-0.73x}\right)^2}$$

$$= \frac{14,000}{1+12.6e^{-0.73x}} + \frac{128,772xe^{-0.73x}}{\left(1+12.6e^{-0.73x}\right)^2}$$

27. a. Price: $P(m) = 0.049m + 1.144$ dollars m months after December

Quantity sold: $Q(m) = -0.946m^2 + 0.244m + 279.911$ units sold m months after December

b. $R(m) = P(m)Q(m) = (0.049m + 1.144)(-0.946m^2 + 0.244m + 279.911)$ dollars of revenue m months after December

c. $R(8) \approx \$340.05$; $R(9) \approx \$325.78$

d. Because the revenue in September is less than that in August, the rate of change in August is probably negative.

e. $P'(m) = 0.049$; $Q'(m) = -1.893m + 0.244$
Rate of change for revenue:
$R'(m) = P'(m) \cdot Q(m) + P(m)Q'(m)$

$$= 0.049(-0.946m^2 + 0.244m + 279.911) + (0.049m + 1.144)(-1.893m + 0.244)$$

dollars of revenue per month m months after December.

f. $R'(2) \approx \$9.17$ per month; $R'(8) \approx -\$12.04$ per month; $R'(9) \approx -\$16.55$ per month

29. a. $P(t) = 0.023t^2 + 1.764t + 205.895$ million people t years after 1970

b. $F(t) = P(t)\dfrac{m(t)}{100}$ (Note that this formula is correct only if the input to P is the number of years since 1970.)

$F(t) = (0.023t^2 + 1.764t + 205.895)\left(\dfrac{6.53(0.941163)^t + 22}{100}\right)$ million people t years after 1970

c. $F'(t) = P'(t) \cdot m(t) + P(t) \cdot m'(t)$

$$= \left[(0.046 + 1.764)\left(\frac{6.53(0.941163)^t + 22}{100}\right) + (0.023t^2 + 1.764t + 205.895)\left(\frac{6.53(\ln 0.941163)(0.941163)^t}{100}\right)\right]$$

million people t years after 1970

d. 1990: $F'(20) \approx 0.35$ million people per year
1995: $F'(25) \approx 0.46$ million people per year
2000: $F'(30) \approx 0.55$ million people per year

31. a. $m(x) = 0.209x + 8.208$ million men 65 years or older x years after 1970

$p(x) = \dfrac{0.0223x^2 - 1.085x + 20.07}{100}$ percentage (expressed as a decimal) of men 65 or older below poverty level x years after 1970

b. $n(x) = m(x)p(x) = (0.209x + 8.208)\left(\dfrac{0.0223x^2 - 1.085x + 20.07}{100}\right)$ million men 65 or older below poverty level

c. $n'(x) = m'(x)p(x) + m(x)p'(x)$

$$= 0.209\left(\frac{0.0223x^2 - 1.085x + 20.07}{100}\right) + (0.209x + 8.208)\left(\frac{0.0446x - 1.085}{100}\right)$$

1990: $n'(20) \approx -0.009$ million men per year
2000: $n'(30) \approx 0.05$ million men per year

33. a. $E(x) = \frac{0.73(1.2912^x) + 8}{100}(-0.026x^2 - 3.842x + 538.868)$ women receiving epidurals at the Arizona hospital x years after 1980

$E'(x) = \frac{0.73(1.2912^x) + 8}{100}(-0.052x - 3.842) +$

$\frac{0.73(\ln 1.2912)(1.2912^x)}{100}(-0.026x^2 - 3.842x + 538.868)$ women per year x years after 1980

b. Increasing by $p'(17) \approx 14.4$ percentage points per year

c. Decreasing by approximately 5 births per year ($b'(17) \approx -4.7$)

d. Increasing by $E'(17) \approx 64$ women per year.

e. Profit $= \$57 \cdot E(17) \approx \$17{,}043$ (using a value of 299 for the number of births) or $17,071 (using an unrounded number of births)

35. a. $E(x) = -151.516x^3 + 2060.988x^2 - 8819.062x + 195{,}291.201$ students enrolled

$D(x) = -14.271x^3 + 213.882x^2 - 1393.655x + 11{,}697.292$ students dropping out x years after the 1980–81 school year

b. $P(x) = \frac{D(x)}{E(x)} \cdot 100$ percent x years after the 1980–81 school year

c. $D'(x) = -42.813x^2 + 427.763x - 1393.655$; $E'(x) = -454.548x^2 + 4121.976x - 8819.062$

Write $P(x) = 100D(x)[E(x)]^{-1}$.

$$P'(x) = 100\frac{d}{dx}\left(D(x)[E(x)]^{-1}\right)$$

$$= 100\left(\frac{d}{dx}D(x)\right)[E(x)]^{-1} + 100D(x)\frac{d}{dx}[E(x)]^{-1}$$

$$= 100D'(x)[E(x)]^{-1} + 100D(x)(-1)[E(x)]^{-2}E'(x)$$

$$= \frac{100D'(x)}{E(x)} - \frac{100D(x) \cdot E'(x)}{[E(x)]^2}$$

$$= \frac{100(-42.813x^2 + 427.763x - 1393.655)}{-151.516x^3 + 2060.988x^2 - 8819.062x + 195{,}291.201} -$$

$$\frac{100(-14.271x^3 + 213.882x^2 - 1393.655x + 11{,}697.292)(-454.548x^2 + 4121.976x - 8819.062)}{(-151.516x^3 + 2060.988x^2 - 8819.062x + 195{,}291.201)^2}$$

percentage points per year x years after the 1980–81 school year

d.

x	$P'(x)$ (percentage point per year)	x	$P'(x)$ (percentage point per year)
0	–0.44	5	–0.19
1	–0.38	6	–0.19
2	–0.32	7	–0.22
3	–0.26	8	–0.29
4	–0.21	9	–0.41

In the 1980–81 school year, the rate of change was most negative with a value of –0.44 percentage point per year. This is the most rapid decline during this time period. The rate of change was least negative in the 1985–86 school year with a value of –0.187 percentage point per year.

e. Negative rates of change indicate that high school attrition in South Carolina was improving during the 1980s.

37. Quill Activity

Chapter 4 Review Test

1. **a.** $x \approx 0.8$

   b. positive slope: $0.8 < x < 2$
   negative slope: $-3 < x < 0, 0 < x < 0.8$

   c. $x = 0$

   d. $f'(-2) \approx -4, f'(1) \approx 1.1$, See Answer Key page A-31 in Text for figure.

   e. See Answer Key page A-31 in Text for figure.

2. **a.** Note that D is a logistic function with $L = 8.101, A = 214.8$, and $B = 0.797$ and derivative of the form $D'(t) = \frac{LABe^{-Bt}}{(1+Ae^{-Bt})^2} = \frac{8.101(214.8)(0.797)e^{-0.797t}}{(1+214.8e^{-0.797t})^2}$ pounds per person per year t years after 1980

   b. $D'(10) \approx 0.4$ pound per person per year
   In 1990 the average annual per capita consumption of turkey in the United States was increasing by 0.4 pound per person per year.

   c. $D'(21) \approx 0.00007$ pound per person per year. There was essentially no growth in the per capita consumption of turkey in 2001.

3. **a.** $\frac{6890-4865}{2000-1996} = 506.3$ billion per year

   b See Answer Key page A-32 in Text for figure.

   c. $A'(18) \approx \$496.4$ billion per year

4. **a.** Rewrite $P(t)$ as $P(t) = 100N(t)[A(t)]^{-1}$.
   Use the Product Rule to find the derivative.

$$P'(t) = 100\frac{d}{dt}\left(N(t)[A(t)]^{-1}\right)$$

$$= 100\left(\left[\frac{d}{dt}N(t)\right][A(t)]^{-1} + N(t)\frac{d}{dt}[A(t)]^{-1}\right)$$

$$= 100\left(N'(t)[A(t)]^{-1} + N(t)(-1)[A(t)]^{-2}A'(t)\right)$$

$$= \frac{100N'(t)}{A(t)} - \frac{100N(t)A'(t)}{[A(t)]^2}$$ percentage points per year t years after 1980

   b. Input units: years
   Output units: percentage points per year

Chapter 5

Section 5.1 Approximating Change

1. $32\% - (4 \text{ percentage points per hour})\left(\frac{1}{3}\text{hour}\right) \approx 30.7\%$

3. $f(3.5) \approx f(3) + f'(3)(0.5) = 17 + 4.6(0.5) = 19.3$

5. **a.** Increasing production from 500 to 501 units will increase cost by approximately \$17.

   b. If sales increase from 150 to 151 units, then profit will increase by approximately \$4.75.

7. A marginal profit of –\$4 per shirt means that the fraternity is actually losing \$4 for each additional shirt sold. The fraternity should consider selling fewer shirts or increasing the sales price.

9. Slope of tangent line $\approx \dfrac{\$8600}{10 \text{ years of age}} = \$860$ per year of age

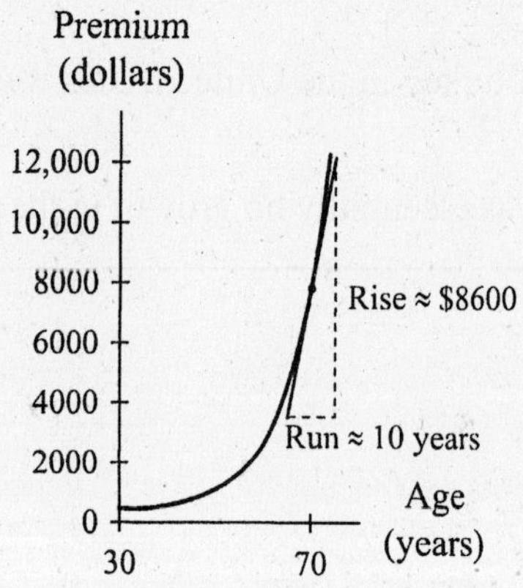

   Annual premium for 70-year-old $\approx \$7850$

   Annual premium for 72-year-old $\approx \$7850 + \left(\dfrac{\$860}{\text{year}}\right)(2 \text{ years}) = \$9570$

   (Answers will vary.)

11. **a.** See Answer Key page A-32 in Text for figure. Slope of tangent line is approximately $\dfrac{97 \text{ billion dollars}}{1 \text{ billion dollars}} = 97$ billion dollars per billion dollars (revenue dollars per sales dollars).
    Revenue is approximately \$614 billion when \$6 billion is spent on advertising.
    Revenue is approximately $614 + 0.5(97) = 662.5$ billion dollars when \$6.5 billion is spent on advertising. (Discussion will vary.)

    b. $R(6.5) \approx \$658$ billion

    c. Answers will vary.

13. **a.** $P(x) = 268.79(1.013087^x)$ thousand people in year x

    $P'(x) = 268.79(\ln 1.013087)(1.013087^x)$ thousand people per year in year x
    In 2000 the population of South Carolina was increasing by 53.6 thousand people per year.

    b. Between 2000 and 2003, the population increased by approximately 160.8 thousand people.

    c. By finding the slope of the tangent line at 2000 and multiplying by 3, we determine the change in the tangent line from 2000 through 2003 and use that change to estimate the change in the population function.

15. a. The population was growing at a rate of 2.53 million people per year in 1990.

b. Between 1998 and 1999, the population of Mexico increased by approximately 2.53 million people.

17. a. In 1998 the amount was increasing by 1.15 million pieces per year.

b. We would expect an increase of approximately 1.15 million pieces between 1998 and 1999.

c. $p(24) - p(23) \approx 1.3$ million pieces

d. $101.9 - 100.4 = 1.5$ million pieces

e. As long as the data in part *d* were correctly reported, the answer to part *d* is the most accurate one.

19. a. $R(x) = \left(-7.032 \cdot 10^{-4}\right)x^2 + 1.666x + 47.130$ dollars when x hot dogs are sold

b. Cost: $c(x) = 0.5x$ dollars per hot dog

Profit: $p(x) = R(x) - c(x) = (-7.032 \cdot 10^{-4})x^2 + 1.166x + 47.130$ dollars per hot dog

c.

x (hot dogs)	$R'(x)$ (dollars per hot dog)	$c'(x)$ (dollars per hot dog)	$p'(x)$ (dollars per hot dog)
200	1.38	0.50	0.88
800	0.54	0.50	0.04
1100	0.12	0.50	–0.38
1400	–0.30	0.50	–0.80

If the number of hot dogs sold increases from 200 to 201, the revenue increases by approximately $1.38 and the profit increases by approximately $0.88. If the number increases from 800 to 801, the revenue increases by 0.54, but the profit sees almost no increase (4 cents). If the number increases from 1100 to 1100, the increase in revenue is only about 12 cents. Because this marginal revenue is less than the marginal cost at a sales level of 1100, the result of the sales increase from 1100 to 1101 is a decrease of $0.38 in profit. If the number of hot dogs increases from 1400 to 1401, revenue declines by approximately 30 cents and profit declines by approximately 80 cents.

d.

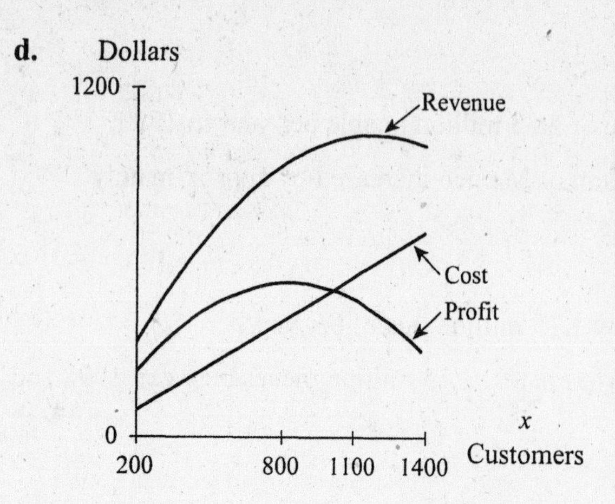

The marginal values in part c are the slopes of the graphs shown here. For example, at $x = 800$, the slope of the revenue graph is \$0.54 per hot dog, the slope of the cost graph is \$0.50 per hot dog, and the slope of the profit graph is \$0.04 per hot dog. We see from the graph that maximum profit is realized when about 800 hot dogs are sold.

Revenue is greatest near $x = 1100$, so the marginal revenue there is small. However, once costs are factored in, the profit is actually declining at this sales level. This is illustrated by the graph.

21. *United States:*

a. $A(t) = 0.109t^3 - 1.555t^2 + 10.927t + 100.320$ t years after 1980

b. $A'(t) = 0.327t^2 - 3.111t + 10.927$ index points per year t years after 1980

$A'(7) \approx 5.2$ index points per year

c. 1988 CPI estimate:

(CPI in 1987) + $A'(7)$ (1 year) $\approx$ 137.9 + (5.2 index points per year)(1 year)

= 143.1

Note: The estimate can also be calculated using the value of $A(7)$ instead of the actual CPI in 1987. Because the model closely agrees with the actual value in 1987, the value of this estimate is not significantly affected by this choice.

Canada:

a. $C(t) = 0.150t^3 - 2.171t^2 + 15.814t + 99.650$ t years after 1980

b. $C'(t) = 0.450t^2 - 4.343t + 15.814$ index points per year t years after 1980

$C'(7) \approx 7.5$ index points per year

c. 1988 CPI estimate:

(CPI in 1987) + $C'(7)$ (1 year) $\approx$ 155.4 + (7.5 index points per year)(1 year)

= 162.9

Note: The estimate can also be calculated using the value of $C(7)$ instead of the actual CPI in 1987. Because the model closely agrees with the actual CPI in 1987, the value of this estimate is not significantly affected by this choice.

Peru:

a. $P(t) = 85.112(2.013252^t)$ t years after 1980

b. $P'(t) = 85.112(\ln 2.013252)(2.013252^t)$

$\approx 59.558(2.013252^t)$ index points per year t years after 1980

$P'(7) \approx 7984$ index points per year

c. 1988 CPI estimate:

(CPI in 1987) + $P'(7)$ (1 year) $\approx$ 11,150 + (7984 index points per year)(1 year)

$= 19{,}134$

Note: The estimate can also be calculated using the value of $P(7)$ instead of the actual CPI in 1987. If this is done, the estimate will be approximately 19,394.

Brazil:

a. $B(t) = 73{,}430(2.615939^t)$ t years after 1980

b. $B'(t) = 73.430(\ln 2.615939)(2.615939^t)$

$\approx 70.612(2.615939^t)$ index points per year t years after 1980

$B'(7) \approx 59{,}193$ index points per year

c. 1988 CPI estimate:

(CPI in 1987) + $B'(7)$ (1 year) $\approx 77{,}258 +$ (59,193 index points per year)(1 year)

$= 136{,}451$

Note: The estimate can also be calculated using the value of $B(7)$ instead of the actual CPI in 1987. If this is done, the estimate will be approximately 120,748.

23. *Note:* This Activity can be solved using either a cubic model or a logistic model. The following solution uses a cubic model.

a. $R(A) = -0.158A^3 + 5.235A^2 - 23.056A + 154.884$ thousand dollars of revenue when A thousand dollars is spent on advertising

b. $R'(A) = -0.473A^3 + 10.471A^2 - 23.056$ thousand dollars of revenue per thousand dollars of advertising when A thousand dollars is spent on advertising

$R'(10) \approx 34.3$ thousand dollars of revenue per thousand dollars of advertising

When \$10,000 is spent on advertising, revenue is increasing by \$34.3 thousand per thousand advertising dollars. If advertising is increased from \$10,000 to \$11,000, the car dealership can expect an approximate increase in revenue of \$34,300.

c. $R'(18) \approx 12.0$ thousand dollars of revenue per thousand of dollars of advertising

When \$18,000 is spent on advertising, revenue is increasing by \$12.0 thousand per thousand advertising dollars. If advertising is increased from \$18,000 to \$19,000, the car dealership can expect an approximate increase in revenue of \$12,000.

25. a. $A(t) = 300\left(1 + \frac{0.065}{12}\right)^{12t}$ dollars after t years

b. $A(t) = 300\left[\left(1 + \frac{0.065}{12}\right)^{12}\right]^t \approx 300(1.066972^t)$ dollars after t years

c. $A(2) \approx \$341.53$

d. $A'(t) \approx 300(\ln 1.066972)(1.066972^t)$ dollars per year

$A'(2) \approx \$22.14$ per year

e. $A'(2)\left(\frac{1}{4}\text{ year}\right) \approx (\$22.14 \text{ per year})\left(\frac{1}{4}\text{ year}\right) \approx \$5.53$

27. Quill Activity

Answers will vary. One possible answer: Close to the point of tangency, a tangent line and a curve are close to one another. The farther away from the point of tangency we move, the more the tangent line deviates from the curve. Thus the tangent line near the point of tangency will usually produce a good estimate, but the tangent line farther away from the point of tangency will produce a poor estimate.

Section 5.2 Relative and Absolute Extreme Points

1. Quadratic, cubic, and many product, quotient, and composite functions could have relative maxima or minima.

3.

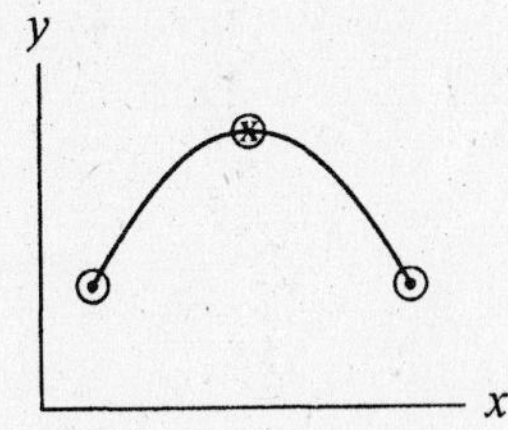

The derivative is zero at the absolute maximum point.

5.

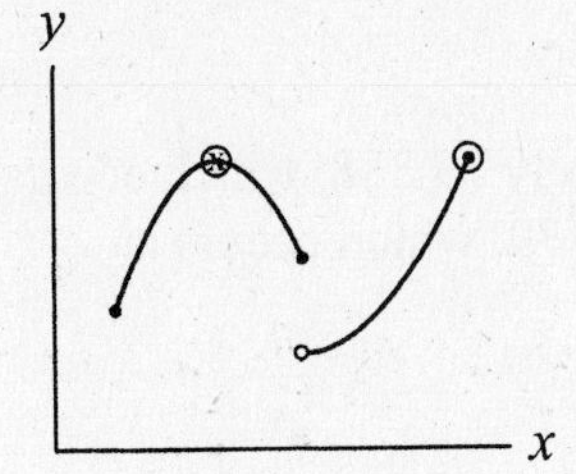

The derivative is zero at the absolute maximum point marked with an X.

7.

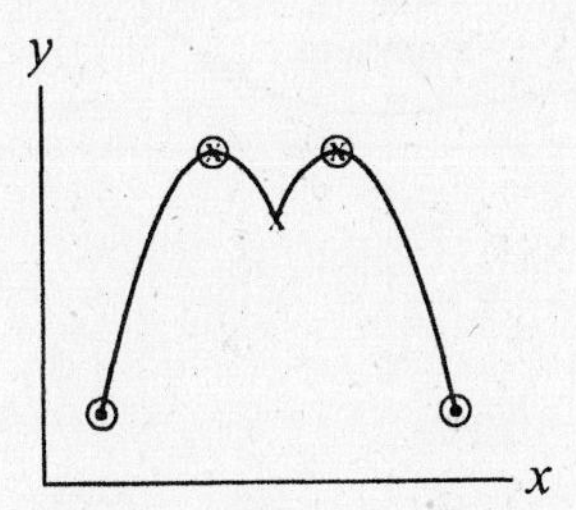

The derivative is zero at both absolute maximum points. The derivative does not exist at the relative minimum point.

9. Answers will vary. One such graph is $y = x^3$, which does not have a relative minimum or maximum at $x = 0$ even though the derivative is zero at this point.

11. **a.** All statements are true.

b. The derivative does not exist at $x = 2$ because f is not continuous there, so the third statement is false.

c. The slope of the graph is negative, $f'(x) < 0$, to the left of $x = 2$ because the graph is decreasing, so the second statement is false.

d. The derivative does not exist at $x = 2$ because f is not smooth there, so the third statement is not true.

13. Answers will vary. One such graph is

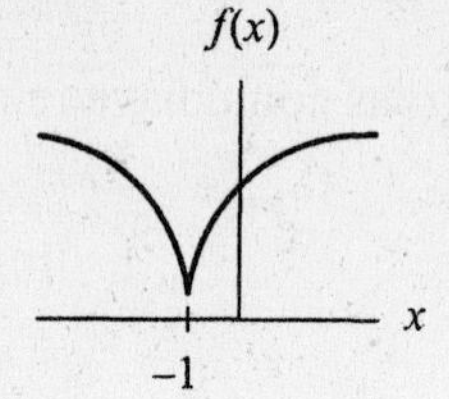

15. Answers will vary. One such graph is

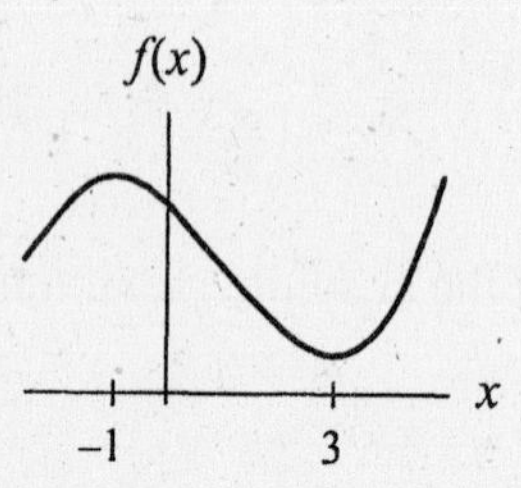

17. a. Using technology, the relative maximum value is approximately 19.888, which occurs at $x \approx 3.633$. The relative minimum value is approximately 11.779, which occurs at $x \approx 11.034$.

b. The absolute maximum and minimum are the relative maximum and minimum found in part *a*.

c.

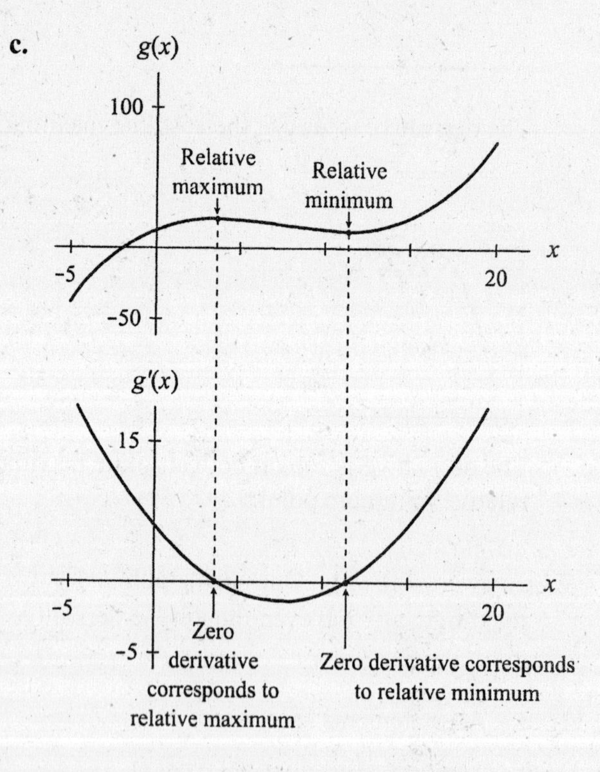

19. a. The greatest percentage of eggs hatching (about 95.6%) occurs at about 9.4°C.

b. The Fahrenheit temperature is about $\frac{9}{5}(9.4)+32 \approx 49°\text{F}$.

21. **a.** $C(0) = -0.865(0)^3 + 12.045(0)^2 - 8.952(0) + 123.02 \approx 123$ cfs

$C(24) = -16.643(24) + 539.429 \approx 140$ cfs

b. The highest flow rate is about 387.6 cfs; it occurs when $h \approx 8.9$ hours. The lowest flow rate is about 121.3 cfs; it occurs when $h \approx 0.4$ hours.

23. **a.** $S(x) = 0.181x^2 - 8.463x + 147.376$ seconds at age x years.

b. The model gives a minimum time of 48.5 seconds occurring at 23.4 years.

c. The minimum time in the table is 49 seconds, which occurs at 24 years of age.

25. **a.** A quadratic or exponential model can be used to model the data, but the exponential model may be a better choice because it does not predict that demand will increase for prices above \$40. An exponential model for the data is $R(p) = 316.765(0.949^p)$ dozen roses when the price per dozen is p dollars.

b. Multiply $R(p)$ by the price, p. The consumer expenditure is $E(p) = 316.765p(0.949^p)$ dollars spent on roses each week when the price per dozen is p dollars.

c. Using technology, we find that $E(p)$ is maximized at $p \approx 19.16$ dollars. A price of \$19.16 per dozen maximizes consumer expenditure.

d. Profit is given by $F(p) = E(p) - 6R(p) = 316.765(p-6)(0.949^p)$.
Using technology, $F(p)$ is maximized at $p \approx 25.16$ dollars. A price of \$25.16 per dozen maximizes profit.

e. Marginal values are with respect to the number of units sold or produced. In this activity, the input is price, so derivatives are with respect to price and are not marginals.

27. **a.** $G(t) = 0.008t^3 - 0.347t^2 + 6.108t + 79.690$ million tons of garbage taken to a landfill t years after 1975

b. $G'(t) = 0.025t^2 - 0.693t + 6.108$ million tons of garbage per year t years after 1975

c. In 2005 the amount of garbage was increasing by $G'(30) \approx 8.1$ million tons per year.

d.

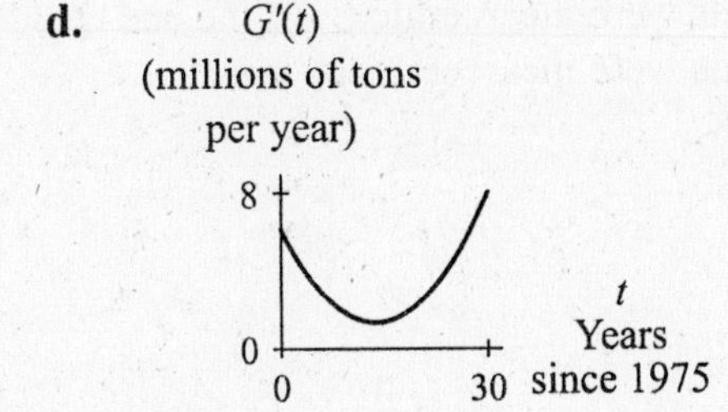

Because the derivative graph exists for all input values and never crosses the horizontal axis, $G(t)$ has no relative maxima.

29. a. Exponential model: $A(p) = 568.074(0.965582^p)$ tickets sold on average when the price is p dollars

Quadratic model: $A(p) = 0.15p^2 - 16.007p + 543.286$ tickets sold on average when the price is p dollars

The exponential model probably better reflects the probable attendance if the price is raised beyond \$35 because attendance is likely to continue to decline. (The quadratic model predicts that attendance will begin to increase around \$53.)

b. Multiply the exponential ticket function by the price, p, to obtain the revenue function.
$R(p) = 568.074p(0.965582^p)$ dollars of revenue when the ticket price is p dollars

c. Using technology, the maximum point on the revenue graph is about (28.55, 5966.86). This corresponds to a ticket price of \$28.55, which results in revenue of approximately \$5967. The resulting average attendance is approximately $A(28.55) \approx 209$.

31. The derivative of y is zero for three values of x: $x = -3.5$, $x \approx -1.049$, and $x \approx 1.549$. A graph indicates that $x = -3.5$ and $x \approx 1.549$ correspond to relative minima and $x \approx -1.049$ corresponds to a relative maximum. There are no places where the derivative is not defined. Observing the end behavior of the graph (rising infinitely on both sides) and comparing the values of y for $x = -3.5$ and $x \approx 1.549$, we conclude that there is no absolute maximum and the absolute minimum is $y = \left[2 - 3(1.549) + (1.549)^2\right](3.5 + 1.549)^2 \approx -6.312$.

33. Excel Activity

a. $N(x) = -0.049x^3 + 1.678x^2 - 2.166x + 82.151$ inmates per 100,000 residents x years after 1977.

b. The absolute minimum point is (0.665, 81.438) and the absolute maximum point is (21.973, 320.487).

c. In the middle of 1978 approximately 81.4 inmates per 100,000 residents were incarcerated. Near the end of 1999, approximately 320.5 inmates per 100,000 residents were incarcerated.

d. For the Midwest, $M(x) = -0.035646x^3 + 1.428806x^2 - 2.553844x + 107.044786$ inmates per 100,000 residents x years after 1977. The absolute minimum point is (0.926, 105.877)and the absolute maximum point is (24, 375.975). Near the end of 1978 approximately 105.9 inmates per 100,000 residents were incarcerated. At the end of 2001, approximately 376.0 inmates per 100,000 residents were incarcerated.

Section 5.3 Inflection Points

1. **a.** One visual estimate of the inflection points is (1982, 25) and (2018, 25). *Note:* There are also "smaller" inflection points at about (1921, 2.5), (1927, 2), (1930, 2), and (1935, 3).

   b. The input values of the inflection points are the years in which the rate of crude oil production is estimated to be increasing and decreasing most rapidly. We estimate that the rate of production was increasing most rapidly in 1982, when production was approximately 25 billion barrels per year, and that it will be decreasing most rapidly in 2018, when production is estimated to be approximately 25 billion barrels per year.

3. For polynomial functions, as these appear to be, you can identify the function and its derivative by noticing the number of inflection points. Because a derivative has a power one less than the original function, it will also have one less inflection point. Thus graph *b* with two inflection points is the function. Graph *a* with one inflection point is the derivative, and graph *c* with no inflection points is the second derivative.

5. Graph *c* appears to have a minimum at –1 and an inflection point at –2. Graph *b* crosses the horizontal axis at –1 and graph *a* crosses it at –2. Thus graph *c* is the function, graph *b* is the derivative, and graph *a* is the second derivative.

7. **a.** $g(x) = 0.04x^3 - 0.88x^2 + 4.81x + 12.11$

   $g'(x) = 0.12x^2 - 1.76x + 4.81$

   $g''(x) = 0.24x - 1.76$

   b. The inflection point on the graph of g is approximately (7.333, 15.834). This is a point of most rapid decline.

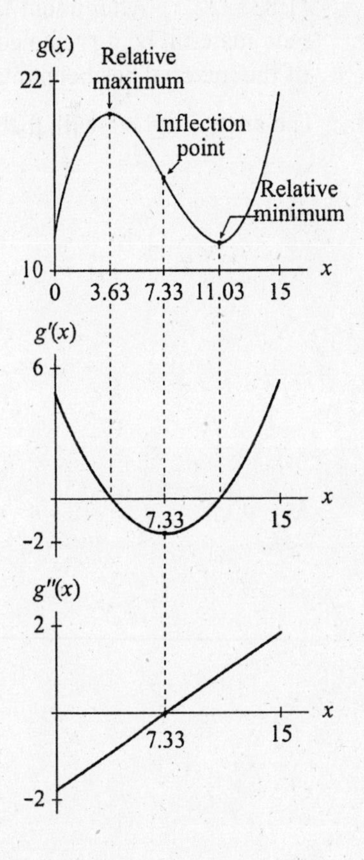

9. **a.** $P(t) = \dfrac{45}{1 + 5.94e^{-0.969125t}} = 45\left(1 + 5.94e^{-0.969125t}\right)^{-1}$ percent after studying for t hours

   $P'(t) = 45(-1)\left(1 + 5.94e^{-0.969125t}\right)^{-2}\left(5.94e^{-0.969125t}\right)(-0.969125)$

$$= 259.0471125e^{-0.969125t}\left(1+5.94e^{-0.969125t}\right)^{-2}$$ percentage points per hour after studying for t hours

$$P''(t) = 259.0471125\left(\frac{d}{dx}\left(e^{-0.969125t}\right)\right)\left(1+5.94e^{-0.969125t}\right)^{-2}$$
$$+ 259.0471125e^{-0.969125t}\left(\frac{d}{dx}\left(1+5.94e^{-0.969125t}\right)^{-2}\right)$$

$$= 259.0471125\left(e^{-0.969125t}\right)(-0.969125)\left(1+5.94e^{-0.969125t}\right)^{-2}$$
$$+ 259.0471125e^{-0.969125t}(-2)\left(1+5.94e^{-0.969125t}\right)^{-3}\left(5.94e^{-0.969125t}\right)(-0.969125)$$

$$\approx -251.049033e^{-0.969125t}\left(1+5.94e^{-0.969125t}\right)^{-2}$$
$$+ 2982.462511e^{-1.93825t}(1+5.94e^{-0.969125t})^{-3}$$ percentage points per hour per hour after studying for t hours

Solving $P''(t) = 0$ for t gives $t \approx 1.838$. The inflection point on P is approximately (1.838, 22.5). After about 1.8 hours of study (1 hour and 50 minutes), the percentage of new material being retained is increasing most rapidly. At that time, approximately 22.5% of the material has been retained.

b. The answer agrees with that given in the discussion at the end of the section.

11. **a.** $P(t) = -0.00645t^4 + 0.488t^3 - 12.991t^2 + 136.560t - 395.154$ percent when the temperature is t°C

$P'(t) = -0.0258t^3 + 1.464t^2 - 25.982t + 136.560$ percentage points per °C when the temperature is t°C

$P''(t) = -0.0774t^2 + 2.928t - 25.982$ percentage points per °C per °C when the temperature is t°C

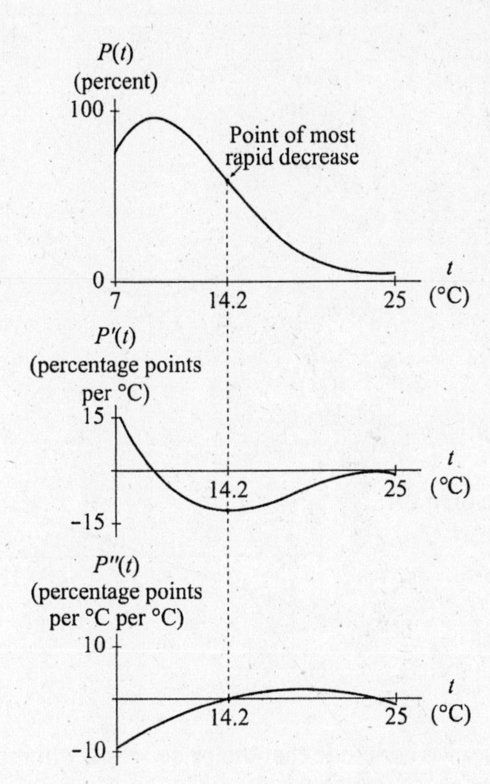

b. Because the graph of P'' crosses the t-axis twice, there are two inflection points. These are approximately (14.2, 59.4) and (23.6, 5.8). The point of most rapid decrease on the graph of P is (14.2, 59.4). (The other inflection point is a point of least rapid decrease.) The most rapid decrease occurs at 14.2°C, when 59.4% of eggs hatch. At this temperature, $P'(14.2) \approx -11.1$, so the percentage of eggs hatching is declining by 11.1 percentage points per °C. A small increase in temperature will result in a relatively large increase in the percentage of eggs not hatching.

13. a. $p(x) = 0.04265x^3 - 0.11816x^2 + 0.9674x + 3.681$ dollars x years after 1980

$p'(x) = 3(0.04265)x^2 - 2(0.11816)x + 0.9674$ dollars per year x years after 1980

$p''(x) = 6(0.04265)x - 2(0.11816)$ dollars per year per year x years after 1980.

On the derivative graph, the minimum point, approximately (9.23, –0.124), corresponds to the inflection point of the original function. On the second derivative graph, the x-intercept, approximately (9.23, 0), corresponds to the inflection point of the original function.

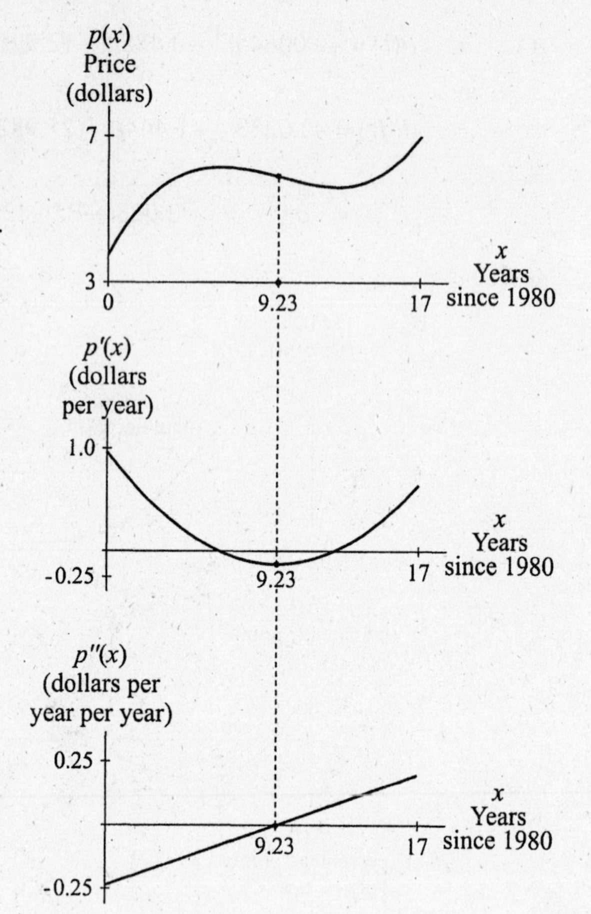

b. The x-intercept of the graph of p'' is $x \approx 9.23$. This is the input of the inflection point of p.

c. This function can be considered essentially continuous without restriction if you consider that it models the average price on a daily basis. If you assume it models a yearly average price, then it must be interpreted discretely. We are not given enough information to know for certain, so either one or the other assumption needs to be made.

In the continuous without restriction case, we conclude that the price was declining most rapidly in March of 1990 ($x \approx 9.23$) at a rate of –\$0.124 per year.

In the case of discrete interpretation, we compare $p'(9)$ and $p'(10)$ conclude that the price was declining most rapidly in 1989 at a rate of –\$0.12 per year.

d. The endpoints are the candidates for the point of most rapid increase. We compare the derivative values at the endpoints: $p'(0) \approx \$0.97$ per year and $p'(17) \approx \$0.65$ per year. We conclude that the price was increasing most rapidly at the end of 1980 at a rate of approximately \$0.97 per year.

15. $D(t) = -10.247t^3 + 208.114t^2 - 168.805t + 9775.035$ donors t years after 1975

$D'(t) = -30.741t^2 + 416.288t - 168.805$ donors per year t years after 1975

a. Using technology, we find that (0.418, 9740.089) is the approximate relative minimum point, and (13.121, 20,242.033) is the approximate relative maximum point on the cubic model.

b. The inflection point occurs where $D'(t)$ has its maximum, at $t \approx 6.770$. The inflection point is approximately (6.8, 14,991.1).

c. i. Because 6.8 is between $t = 6$ (the end of 1981) and $t = 7$ (the end of 1982), the inflection point occurs during 1982, shortly after the team won the National Championship. This is when the number of donors was increasing most rapidly.

ii. The relative maximum occurred around the same time that a new coach was hired. After this time, the number of donors declined.

17. a. $N(h) = \dfrac{62}{1+11.49e^{-0.654h}}$ components after h hours. Using the formula $\dfrac{LABe^{-Bx}}{(1+Ae^{-Bx})^2}$ for the derivative, we have

$$N'(h) = \frac{62(-0.654)(11.49)\left(e^{-0.654h}\right)}{\left(1+11.49e^{-0.654h}\right)^2} = 465.89652e^{-0.654h}\left(1+11.49e^{-0.654h}\right)^{-2}$$

components per hour after h hours.

The greatest rate occurs when $N'(h)$ is maximized, at $h \approx 3.733$ hours, or approximately 3 hours and 44 minutes after she began working.

b. Her employer may wish to give her a break after 4 hours to prevent a decline in her productivity.

19. a. $H(w) = \dfrac{10{,}111.102}{1+1153.222e^{-0.727966w}}$ total labor-hours after w weeks

b. $$H'(w) = 10{,}111.102(-1)\left(1+1153.222e^{-0.727966w}\right)^{-2}\left(1153.22e^{-0.727966w}\right)(-0.727966)$$

$$\approx 8{,}488{,}330.433e^{-0.727966w}\left(1+1153.222e^{-0.727966w}\right)^{-2}$$ labor-hours per week after w weeks

c.

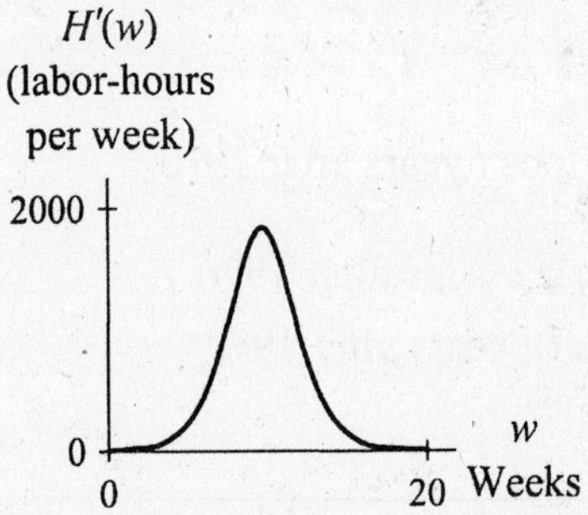

$H'(w)$ (labor-hours per week)

2000

0

0 20 Weeks w

The derivative gives the manager information about the number of labor-hours spent each week

d. The maximum point on the graph of H' is about (9.685, 1840.134). Keeping in mind that the model must be discretely interpreted, we conclude that in the tenth week, the most labor-hours are needed. That number is $H'(10) \approx 1816$ labor-hours.

e. $$H''(w) = 8{,}488{,}330.433\left(\frac{d}{dx}\left(e^{-0.727966w}\right)\right)\left(1+1153.222e^{-0.727966w}\right)^{-2}$$

$$+8{,}488{,}330.433\left(e^{-0.727966w}\right)\left(\frac{d}{dx}\left(1+1153.222e^{-0.727966w}\right)^{-2}\right)$$

$$= 8{,}488{,}330.433\left(e^{-0.727966w}\right)(-0.727966)\left(1+1153.222e^{-0.727966w}\right)^{-2}$$

$$+8{,}488{,}330.433\left(e^{-0.727966w}\right)(-2)\left(1+1153.222e^{-0.727966w}\right)^{-3}$$
$$\left(1153.222e^{-0.727966w}\right)(-0.727966)$$
$$\approx -6{,}179{,}214.512\left(e^{-0.727966w}\right)\left(1+1153.222e^{-0.727966w}\right)^{-2}$$
$$+\left(1.4252\cdot 10^{10}\right)\left(e^{-1.455932w}\right)\left(1+1153.222e^{-0.72766w}\right)^{-3}$$

Use technology to find the maximum of $H''(w)$, which occurs at $w \approx 7.876$. The point of most rapid increase on the graph of H' is (7.876, 1226.756). This occurs approximately 8 weeks into the job, and the number of labor-hours per week is increasing by approximately $H''(8) \approx 513$ labor-hours per week per week.

f. Use technology to find the minimum of the graph of H'', which occurs at $w \approx 11.494$. The point of most rapid decrease on the graph of H' is (11.494, 1226.756). This occurs approximately 12 weeks into the job when the number of labor-hours per week is changing by about $H''(12) = -486$ labor-hours per week per week.

g. By solving the equation $H'''(w) = 0$, we can find the input values that correspond to a maximum or minimum point on the graph of H'', which corresponds to inflection points on the graph of H', the weekly labor-hour curve.

h. Since the minimum of $H''(w)$ occurs about 4 weeks after the maximum of $H''(w)$, the second job should begin about 4 weeks into the first job.

21. a. Between 1980 and 1985, the average rate of change was smallest at $\dfrac{122-117}{1985-1980} = 1$ million tons per year.

b. $g(t) = 0.008t^3 - 0.347t^2 + 6.108t + 79.690$ million tons t years after 1970

c. $g'(t) = 0.025t^2 - 0.693t + 6.108$ million tons per year t years after 1970

$g''(t) = 0.051t - 0.693$ million tons per year per year t years after 1970

d.

$$g''(t) = 0$$
$$0.0507t - 0.693 = 0$$
$$0.0507t = 0.693$$
$$t \approx 13.684$$

Solving $g''(t) = 0$ gives $t \approx 13.684$, which corresponds to mid-1984. The corresponding amount of garbage is $g(13.684) \approx 120$ million tons and the corresponding rate of increase is $g'(13.684) \approx 1.4$ million tons per year.

e.

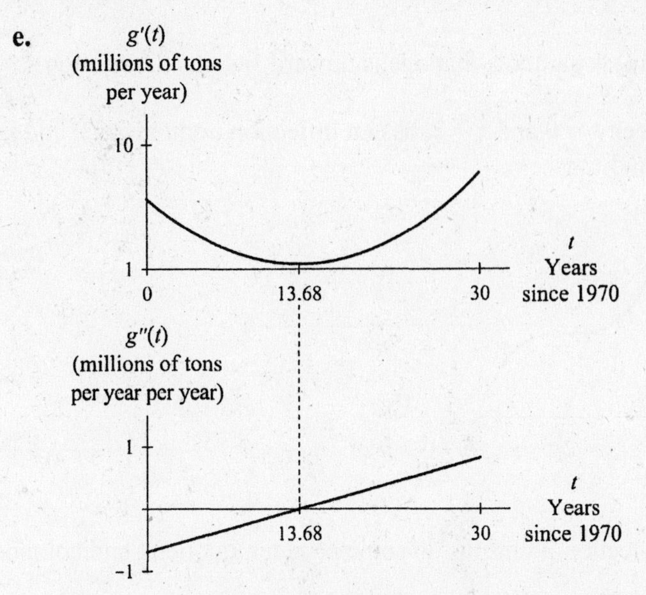

Because the graph of g'' crosses the t-axis at 13.68, we know that input corresponds to an inflection point of the graph of g. Because the graph of g' has a minimum at that same value, we know that it corresponds to a point of slowest increase on the graph of g.

f. The year with the smallest rate of change is 1984, with $g(14) \approx 120.4$ million tons of garbage, increasing at a rate of $g'(14) \approx 1.37$ million tons per year.

23. a. The first differences are greatest between 6 and 10 minutes, indicating the most rapid increase in activity.

b. $A(m) = \dfrac{1.930}{1 + 31.720e^{-0.439118m}}$ U / 100 μL m minutes after the mixture reaches 95°C

The inflection point, whose input is found using technology to locate the maximum point on a graph of A', is approximately (7.872, 0.965). After approximately 7.9 minutes, the activity was approximately 0.97 U/100 μL and was increasing most rapidly at a rate of approximately 0.212 U/100 μL/min.

25. a. $L(x) = 251.3\left(1 + 0.1376e^{0.2854x}\right)^{-1}$ million tons x years after the end of 1970

Using the formula $\dfrac{LABe^{-Bx}}{(1 + Ae^{-Bx})^2}$ for the derivative, we have

$$L'(x) = \frac{251.3(0.1376)(-0.2854)e^{0.2854x}}{\left(1 + 0.1376e^{0.2854x}\right)^2} = \frac{-9.8688e^{0.2854x}}{\left(1 + 0.1376e^{0.2854x}\right)^2} \text{ million tons per year}$$

x years after the end of 1970

b. 1970: $L'(0) \approx -7.63$ million tons per year
1995: $L'(25) \approx 0.41$ million tons per year

c. Using technology to find the input of the minimum point on the graph of L' gives $x \approx 6.95$. Because L must be discretely interpreted, we check values of L' at $x = 6$ and $x = 7$ to determine that emissions were declining most rapidly in 1977 at a rate of $L'(7) \approx 17.9$ million tons per year. At that time, yearly emissions were $L(7) \approx 124.7$ tons.

27. The graph of f is always concave up. A parabola that opens upward fits this description.

29. a. The graph is concave up between $x = 0$ and $x = 2$, has an inflection point at $x = 2$ and is concave down between $x = 2$ and $x = 4$.

b.

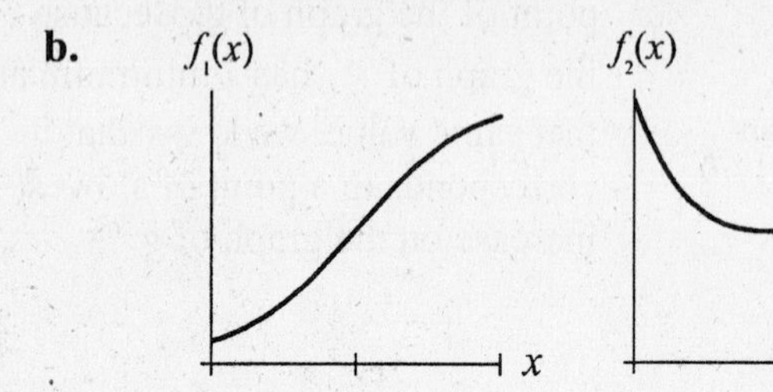

31. Quill Activity
Cubic and logistic models have inflection points, as do some product, quotient, and composite functions.

33. Excel Activity

a. $N(x) = -0.049x^3 + 1.678x^2 - 2.166x + 82.151$ inmates per 100,000 residents x years after 1977.

b. $N'(x) = -0.148x^2 + 3.356x - 2.166$ inmates per 100,000 residents per year x years after 1977.
$N''(x) = -0.297x + 3.356$ inmates per 100,000 residents per year squared x years after 1977.

c. The inflection point is approximately (11.319, 200.963). That is the point of most rapid growth occurred in 1989. The incarceration rate near the beginning of 1989 was approximately 201 inmates per 100,000 residents and was growing at a rate of 16.8 inmates per 100,000 residents per year.

d. For the Midwest, $M(x) = -0.036x^3 + 1.429x^2 - 2.554x + 107.045$ inmates per 100,000 residents x years after 1977.
$M'(x) = -0.107x^2 + 2.858x - 2.55$ inmates per 100,000 residents per year x years after 1977.
$M'(x) = -0.214x + 2.858$ inmates per 100,000 residents per year squared x years after 1977.
The inflection point is approximately (13.361, 242.968). That is the point of most rapid growth occurred in 1991. The incarceration rate near the beginning of 1991 was approximately 243 inmates per 100,000 residents and was growing at a rate of 16.5 inmates per 100,000 residents per year.

Section 5.4 Derivatives in Action

1. ***Step 1:*** Output quantity to be minimized: amount of material
   Input quantities: dimension of the tin

   Step 2:

   Bottom: $\frac{d}{2}$ inches

   Top: $\frac{d}{2}+\frac{9}{8}$ inches

   Side: πd inches by h inches

   Step 3: The equation for surface area in terms of diameter d inches and height h inches is $S=\pi dh+\pi\left(\frac{d}{2}\right)^2+\pi\left(\frac{d}{2}+\frac{9}{8}\right)^2$ square inches.

   Step 4: The volume equation is $V=\pi\left(\frac{d}{2}\right)^2 h=808.5$ cubic inches. Solving the volume equation for h in terms of d gives $h=\frac{808.5}{\pi\left(\frac{d}{2}\right)^2}=\frac{3234}{\pi d^2}$ inches. Substituting this expression into the surface area equation results in the equation we seek to optimize: $S=\frac{3234}{d}+\pi\left(\frac{d}{2}\right)^2+\pi\left(\frac{d}{2}+\frac{9}{8}\right)^2$ square inches when d is the diameter in inches.

   Step 5: Input interval: $d>0$

   Steps 6 and 7: Setting the derivative equal to zero and solving for d gives $d\approx 9.73568$, which is twice the optimal radius found in Example 3.

3. ***Step 1:*** Output quantity to be minimized: area
   Input quantities: length and width

   Step 2:

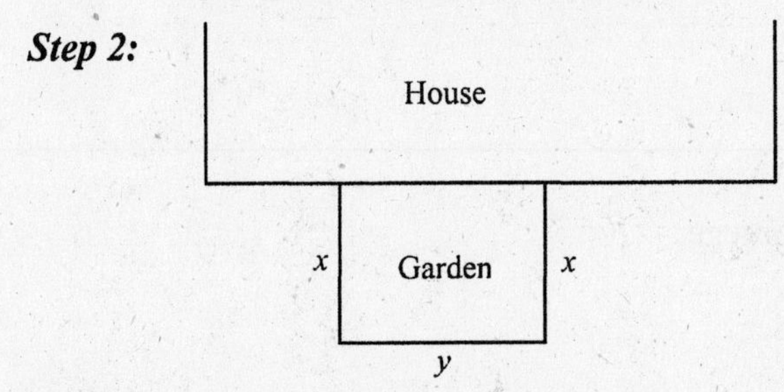

   Step 3: Area $=xy$ square feet for a width of x feet and a length of y feet.

   Step 4: The total amount of fencing is 60 feet. Thus $2x+y=60$, and $y=60-2x$.

   Substituting into the equation in Step 3 gives $A=xy=x(60-2x)=60x-x^2$ square feet where x is the width in feet.

   Step 5: Input interval: $0<x<60$

***Step 6*:** Setting the derivative equal to zero and solving for x gives $x = 15$ feet.

***Step 7*:** The maximum area is $A = 60(15) - (15)^2 = 450$ square feet, occurring when two sides are each 15 feet long and the other side is $y = 60 - 2(15) = 30$ feet long.

5. a. ***Step 1:*** Output quantity to be maximized: volume
Input quantity: length of corner cut

Step 2:

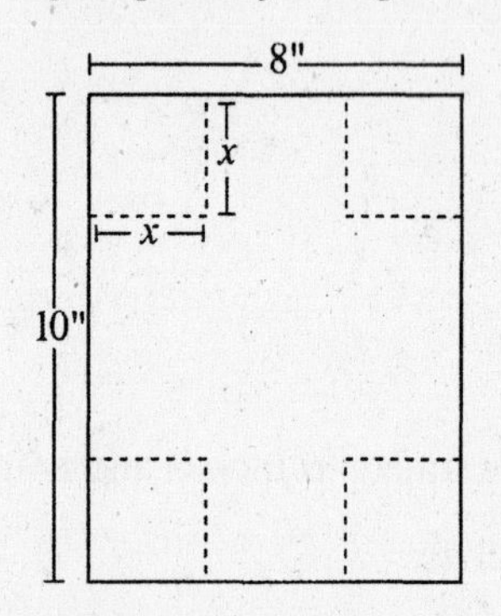

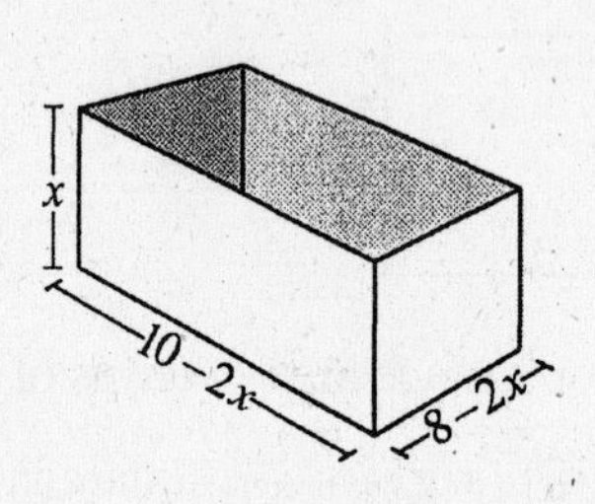

Step 3: Volume = lwh cubic inches for a length of l inches, a width of w inches, and a height of h inches

Step 4: If the length of the corner cut is x inches, then x is the height of the box, the width is $8 - 2x$, and the length is $10 - 2x$ (all dimensions in inches). Thus the volume function becomes $V = x(8-2x)(10-2x) = 80x - 36x^2 + 4x^3$ cubic inches.

Step 5: Input interval: $0 < x < 4$

Step 6: Setting the derivative equal to zero and solving for x gives $x \approx 1.47$ inches and $x \approx 4.53$ inches. The second solution is outside of the interval in Step 5.

Step 7: A corner cut of approximately 1.47 inches will result in the maximum volume of 52.5 cubic inches of sand.

b. Solving $50 = 80x - 36x^2 + 4x^3$ yields two solutions: $x \approx 1.12$ and $x \approx 1.86$. Corner cuts of approximately 1.12 inches and 1.86 inches will result in boxes with volume 50 cubic inches.

c. Answers will vary.

7. a. 42 cases

b. $\frac{42}{x}$ orders rounded up to the nearest integer

c. $\frac{42}{x}$(\$12 per order) = $\frac{42}{x}$ dollars to order

d. ($\frac{x}{2}$ cases to store)(\$4 per case to store) = $2x$ dollars to store

e. $C(x) = \frac{504}{x} + 2x$ dollars when x cases are ordered
$C(x)$ is a minimum when $x \approx 15.9$ cases. The manager must order 3 times a year. Because the order sizes are not all the same (16 cases ordered 3 times a year is more cases than the cafeteria needs), we cannot just substitute the optimal value of x into the cost equation to

obtain the total cost. There are two most cost-effective ways to order: 15, 15, and 12 cases, or 16, 16, and 10 cases (assuming that the storage fees can be calculated as $1.33 per case for 4 months). The total cost associated with each option is $63.93.

9. ***Step 1:*** Output quantity to be minimized: cost
Input quantities: height and radius

Step 2:

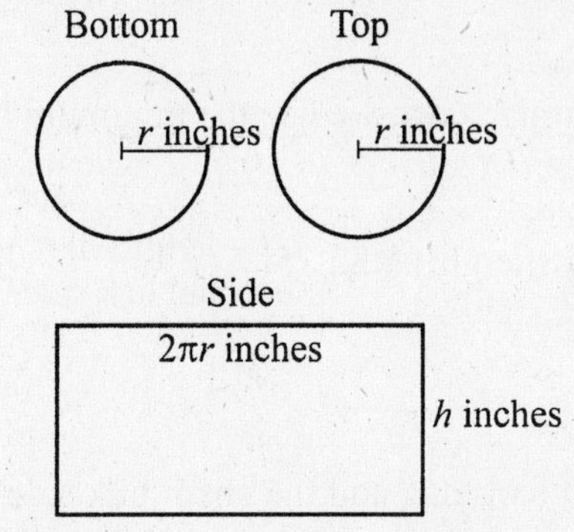

Step 3: The cost is some fixed amount multiplied by the area of the side and twice the combined area of the top and bottom. Although we do not know the fixed amount, we can still minimize the cost by minimizing the area of the side and twice the combined area of the top and bottom. We call this the cost function, but understand that the actual cost is a constant multiplied by this function. Thus we can determine the dimensions that result in minimum cost, but we cannot determine the actual cost to produce such a can.

Cost = 2(area of top and bottom) + area of side = $2(2\pi r^2) + 2\pi rh$ where r is the radius of the can and h is the height, both measured in inches.

Step 4: The volume of the can is 12 fluid ounces. Using the fact that 1 fluid ounce is 1.8047 cubic inches, we have

Volume = (12 fluid ounces)(1.8047 cubic inches per fluid ounce)

$$\pi r^2 h = 21.6564 \text{ or } h = \frac{21.6564}{\pi r^2}$$

Substituting this expression for h into the equation in Step 3 gives

$$\begin{aligned}\text{Cost} &= 2(2\pi r^2) + 2\pi rh \\ &= 4\pi r^2 + 2\pi r\frac{21.6564}{\pi r^2} \\ &= 4\pi r^2 + \frac{43.3128}{r}\end{aligned}$$

Step 5: Input interval: $r > 0$

Step 6: Setting the derivative equal to zero and solving for r gives $r = 1.199$ inches.

Step 7: A radius of approximately 1.2 inches and a height of approximately $h = \dfrac{21.6564}{\pi(1.199^2)} \approx 4.8$ inches will minimize the cost.

11. a. ***Step 1:*** Output quantity to be minimized: cost
Input quantities: length of cinderblock sides, length of chain link side

Step 2:

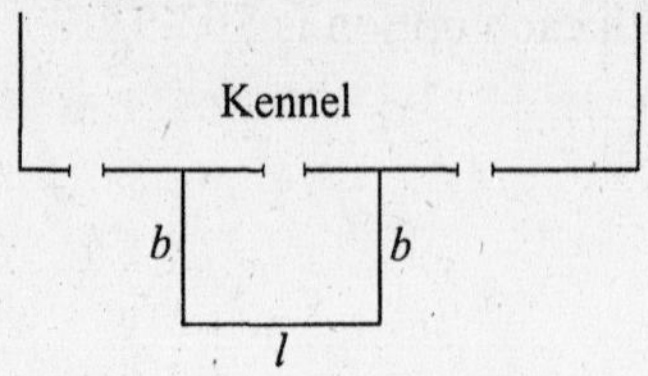

Kennel

b b

l

Step 3: Note that because we seek to minimize cost, we use the minimum height allowed for the walls and fence, in this case, 7 feet.

$$\begin{aligned}\text{Cost} &= \$0.50(2)\left(\begin{matrix}\text{length of}\\ \text{cinder block side}\end{matrix}\right)(\text{height}) + \$2.75\left(\begin{matrix}\text{length of}\\ \text{chain link side}\end{matrix}\right)(\text{height})\\ &= \$1.00(b)(7) + \$2.75(l)(7)\\ &= 7b + 19.25l \text{ dollars}\end{aligned}$$

where the cinder block walls are b feet long and the chain link side is l feet long.

Step 4: A minimum cost requires that we use the minimum area allowed, in this case, 120 square feet. Thus we have the equation $bl = 120$ or $b = \frac{120}{l}$. Substituting this expression for b in the equation in Step 3 gives

$$\text{Cost} = 7\left(\tfrac{120}{l}\right) + 19.25l = \frac{840}{l} + 19.25l$$

Step 5: Input interval: $l > 0$

Step 6: Setting the derivative equal to zero and solving for l gives $l \approx \pm 6.6$ feet. The negative solution is outside of the interval in Step 5.

Step 7: A chain link side 6.6 feet long and a cinder block side $b = \frac{120}{6.6} \approx 18.2$ feet long will result in the smallest cost.

b. If two dog runs are built side by side, there are 3 cinder block walls and two chain link fence segments. The equation in Step 3 becomes

$$\begin{aligned}\text{Cost} &= \$0.50(3)\left(\begin{matrix}\text{length of}\\ \text{cinder block side}\end{matrix}\right)(\text{height}) + \$2.75(2)\left(\begin{matrix}\text{length of}\\ \text{chain link side}\end{matrix}\right)(\text{height})\\ &= \$1.50(b)(7) + \$5.50(l)(7)\\ &= 10.5b + 38.5l \text{ dollars}\end{aligned}$$

The equation in Step 4 becomes

$$\text{Cost} = 10.5\left(\tfrac{120}{l}\right) + 38.5l = \frac{1260}{l} + 38.5l$$

Setting the derivative of this cost equation equal to zero and finding the positive solution for l gives $l \approx 5.7$ feet. Thus $b = \frac{120}{5.7} \approx 21.0$ feet. If two side-by-side dog runs are to be built for the least amount of money, each run should have cinder block sides about 21 feet long and a chain link end that is about 5.7 feet wide.

c. The answer to part *a* does not change. The answer to part *b* becomes: The chain link side should be 6 feet and the cinder block sides should be 20 feet.

13. a. Revenue = \$350 + (number of passengers)(cost per passenger)

If there are n passengers, then the cost per passenger is

\$35 + \$2(number of passengers under 44) = \$35 + \$2(44 – n)

$$R(n) = 350 + n[35 + 2(44 - n)] = 350 + 2n(44 - n) + 35n$$

$$= 350 + 123n - 2n^2 \text{ dollars when } n \text{ students go on the trip}$$

b. Sorority pays \$350 + \$2(number of passengers under 44)(number of passengers)

$S(n) = 350 + 2(44 - n)n = 350 + 88n - 2n^2$ dollars when n students go on the trip

c. Because the graph of R is a concave-down parabola, the maximum occurs where the derivative is zero, and the minimum occurs at an endpoint of the interval under consideration. In this case, the interval is $10 \le n \le 44$. Solving $R'(n) = 123 - 4n = 0$ gives $n = 30.75$. Thus either 30 or 31 passengers will maximize the bus company's revenue. We evaluate the bus company's revenue function at 10, 30, 31, and 44:

$$R(10) = \$1380,\ R(30) = \$2240,\ R(31) = \$2241,\ R(44) = \$1890$$

The bus company makes the least revenue when there are 10 passengers and the most revenue when there are 31 passengers.

d. Because the graph of S is a concave-down parabola, the maximum occurs where the derivative is zero, and the minimum occurs at an endpoint of the interval under consideration. In this case, the interval is $10 \le n \le 44$. Solving $S'(n) = 88 - 4n = 0$ gives $n = 22$. Thus 22 passengers will maximize the amount the sorority pays. We evaluate the sorority's cost function at 10 and 44 to determine the least amount the sorority pays: $S(10) = \$1030$ and $S(44) = \$350$. The sorority pays the least when the bus is full with 44 passengers.

15. a. $m(s) = -0.0015s^2 + 0.1043s + 3.5997$ mpg at a speed of s mph

b. i. $\dfrac{400 \text{ miles}}{s \text{ miles per hour}} = \dfrac{400}{s}$ hours

ii. $(\$15.50 \text{ per hour})\left(\frac{400}{s} \text{ hours}\right) = \frac{6200}{s}$ dollars

iii. $\dfrac{400 \text{ miles}}{m(s) \text{ miles per gallon}} = \dfrac{400}{m(s)} = \dfrac{400}{-0.0015s^2 + 0.0143s + 3.5997}$ gallons

iv. $(\$1.15 \text{ per gallon})\left(\dfrac{400}{m(s)} \text{gallons}\right) = \dfrac{460}{m(s)} = \dfrac{460}{-0.0015s^2 + 0.0143s + 3.5997}$ dollars

(assuming a price of \$1.15 per gallon)

v. $C(s) = \dfrac{6200}{s} + \dfrac{460}{m(s)} = \dfrac{6200}{s} + \dfrac{460}{-0.0015s^2 + 0.1043s + 3.5997}$ dollars

when the speed is s mph

c. Solving $C'(s) = \dfrac{-6200}{s^2} - \dfrac{460(-0.003s + 0.1043)}{\left(-0.0015s^2 + 0.1043s + 3.5997\right)^2} = 0$

for positive s gives $s \approx 60.3$ mph. A graph of C confirms that this values corresponds to a minimum.

d. 700 mile trip:

i. $\frac{700}{s}$ hours

ii. $\frac{10,850}{s}$ dollars

iii. $\frac{700}{m(s)}$ gallons

iv. $\frac{805}{m(s)}$ dollars

(assuming a price of $1.15 per gallon)

v. $C(s) = \frac{10,850}{s} + \frac{805}{m(s)}$ dollars

Optimal speed: 60.3 miles per hour

1200 mile trip:

i. $\frac{1200}{s}$ hours

ii. $\frac{18,600}{s}$ dollars

iii. $\frac{1200}{m(s)}$ gallons

iv. $\frac{1380}{m(s)}$ dollars

v. $C(s) = \frac{18,600}{s} + \frac{1380}{m(s)}$ dollars

Optimal speed: 60.3 miles per hour

The optimal speed remains constant. It does not depend on trip length.

e. When the price of gas is $1.35 per gallon, the total cost of fuel becomes ($1.35 per gallon)$\left(\frac{400}{m(s)}\text{gallons}\right) = \frac{540}{m(s)}$ dollars and the cost function is $C(s) = \frac{6200}{s} + \frac{540}{m(s)}$ dollars. Similar changes in the cost function can be made for other fuel prices, giving the following optimal speeds:

Price of gas:	Optimal speed for 400 mile trip:
$1.35 / gallon	58.8 mph
$1.55 / gallon	57.5 mph
$1.75 / gallon	56.4 mph

As the gas price increases, the optimal speed decreases.

f. When the driver's hourly wages change to $17.50 per hour, the total amount paid changes to ($17.50 per hour)$\left(\frac{400}{s}\text{hours}\right) = \frac{7000}{s}$ dollars and the cost function becomes $C(s) = \frac{7000}{s} + \frac{460}{m(s)}$ dollars. Similar changes in the cost function can be made for other hourly wages, giving the following optimal speeds:

Wages of driver:	Optimal speed for 400 mile trip:
$17.50 / hour	61.4 mph
$20.50 / hour	62.9 mph
$25.50 / hour	64.9 mph

As the wages increase, the optimal speed increases

17. Quill Activities

Section 5.5 Interconnected Change: Related Rates

1. $\frac{df}{dt} = 3\frac{dx}{dt}$

3. $\frac{dk}{dy} = 12x\frac{dx}{dy}$

5. $\frac{dg}{dt} = 3e^{3x}\frac{dx}{dt}$

7. $\frac{df}{dt} = 62(\ln 1.02)(1.02^x)\frac{dx}{dt}$

9.
$$\frac{dh}{dy} = 6\frac{da}{dy} + 6\ln a\frac{da}{dy}$$
$$= 6(1+\ln a)\frac{da}{dy}$$

11.
$$\frac{ds}{dt} = \pi r\frac{1}{2}\left(r^2+h^2\right)^{-1/2}(2h)\frac{dh}{dt}$$
$$= \frac{\pi rh}{\sqrt{r^2+h^2}}\frac{dh}{dt}$$

13. Use the Product Rule with πr as the first term and $\sqrt{r^2+h^2}$ as the second term.
$$0 = \pi r\frac{1}{2}\left(r^2+h^2\right)^{-1/2}\left(2r\frac{dr}{dt}+2h\frac{dh}{dt}\right)+\left(\pi\frac{dr}{dt}\right)\sqrt{r^2+h^2}$$
$$0 = \frac{\pi r}{\sqrt{r^2+h^2}}\left(r\frac{dr}{dt}+h\frac{dh}{dt}\right)+\pi\sqrt{r^2+h^2}\,\frac{dr}{dt}$$

15. **a.** $w = 31.54 + 12.97\ln 5 \approx 52.4$ gallons per day

b. $\frac{dw}{dt} = \frac{12.97}{g}\cdot\frac{dg}{dt} = \frac{12.97}{5}\left(\frac{2}{12}\text{ inches per year}\right) \approx 0.43$ gallon per day per year

The amount of water transpired is increasing by approximately 0.43 gallon per day per year. In other words, in one year, the tree will be transpiring about 0.4 gallon more each day than it currently is transpiring.

17. **a.** $B = \frac{0.45(100)}{0.00064516h^2} = \frac{45}{0.00064516h^2}$ points

b. $\frac{dB}{dt} = \frac{45}{0.00064516}\left(-2h^{-3}\right)\frac{dh}{dt} = \frac{45}{0.00064516h^3}\frac{dh}{dt}$

c. Evaluate the equation in part *b* at $h = 63$ inches and $\frac{dh}{dt} = 0.5$ inch per year to obtain $\frac{dB}{dt} \approx -0.2789$ point per year.

19. a. We know $h = 32$ feet, $d = \frac{10}{12}$ foot, $\frac{dh}{dt} = 0.5$ foot per year, and we wish to find $\frac{dV}{dt}$. We treat d as a constant and find the derivative with respect to time t to obtain the related rates equation $\frac{dV}{dt} = 0.002198d^{1.739925}1.133187h^{0.133187}\frac{dh}{dt}$.

Substituting the values given above results in $\frac{dV}{dt} \approx 0.0014$ cubic foot per year.

b. We know $h = 34$ feet, $d = 1$ foot, $\frac{dd}{dt} = \frac{2}{12}$ foot per year, and we wish to find $\frac{dV}{dt}$. We treat h as a constant and find the derivative with respect to time t to obtain the related rates equation $\frac{dV}{dt} = 0.002198(1.739925d^{0.739925})h^{1.133187}\frac{dd}{dt}$.

Substituting the values given above results in $\frac{dV}{dt} \approx 0.0347$ cubic foot per year.

21. a. $L = \left(\frac{M}{48.10352K^{0.4}}\right)^{5/3} = \left(\frac{M}{48.10352}\right)^{5/3} K^{-2/3}$

b. $\frac{dL}{dt} = \left(\frac{M}{48.10352}\right)^{5/3}\left(\frac{-2}{3}K^{-5/3}\right)\frac{dK}{dt}$

c. We are given $K = 47$ and $\frac{dK}{dL} = 0.5$. Using the fact that $L = 8$ and the original equation, we can find the value of M corresponding to the current levels of labor and capital: $M \approx 781.39$. Substituting the known values into the equation in part *b* gives $\frac{dL}{dt} \approx -0.057$ thousand worker-hours per year.

23.

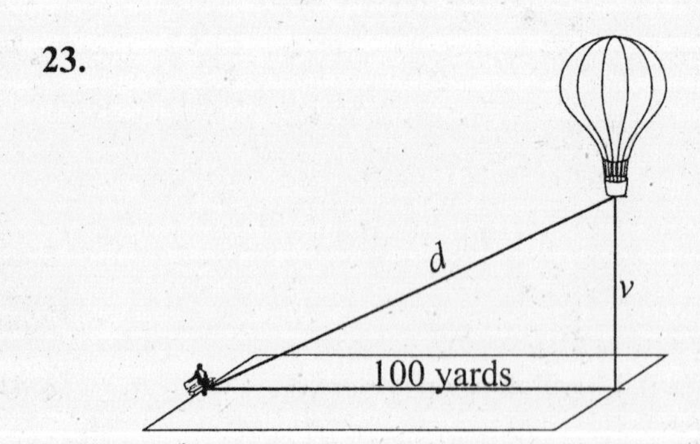

We are told that $\frac{dV}{dt} = 2$ feet per second and v = (500 yards)(3 feet per yard) = 1500 feet, and we need to know $\frac{dd}{dt}$. Converting 100 yards to feet and using the Pythagorean Theorem, we know that $v^2 + 300^2 = d^2$. Taking the derivatives of both sides with respect to time gives $2v\frac{dv}{dt} + 0 = 2d\frac{dd}{dt}$.

To solve for $\frac{dd}{dt}$, we need to know the value of d when v = 1500 feet. Use the Pythagorean Theorem: $1500^2 + 300^2 = d^2$ to find that $d \approx 1529.71$ feet. Thus we have

$$2v\frac{dv}{dt} = 2d\frac{dd}{dt}$$

$$9(1500)(2) = 2(1529.71)\frac{dd}{dt}$$

$$\frac{dd}{dt} \approx 1.96 \text{ feet per second}$$

The balloon is approximately 1529.7 feet from the observer, and that distance is increasing by approximately 1.96 feet per second.

25.

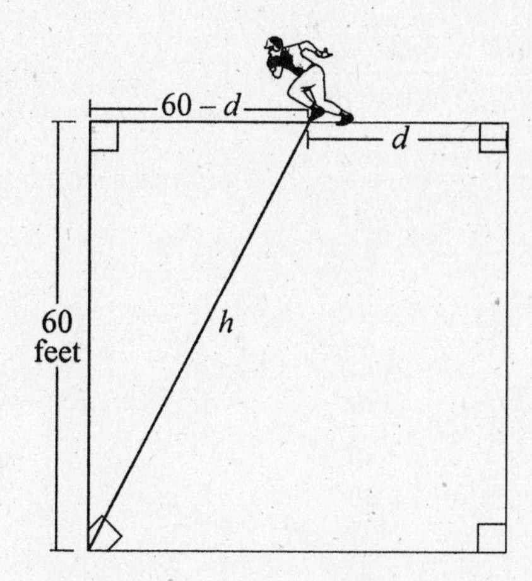

We are told that $\frac{dd}{dt} = 22$ feet per second and $d = 30$ feet. We wish to find $\frac{dh}{dt}$. We use the Pythagorean Theorem: $h^2 = 60^2 + (60-d)^2$ to obtain the related rates equation:

$$2h\frac{dh}{dt} = 0 + 2(60-d)(-1)\frac{dd}{dt}$$

To find the value of h, we substitute $d = 30$ into the Pythagorean Theorem: $h^2 = 60^2 + 30^2$

$$h \approx 67.08$$

Thus we have

$$2h\frac{dh}{dt} = -2(60-d)\frac{dd}{dt}$$

$$2(67.08)\frac{dh}{dt} = -2(60-30)(22)$$

$$\frac{dh}{dt} \approx \frac{1320}{2(67.08)} \approx 9.8 \text{ feet per second}$$

The runner is approximately 67.1 feet from home plate, and that distance is decreasing by about 9.84 feet per second.

27. a. The volume of a sphere with radius r centimeters is given by the formula $V = \frac{4}{3}\pi r^3$ cubic centimeters. When $r = 10$, $V \approx 4188.79$ cm^3.

b. Differentiating the volume equation with respect to time t yields $\frac{dV}{dt} = \frac{4}{3}(3\pi r^2)\frac{dr}{dt}$.

We find $\frac{dr}{dt}$ as the average rate of change between the points (0, 12) and (30, 8):

$$\frac{dr}{dt} = \frac{8-12}{30-0} = \frac{-4}{30} \text{ cm per minute}$$

Substituting $r = 10$ and $\frac{dr}{dt} = \frac{-4}{30}$ into the related rates equation, we find that

$$\frac{dV}{dt} = \frac{4}{3}(3\pi(10^2))\frac{-4}{30} \approx -167.6 \text{ cm}^3 \text{ per minute}$$

29.

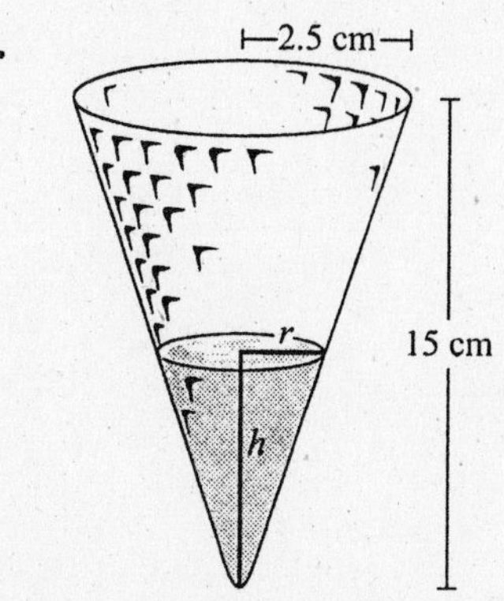

We are told $\frac{dV}{dt} = \frac{1\text{T}}{\text{sec}} = \frac{1\text{T}}{\text{sec}}\left(\frac{1\text{ cm}^3}{0.06\text{T}}\right) = \frac{1}{0.06}\text{cm}^3$ per second

We wish to find $\frac{dh}{dt}$. The volume of a cone with radius r and height h, both in centimeters is $V = \frac{\pi r^2 h}{3}\text{cm}^3$. Because of similar triangles, we know that $\frac{15}{2.5} = \frac{h}{r}$ or $r = \frac{h}{6}$. Substituting this expression into the volume equation gives volume in terms of height: $V = \dfrac{\pi\left(\frac{h}{6}\right)^2 h}{3} = \dfrac{\pi h^3}{108}$

Differentiating both sides with respect to time t gives $\dfrac{dV}{dt} = \dfrac{3\pi h^2}{108}\dfrac{dh}{dt}$.

When $h = 6$ cm and $\dfrac{dV}{dt} = \dfrac{1}{0.06}$, $\dfrac{1}{0.06} = \dfrac{3\pi(6^2)}{108}\dfrac{dh}{dt}$ which gives $\dfrac{dh}{dt} \approx 4.34$ cm per second

31. Begin by solving for h: $h = \dfrac{V}{\pi r^2} = \dfrac{V}{\pi} r^{-2}$

Differentiate with respect to t (V is a constant): $\dfrac{dh}{dt} = \dfrac{V}{\pi}\left(-2r^{-3}\right)\dfrac{dr}{dt}$

Substitute $\pi r^2 h$ for V: $\dfrac{dh}{dt} = \dfrac{\pi r^2 h}{\pi}\left(-2r^{-3}\right)\dfrac{dr}{dt}$

Simplify: $\dfrac{dh}{dt} = \dfrac{-2h}{r}\dfrac{dr}{dt}$

Rewrite: $\dfrac{dr}{dt} = \dfrac{r}{-2h}\dfrac{dh}{dt}$

Chapter 5 Review Test

1. a. T has a relative maximum point at (0.682, 143.098) and a relative minimum point at (3.160, 120.687). These points can be determined by finding the values of x between 0 and 6 at which the graph of T' crosses the x-axis. (There is also a relative maximum to the right of $x = 6$.)

b. T has two inflection points: (1.762, 132.939) and (5.143, 149.067). These points can be determined by finding the values of x between 0 and 6 at which the graph of T'' crosses the x-axis. These are also the points at which T' has a relative maximum and relative minimum.

c.

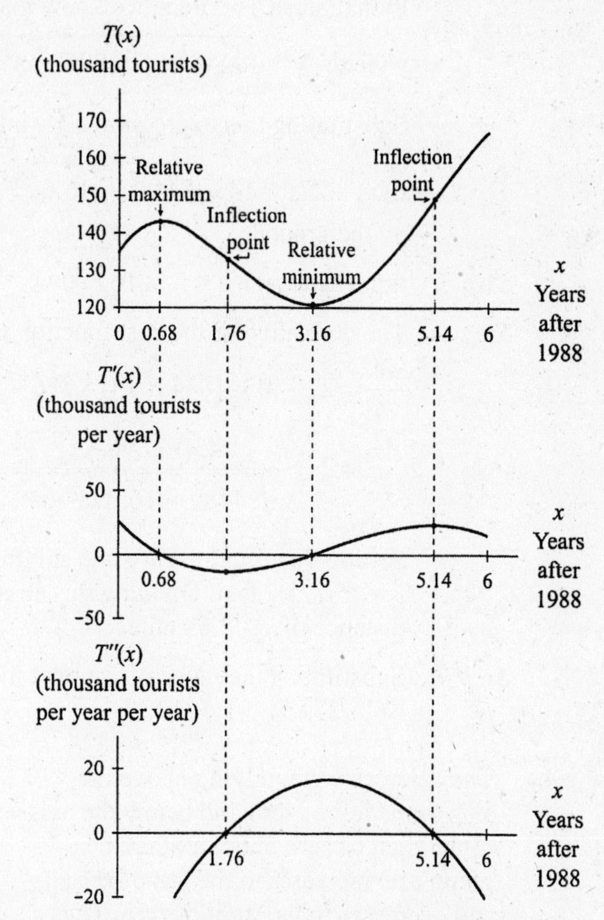

d. To determine the absolute maximum and minimum, we compare the outputs of the relative extrema with the outputs at the endpoints $x = 0$ and $x = 6$. The number of tourists was greatest in 1994 at 166.8 thousand tourists. The number was least in 1991 at 120.9 thousand.

e. To determine the greatest and least slopes, we compare the slopes at the inflection points with the slopes at the endpoints $x = 0$ and $x = 6$. The number of tourists was increasing most rapidly in 1993 at a rate of 23.1 thousand tourists per year. The number of tourists was decreasing most rapidly in 1990 at a rate of 13.3 thousand tourists per year.

2. a. (4.5 thousand people per year)$\left(\frac{1}{4}\text{ year}\right)$= 1.125 thousand people

b. $202 + \frac{1}{2}(4.5) = 204.25$ thousand people

3. ***Step 1:*** Output quantity to be minimized: cost

Input quantities: distances x and y

Step 2: See Figure 5.27 in the Chapter 5 Review Test.

Step 3: Cost = $27x + 143y$ dollars for distances of x feet and y feet.

Step 4: Convert the distances in miles in the figure to distances in feet using the fact the

1 mile = 5280 feet: 3.2 miles = 16,896 feet and 1.6 miles = 8448 feet. Using the Pythagorean Theorem, we know that $8448^2 + (16{,}890 - x)^2 = y^2$. Solving for positive y yields $y = \sqrt{8448^2 + (16{,}896 - x)^2}$.

Substituting this expression for y into the equation in Step 3 gives

$C(x) = 27x + 143\sqrt{8448^2 + (16{,}896 - x)^2}$ dollars where x is the distance the pipe is run on the ground.

Step 5: Input interval: $0 < x < 16{,}890$

Step 6: The derivative of the cost function is

$$\frac{dC}{dx} = 27 + 143\tfrac{1}{2}[8448^2 + (16{,}896 - x)^2]^{-1/2}2(16{,}896 - x)(-1)$$

$$= 27 - \frac{143(16{,}896 - x)}{\sqrt{8448^2 + (16{,}896 - x)^2}}$$

Setting this equal to zero and solving for x between 0 and 16,890 gives $x = 15{,}271.7$ feet. Dividing this answer by 5280 feet per mile, we have an optimal distance of $x \approx 2.89$ miles

Step 7: Substituting the value of x in feet into the cost equation gives a cost of $C(15{,}271.7) \approx 1{,}642{,}527$.

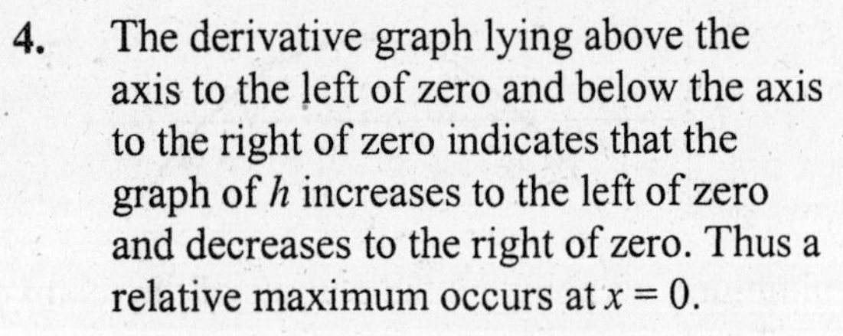

4. The derivative graph lying above the axis to the left of zero and below the axis to the right of zero indicates that the graph of h increases to the left of zero and decreases to the right of zero. Thus a relative maximum occurs at $x = 0$.

The derivative graph indicates a maximum slope of h between $x = a$ and $x = 0$ and a minimum slope of between $x = 0$ and $x = b$. These points of extreme slope are inflection points on the graph of h.

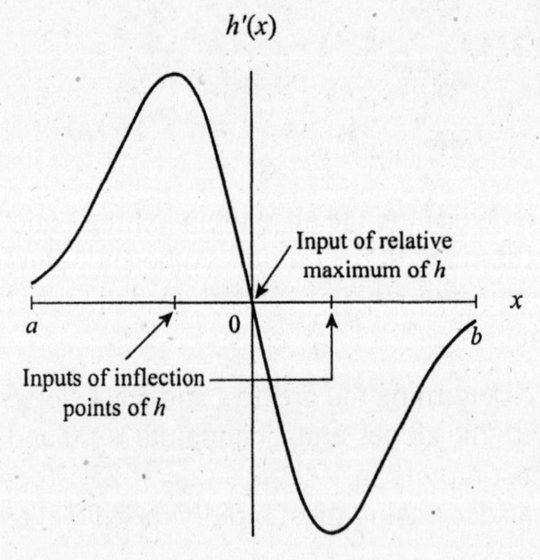

5. Treating w as a constant and differentiating the function S with respect to time yields $\frac{dS}{dt} = 0.000013w(2v)\frac{dv}{dt}$. We have $w = 4000$ pounds, $v = 60$ mph, and $\frac{dv}{dt} = -5$ mph per second. Substituting these values into the related rates equation gives

$\frac{dS}{dt} = 0.000013(4000)(2 \cdot 60)(-5) = -31.2$ feet per second.

The length of the skid marks is decreasing by 31.2 feet per second.

Chapter 6

Section 6.1 Results of Change and Area Approximations

1. **a.** The heights are in thousands of bacteria per hour.

   b. The widths are in hours.

   c. Because the area of each rectangle is a product of thousands of bacteria per hour and hours, its units are thousands of bacteria.

   d. The area has the same units as the areas of the rectangles, that is, thousands of bacteria.

   e. The accumulated change is a number of bacteria. Its units will be thousands of bacteria.

3. **a.** The area would represent how much farther a car going 60 mph will travel before stopping than a car going 40 mph.

   b. i. The heights are in feet per mile per hour, and the widths are in miles per hour.

   ii. Because the area of each rectangle is a product of feet per mph and mph, each rectangle area (and hence the total area) is in feet.

5. **a.** On the horizontal axis, mark integer values of x between 0 and 8. Construct a rectangle with width from $x = 0$ to $x = 1$ and height $f(1)$. Because the width of the rectangle is 1, the area will be the same as the height. Repeat the rectangle constructions between each pair of consecutive integer input values. Note that for the rectangles that lie below the horizontal axis, the heights are the absolute values of the function values. Also note that the fourth rectangle has height 0.

   Sum the areas (heights) of the rectangles to obtain the area estimate.

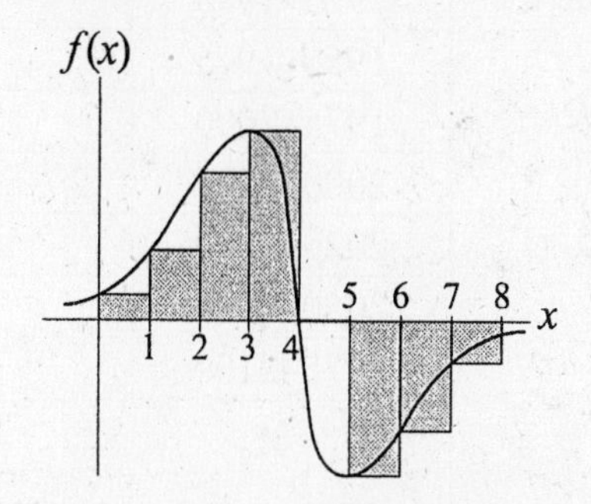

   b. Repeat part a, except the height of each rectangle is determined by the function value corresponding to the left side of the interval. In the case of the first rectangle, the height is $f(0)$. When we use left rectangles, the fifth rectangle has height 0.

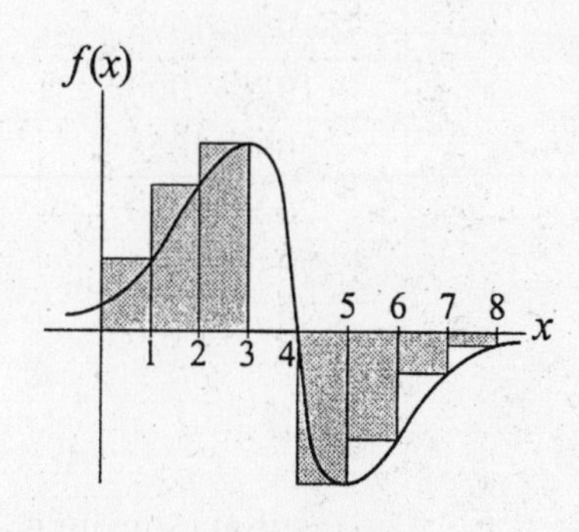

7. Divide the interval from a to b into four equal subintervals. Determine the midpoint of each subinterval and substitute the midpoints into the function to find the heights of the rectangles. Determine the area of each rectangle by multiplying the height by the width of the subintervals. Add the four areas to obtain the midpoint-rectangle estimate.

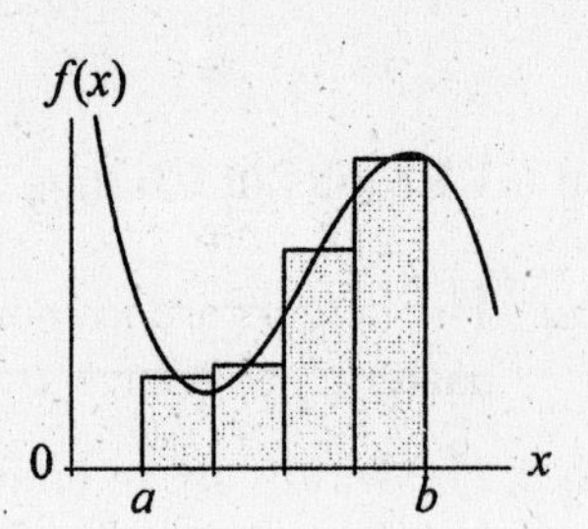

9. Possible solution:

a. Each rectangle has a width of 25 years.

Interval	Midpoint height (billion barrels per year)	Midpoint-rectangle area (25 years)(height)→ (billions of barrels)
1900–1925	0	0.0
1925–1950	3	75.0
1950–1975	9	225.0
1975–2000	33	825.0
2000–2025	33	825.0
2025–2050	9	225.0
2050–2075	3	75.0
2075–2100	1	25.0
		Total area of rectangles ≈ 2275 billion barrels **Total oil production ≈ 2275 billion barrels**

b. Each rectangle has a width of 25 years.

Interval	Midpoint height (billion barrels per year)	Midpoint-rectangle area (25 years)(height)→ (billions of barrels)
1900–1925	0	0.0
1925–1950	3	75.0
1950–1975	9	225.0
1975–2000	24	825.0
2000–2025	13	825.0
2025–2050	4	225.0
2050–2075	1	75.0
2075–2100	0	25.0
		Total area of rectangles ≈ 1625 billion barrels **Total oil production ≈ 1625 billion barrels**

c. The graph A estimate is 175 billion barrels above the journal's estimate. The graph B estimate is 275 billion barrels above the journal's estimate.

d. Oil production will decline as the world's resources are used up.

11. a.

Velocity (feet per minute)

880

0

0 1 3 3.5

Time (minutes)

b. The region can be divided into two triangles and one rectangle. The area is calculated as

Area of triangle + area of rectangle + area of triangle

$= \frac{1}{2}(\text{base})(\text{height}) + (\text{length})(\text{width}) + \frac{1}{2}(\text{base})(\text{height})$

$= \frac{1}{2}(1 \text{ min})(880 \text{ feet/min}) + (2 \text{ min})(880 \text{ feet/min}) +$

$\frac{1}{2}(0.5 \text{ min})(880 \text{ feet/min})$

$= 440 \text{ feet} + 1760 \text{ feet} + 220 \text{ feet} = 2420 \text{ feet}$

c. The area in part *b* is the distance the robot traveled. The robot traveled 2420 feet during the $3\frac{1}{2}$-minute experiment.

13. a.

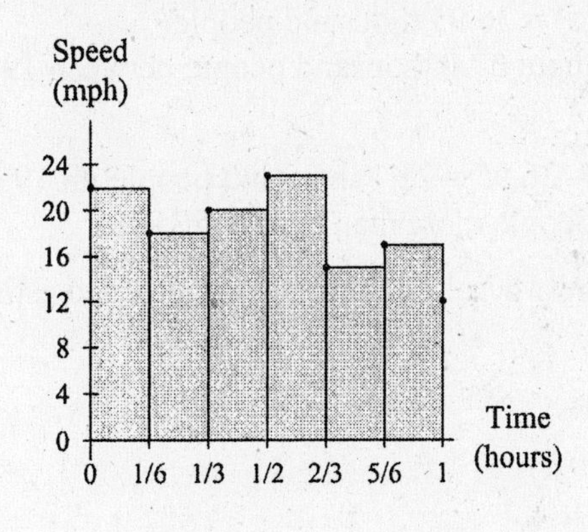

Note that we convert the times to hours to correspond with the speeds given in miles per hour: $10 \text{ minutes}\left(\frac{1 \text{ hour}}{60 \text{ minutes}}\right) = \frac{1}{6} \text{ hour}$. The heights of left rectangles are simply the output data values except the last one. You can calculate the left-rectangle sum by adding first six speed values and multiplying by $\frac{1}{6}$:

Distance traveled

$\approx \frac{1}{6}(22 + 18 + 20 + 23 + 15 + 17) = \left(\frac{1}{6} \text{ hour}\right)(115)$

≈ 19.2 miles

b.

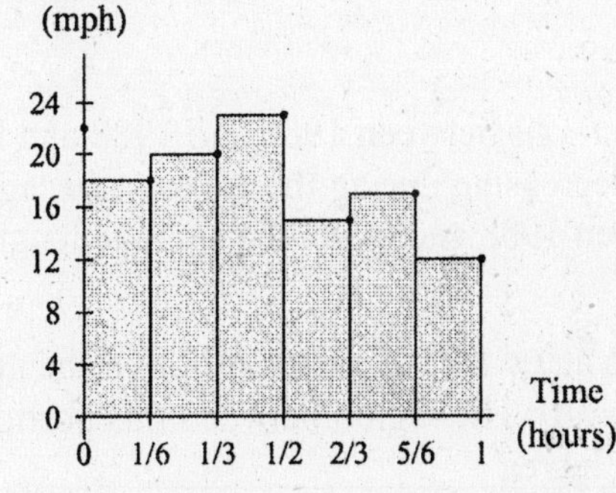

The heights of right rectangles are simply the output data values except the first one. You can calculate the right-rectangle sum by adding last six speed values and multiplying by $\frac{1}{6}$:

Distance traveled

$\approx \frac{1}{6}(18 + 20 + 23 + 15 + 17 + 12) = \left(\frac{1}{6} \text{ hour}\right)(105)$

≈ 17.5 miles

15. a.

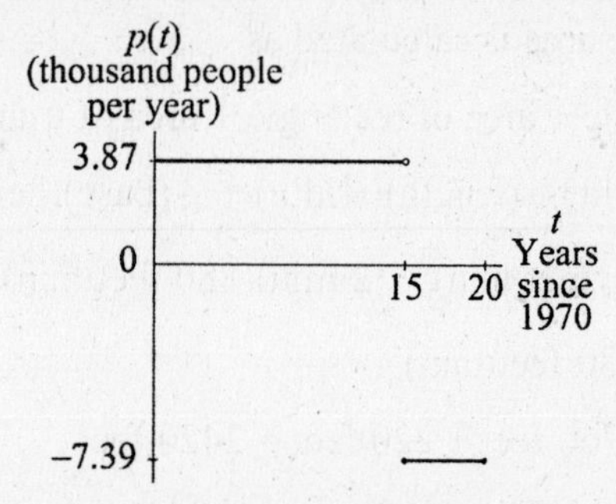

b. Area = (3.87 thousand people per year)(15 years) = 58.05 thousand people
The population of North Dakota grew by about 58,100 people between 1970 and 1985.

c. Area = (7.39 thousand people per year)(5 years) = 36.95 thousand people
The population of North Dakota declined by about 37.0 thousand people between 1985 and 1990.

d. The net change from 1970 to 1990 was 58.05 – 36.95 = 21.1 thousand people. In 1990, the population was approximately 21.1 thousand people more than it was in 1970.

e. You would need the population in some year between 1970 and 1990 in order to estimate the 1990 population.

17. a.

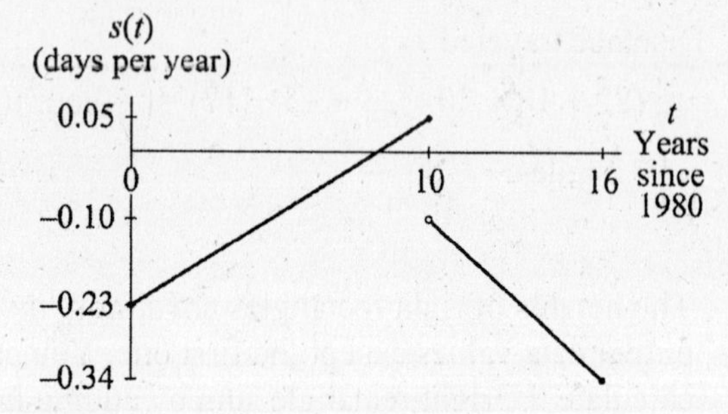

b. Because the rate-of-change graph lies below the t-axis between 1980 and 1988 and between 1990 and 1996, the average length of stay was decreasing during those years. Because the rate-of-change graph lies above the t-axis between 1988 and 1996, the average length of stay was increasing during those years.

c. The left portion of the function crosses the t-axis at $t \approx 8.2143$. The portion of the graph above the axis begins at this value and ends at $t = 10$. The region between this portion of the graph and the horizontal axis is a triangle with area

$$\text{Area} = \tfrac{1}{2}(\text{base})(\text{height})$$
$$\approx \tfrac{1}{2}(10 - 8.2143)(0.028(10) - 0.23) = \tfrac{1}{2}(1.7857)(0.05) \approx 0.0446$$

Between early 1989 and the end of 1990, the average hospital stay increased by about 0.04 day.

d. The region lying below the axis is comprised of a triangle and a trapezoid. The height of the triangle is the absolute value of $s(0)$. The heights of the trapezoid are the absolute values of $s(16)$ and the right function evaluated at $t = 10$, even though this is not the function value at $t = 10$. The combined area is

$$\text{Area} = \tfrac{1}{2}(\text{base})(\text{height}) + \frac{\text{left height} + \text{right height}}{2}(\text{width})$$

$$\approx \tfrac{1}{2}(8.2143)\left|0.028(0) - 0.23\right| + \frac{\left|-0.0408(10) + 0.30833\right| + \left|-0.0408(16) + 0.30833\right|}{2}(6)$$

$$= \tfrac{1}{2}(8.2143)\left|-0.23\right| + \frac{\left|-0.1\right| + \left|-0.3448\right|}{2}(6)$$

$$= \tfrac{1}{2}(8.2143)(0.23) + \frac{0.1 + 0.3448}{2}(6) \approx 2.28$$

Between 1980 and early 1989 and between 1990 and 1996, the average hospital stay declined by about 2.28 days.

e. The next change in hospital stay between 1980 and 1996 was about $-2.28 + 0.04 = -2.24$. The average hospital stay in 1996 was about 2.24 days shorter than the average stay in 1980. We cannot use this information to determine the average hospital stay in 1996 because we do not know the average stay in 1980.

19. a. The total number of births in 1985, 1986, 1987, and 1988 can be found by summing the yearly totals for each of those year:

$$1162 + 1257 + 1375 + 1427 = 5221$$

From the beginning of 1985 through the end of 1988, there were 5221 live births to U.S. women 45 years of age and older. Disregarding reporting error, this answer is exact.

Graphically, this answer is the area of right rectangles.

b. $B(x) = 0.303x^3 - 17.497x^2 + 119.958x + 5343.469$ births x years after 1950

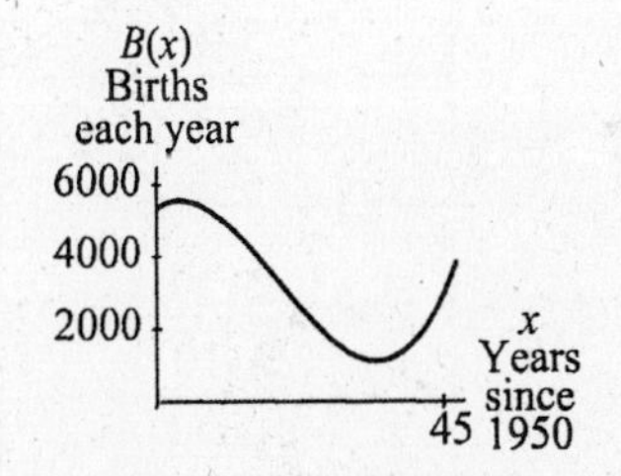

c. Because the time period is 30 years, 30 rectangles are necessary. Because the width of each rectangle is 1 year, the areas of the rectangles are $B(15), B(16), B(17), \ldots, B(44)$.

$B(15) \approx 4229.4$	$B(23) \approx 2535.9$	$B(31) \approx 1280.9$	$B(39) \approx 1395.8$
$B(16) \approx 4025.5$	$B(24) \approx 2335.9$	$B(32) \approx 1201.2$	$B(40) \approx 1552.9$
$B(17) \approx 3815.8$	$B(25) \approx 2144.6$	$B(33) \approx 1144.8$	$B(41) \approx 1747.8$
$B(18) \approx 3602.1$	$B(26) \approx 1963.8$	$B(34) \approx 1113.4$	$B(42) \approx 1982.3$
$B(19) \approx 3386.0$	$B(27) \approx 1795.3$	$B(35) \approx 1108.9$	$B(43) \approx 2258.2$
$B(20) \approx 3169.6$	$B(28) \approx 1641.0$	$B(36) \approx 1133.0$	$B(44) \approx 2577.3$
$B(21) \approx 2954.5$	$B(29) \approx 1502.6$	$B(37) \approx 1187.7$	
$B(22) \approx 2742.7$	$B(30) \approx 1381.9$	$B(38) \approx 1274.7$	

Adding these values, we estimate the number of births to be 62,898. (Note that technology can be used to perform this calculation automatically.)

d. To find the exact answer to the question in part *b*, we need exact data for each year between 1965 and 1995.

21. a. To convert the air speeds to miles per second, multiply by 1.15 mph per knot and divide by 3600 seconds per hour.

Time (seconds)	Air speed (miles per second)
0	0.00000
10	0.01278
15	0.01917
20	0.02236
25	0.02556

b. $S(t) = -\left(2.108 \cdot 10^{-5}\right)t^2 + \left(1.554 \cdot 10^{-3}\right)t - 9.840 \cdot 10^{-5}$ miles per second t seconds after the plane began to taxi

c. We use 18 midpoint rectangles. Answers may vary.

Interval (sec)	Midpoint t (sec)	Height $S(t)$ (mi/sec)	Area (1 sec)(miles/sec) → (miles)
0–1	0.5	0.000673	0.000673
1–2	1.5	0.002185	0.002185
2–3	2.5	0.003656	0.003656
3–4	3.5	0.005084	0.005084
4–5	4.5	0.006469	0.006469
5–6	5.5	0.007813	0.007813
6–7	6.5	0.009114	0.009114
7–8	7.5	0.010374	0.010374
8–9	8.5	0.011591	0.011591
9–10	9.5	0.012766	0.012766
10–11	10.5	0.013899	0.013899
11–12	11.5	0.014990	0.014990
12–13	12.5	0.016038	0.016038
13–14	13.5	0.017045	0.017045
14–15	14.5	0.018009	0.018009
15–16	15.5	0.018931	0.018931
16–17	16.5	0.019811	0.019811
17–18	17.5	0.020649	0.020649
		Sum of midpoint rectangles	**≈ 0.209095 mile**

The area is about 0.209 mile.

d. It took approximately 0.2 mile of runway for the Cessna to taxi for takeoff (assuming no headwind).

23. a. See figure in Answer Key page A-43 of Text.

b. The percentage of low birthweight babies was declining as the mother's weight gain increased from 18 to 43 pounds.

c. The signed area is approximately -5.3. When the mother gained between 18 and 43 pounds during pregnancy, the percentage of low birthweight babies decreased by approximately 5.3 percentage points.

d. We cannot answer this question with the information given.

25. a. The life expectancies were always rising. Declining positive rate-of-change data indicate that life expectancies were increasing at a slower and slower rate.

b. $E(t) = (4.199 \cdot 10^{-4})t^2 - 0.022t + 0.359$ years of life expectancy per year t years after 19

c. The width of each rectangle is $\frac{2010-1970}{8} = 5$ years.

Interval (years)	Midpoint t (years)	Height $E(t)$ (years of life expectancy per year)	Area (5 years)$E(t)$ → (years of life expectancy)
0 to 5	2.5	0.30722	1.53612
5 to 10	7.5	0.21924	1.09619
10 to 15	12.5	0.15225	0.76123
15 to 20	17.5	0.10625	0.53125
20 to 25	22.5	0.08125	0.40625
25 to 30	27.5	0.07725	0.38623
30 to 35	32.5	0.09424	0.47119
35 to 40	37.5	0.13222	0.66112
		Total midpoint area ≈	**5.8 years**

From 1970 to 2010, life expectancy for women is expected to have increased by approximately 5.8 years.

27. a. Using technology, the curve intersects the horizontal axis at $A \approx 43.799$ days and $B \approx 273.382$ days after September 30, 1995.

b. The width of each rectangle is $\frac{43.799-0}{5} = 8.760$ days.

Interval (days)	Midpoint d (days)	Height $\lvert r(d)\rvert$ (feet per day)	Area $8.760\cdot\lvert r(d)\rvert$ (feet)
0 to 8.76	4.380	0.01700	0.14894
8.76 to 17.519	13.140	0.01279	0.11207
17.519 to 26.279	21.899	0.00883	0.07736
26.279 to 35.039	30.659	0.00511	0.04480
35.039 to 43.799	39.419	0.00164	0.01439
		Total midpoint area ≈	**0.39756**

In about 44 days after September 30, 1995, the level of the lake fell by approximately 0.398 foot.

c. The width of each rectangle is $\frac{273.382 - 43.799}{10} \approx 22.958$ days.

Interval (days)	Midpoint d (days)	Height $r(d)$ (feet per day)	Area $22.958r(d)$ (feet)
43.799 to 66.757	55.278	0.00401	0.09217
66.757 to 89.715	78.236	0.01078	0.24740
89.715 to 112.674	101.195	0.01585	0.36383
112.674 to 135.632	124.153	0.01923	0.44144
135.632 to 158.591	147.111	0.02092	0.48025
158.591 to 181.549	170.070	0.02092	0.48025
181.549 to 204.507	193.028	0.01923	0.44144
204.507 to 227.466	215.987	0.01585	0.36383
227.466 to 250.424	238.945	0.01078	0.24740
250.424 to 273.382	261.903	0.00401	0.09217
		Total midpoint area ≈	**3.25018 feet**

Between about 44 and 273 days after September 30, 1995, the level of the lake rose by approximately 3.250 feet.

d. The lake level was approximately $3.25018 - 0.39756 \approx 2.853$ feet higher 273 days after September 30, 1995.

29. a.

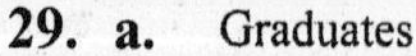

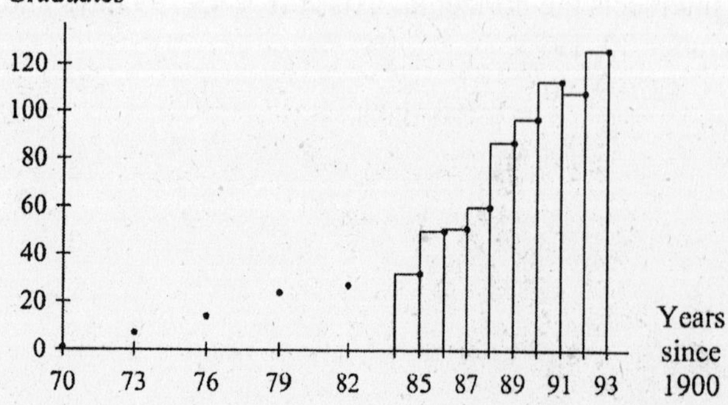

The number of new female PhDs from the beginning of 1985 through the end of 1993 was 32 + 50 + 51 + 60 + 87 + 97 + 113 + 108 + 126 = 724 graduates. This number is exact, as long as the data were correctly reported.

b. $g(t) = 0.016t^3 - 0.233t^2 + 2.669t + 1.633$ graduates t years after 1970

c. If rectangles of width 1 year are used to estimate the area, the areas of the rectangles are $g(0), g(1), g(2), \ldots, g(23)$.

$g(0) \approx 1.6$	$g(5) \approx 11.1$	$g(10) \approx 20.8$	$g(15) \approx 42.5$	$g(20) \approx 88.2$
$g(1) \approx 4.1$	$g(6) \approx 12.7$	$g(11) \approx 23.8$	$g(16) \approx 49.4$	$g(21) \approx 101.2$
$g(2) \approx 6.2$	$g(7) \approx 14.3$	$g(12) \approx 27.4$	$g(17) \approx 57.3$	$g(22) \approx 115.8$
$g(3) \approx 8.0$	$g(8) \approx 16.1$	$g(13) \approx 31.6$	$g(18) \approx 66.3$	$g(23) \approx 131.9$
$g(4) \approx 9.6$	$g(9) \approx 18.3$	$g(14) \approx 36.7$	$g(19) \approx 76.6$	

Adding these values gives an estimate of about 971 graduates. (Note that technology can be used to perform this calculation automatically.)

d. The estimate of 971 graduates is close to the actual number of 987.

31. Quill Activity

Section 6.2 Limit of Sums, Accumulated Change, and The Definite Integral

1. a. (thousand people per year)(years) = thousand people

b, c. Thousand people

3. a. This is the change in the number of organisms when the temperature increases from 25°C to 35°C.

b. Assuming the graph of A does not cross the c-axis between $c = 30$ and $c = 40$, this is the change in the number of organisms when the temperature increases from 30°C to 40°C.

5. a, b. Profit is increasing when the rate-of-change function is positive: between 0 and 300 boxes, and between 400 and 600 boxes.

c. NA

d. Profit reaches a relative maximum when the rate-of-change graph passes from positive to negative: at 300 boxes.

e. Profit reaches a relative minimum when the rate-of-change graph passes from negative to positive: at 400 boxes.

f. Profit is decreasing most rapidly when the rate-of-change graph reaches a minimum: at about 350 boxes.

g. (dollars per box)(boxes) = dollars

h. Note that p' is the function whose graph is shown in the text. Because $p'(b) < 0$ between $b = 300$ and $b = 400$, $\int_{300}^{400} p'(b)db < 0$. Because $p'(b) > 0$ between $b = 100$ and $b = 200$, $\int_{100}^{200} p'(b)db > 0$. Therefore, $\int_{300}^{400} p'(b)db$ is less than $\int_{100}^{200} p'(b)db$.

7. a.

Number of rectangles	Approximation of area
4	17.6908
8	17.8527
16	17.8959
32	17.9069
64	17.9097
128	17.9104
256	17.9106
512	17.9106
1024	17.9106
	Trend ≈ 17.9106

$\int_3^{11} w(t)dt \approx 17.91$

b. Between 3 and 11 weeks of age, the mouse gained 17.91 grams.

c. The mouse's weight at 11 weeks of age is about 4 grams + 17.9 grams = 21.9 grams.

9. a.

Number of rectangles	Approximation of area
5	10.671
10	10.662
20	10.656
40	10.654
80	10.653
160	10.653
320	10.653
	Trend ≈ 10.653

During the first 5 years of production, this oil field produced about 10.65 thousand barrels of oil.

b.

Number of rectangles	Approximation of area
5	12.272
10	12.435
20	12.453
40	12.454
80	12.454
	Trend ≈ 12.454

The yield from the oil field during the first 10 years is approximately 12.45 thousand barrels.

c. $\int_0^5 r(t)dt$ and $\int_0^{10} r(t)dt$

d. Since 10.65 is about $\frac{10.65}{12.45} \cdot 100\% \approx 85.6\%$ of 12.45, the first 5 years account for approximately 85.6% of the first 10 years' production.

11. a. Using technology, the curve intersects the horizontal axis at $A \approx 17.308$ seconds.

b.

Number of rectangles	Approximation of area
4	174.02
8	174.50
16	174.63
32	174.66
64	174.66
128	174.66
	Trend ≈ 174.66

$\int_0^A a(t)dt \approx 174.7$. From 0 to 17.3 seconds, the car's speed increased by approximately 174.7 feet per second (or $\frac{174.7 \text{ feet}}{\text{second}} \cdot \frac{3600 \text{ seconds}}{\text{hour}} \cdot \frac{1 \text{ mile}}{5280 \text{ feet}} \approx 119.1$ mph).

c.

Number of rectangles	Approximation of area
4	95.56
8	95.04
16	94.91
32	94.88
64	94.87
128	94.87
256	94.87
	Trend ≈ 94.87

$\int_A^{35} a(t)dt \approx -94.9$. From 17.3 to 35 seconds, the car's speed decreased by approximately 94.9 feet per second (or 64.7 mph).

d. Since 174.66 – 94.87 ≈ 79.8, the car's speed after 35 seconds was approximately 79.8 feet per second (or 54.4 mph) faster than it was at 0 seconds.

13. a. The heights will be in meters per second, and the widths will be in microseconds.

b. Sample calculation:

$$\frac{148.2 \text{ meters}}{\text{second}} \cdot \frac{1000 \text{ millimeters}}{1 \text{ meter}} \cdot \frac{1 \text{ second}}{1{,}000{,}000 \text{ microseconds}} = 0.148 \text{ millimeters per microsecond}$$

The new values for velocity are 0.1482, 0.1593, 0.1695, 0.1807, 0.1898, and 0.2000 millimeters per microsecond. The area units will be (millimeters per microsecond)(microseconds) = millimeters.

c. $V(m) = -\left(1.589 \cdot 10^{-6}\right)m^2 + \left(1.145 \cdot 10^{-3}\right)m + 0.137$ millimeters per microsecond after m microseconds

Number of rectangles	Approximation of area
4	10.1636
8	10.1622
16	10.1619
32	10.1618
64	10.1618
128	10.1618
	Trend ≈ 10.16

The crack traveled approximately 10.2 millimeters.

e. $\int_0^{60} V(m)dm$

15. a.

Number of rectangles	Approximation of area
5	9859.0
10	10,096.7
20	10,100.0

The estimates are about 9859 labor hours, 10,097 labor hours, and 10,100 labor hours.

b. Because the activity specifies that the function evaluated at the ends of weeks (integers) gives the exact number of labor hours needed, simply adding the function at integer values will give an exact answer for the total number of worker hours. This is the sum of right rectangles (i). Of course, this exact answer doesn't give the exact area beneath the curve.

17. a.

Number of rectangles	Approximation of area
4	2657.37
8	2616.18
16	2605.89
32	2603.31
64	2602.67
128	2602.51
256	2602.47
512	2602.46
1024	2602.46
	Trend ≈ 2602.46

$\int_0^{720} c(m)dm \approx 2602$

b. Between 9:00 a.m. and 9:00 p.m., about 2602 customers entered the store.

19. a. The patient's diastolic blood pressure is rising when the rate of change is positive. Using the table, this is from about 2 a.m. until almost 2 p.m., assuming that the data represent daily averages and "wrap around." Blood pressure falls when the rate of change is negative, from about 2 p.m. until almost 2 a.m.

b. In the table, the greatest rate of change (pressure rising most rapidly) occurs at 8 a.m., and the most negative rate of change (pressure falling most rapidly) occurs at 8 p.m.

c. $B(t) = 0.030t^2 - 0.718t + 3.067$ mm Hg per hour, where t is the number of hours since 8 a.m.

d. The model is zero at $t \approx 5.59$ hours and at $t \approx 18.13$ hours. These are the times when the blood pressure indicated by the model is highest and lowest, respectively.

e. Note that although $B(t)$ changes signs during the interval from $t = 0$ to $t = 12$. We may still use a single sum (instead of calculating the areas above and below the horizontal axis separately) because the question is asking for net change.

Number of rectangles	Approximation of area
6	2.417
12	2.508
24	2.531
48	2.537
96	2.538
192	2.538
	Trend ≈ 2.538

From 8 a.m. to 8 p.m., diastolic pressure rose by about 2.54 mm Hg.

f. $\int_0^{12} B(t)dt$

g. We do not have enough information to answer this question because we do not know the patient's blood pressure at 8 a.m.

21. Excel Activity

a. $M(t) = -0.286t^3 + 15.312t^2 - 224.973t + 682.451$ thousand males t years after 1974.

b. $A \approx 4.080$, $B \approx 19.548$

c. $\int_0^{4.080} M(t) \approx 1238.7$, $\int_{4.080}^{19.548} M(t) \approx -3194.0$, $\int_{19.548}^{27} M(t) \approx 853.7$

d. $\int_0^{27} M(t) \approx -1101.6$ thousand males

23. Quill Activity

Section 6.3 Accumulation Functions

1. a.

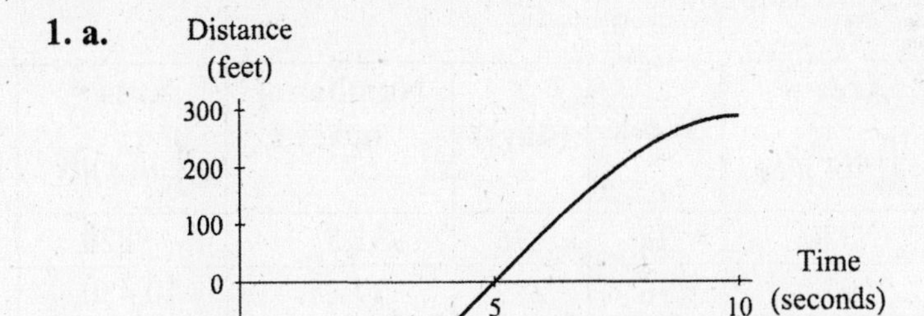

b. $D(x) = \int_5^x f(t)dt$

c. The accumulation function gives the distance traveled between 5 seconds and x seconds. For times before 5 seconds, the accumulation function is the negative of the distance traveled, because we are looking back in time.

3. a. The area of the region between days 0 and 18 represents how much the price of the technology stock declined ($15.40 per share) during the first 18 trading days of 2003.

b. The area of the region between days 18 and 47 represents how much the price of the technology stock rose ($55.80) between days 18 and 47.

c. The price on day 47 was $40.40 more than the price on day 0.

d. The price was $11.10 less on day 55 than it was on day 47.

e.

x	0	8	18	35	47	55
$\int_0^x r(t)dt$	0	−7.1	$-(7.1 + 8.3)$ $= -15.4$	$-15.4 + 30.4$ $= 15.0$	$15 + 25.4$ $= 40.4$	$40.4 - 11.1$ $= 29.3$

f.

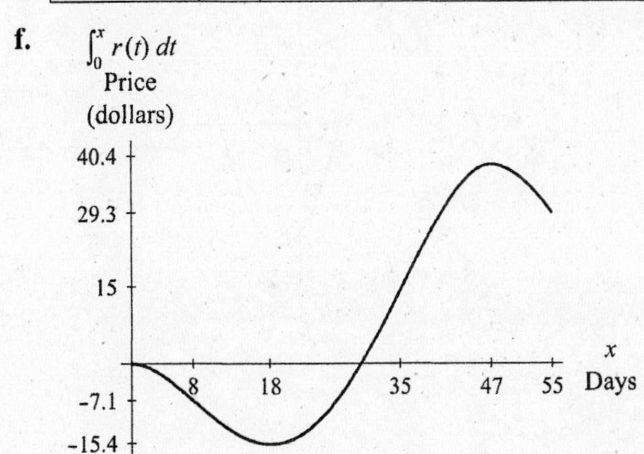

g. On day 55 the stock price was

$127 + $29.30 = $156.30

5. a. Because the rate-of-change graph is always positive, lying above the horizontal axis, the number of subscribers never declined during the first year.

b. The peak corresponds to the time when the number of subscribers was increasing most rapidly, that is, when the rate of change was greatest.

c. The accumulation function represents the change in the number of subscribers between day 140 and day t, in other words, how many new subscribers are added t days after the end of the twentieth week.

d. Each box has width 4 weeks or 28 days and height 10 subscribers per day. Multiplying these values gives an area of (28 days)(10 subscribers per day) = 280 subscribers.

e. Estimates will vary. Multiply the number of boxes by 280 subscribers per box to obtain the area estimate. Possible estimates are as follows:

Week	t (days)	Number of boxes	Area = $\int_0^t n(x)dx$	Week	t (days)	Number of boxes	Area = $\int_0^t n(x)dx$
4	28	1.25	350	28	196	31.5	8820
8	56	3.3	924	36	252	37.25	10,430
12	84	7	1960	44	308	39	10,920
16	112	12.5	3500	52	364	39.5	11,060
20	140	19.25	5390				

f. $\int_{140}^{t} n(x)\,dx$

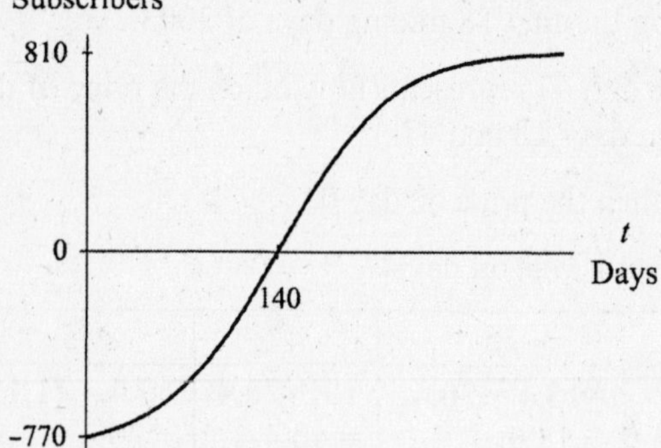

7. Rainfall (inches)

5.5

3

0 3 6 Hours

b. $\int_{B}^{x} f(t)\,dt$

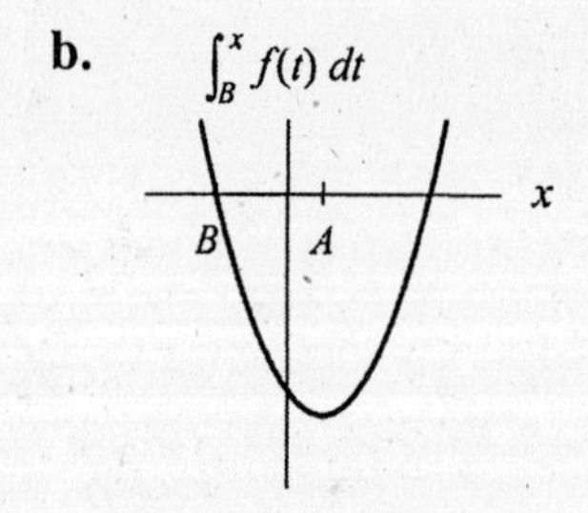

9. a. $\int_{A}^{x} f(t)\,dt$

x

A

11. a.

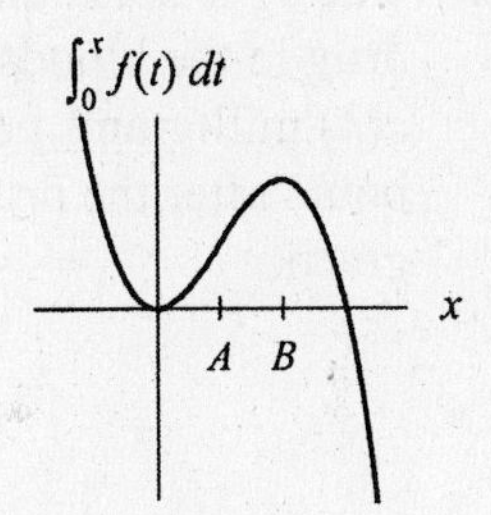

b.

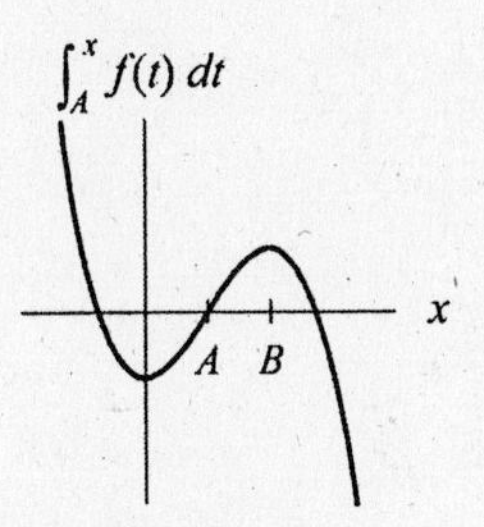

c.

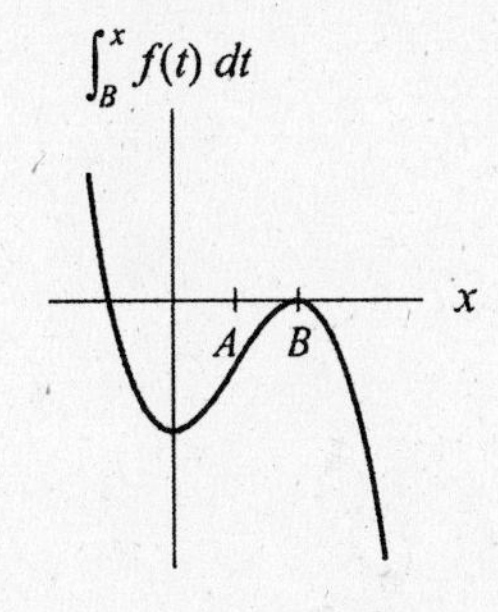

13.

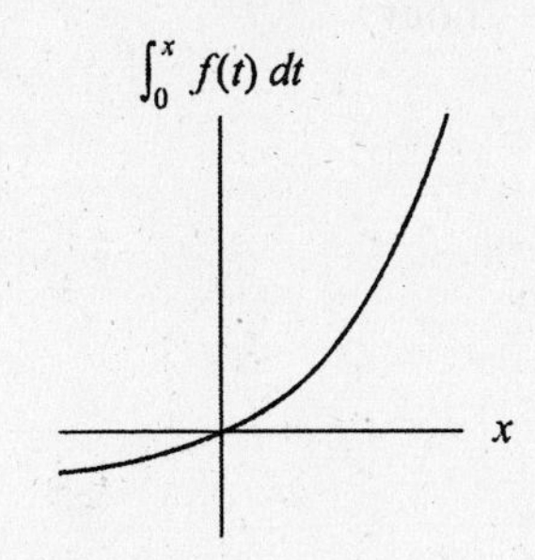

15. Derivative graph: b
Accumulation graph: f

17. Derivative graph: f
Accumulation graph: e

19. Because the table values indicate that a graph of f decreases between $x = 0$ and $x = 2$ and increases between $x = 2$ and $x = 5$, we expect the derivative values to be non-positive between $x = 0$ and $x = 2$ and non-negative between $x = 2$ and $x = 5$. The left table fits this description and is thus the derivative table. The accumulation function is the table on the right, with values increasingly negative between $x = 0$ and $x = 3$ as the area accumulates below the x-axis.

21. a. $\dfrac{\text{million dollars of revenue}}{\text{thousand advertising dollars}}$

b.

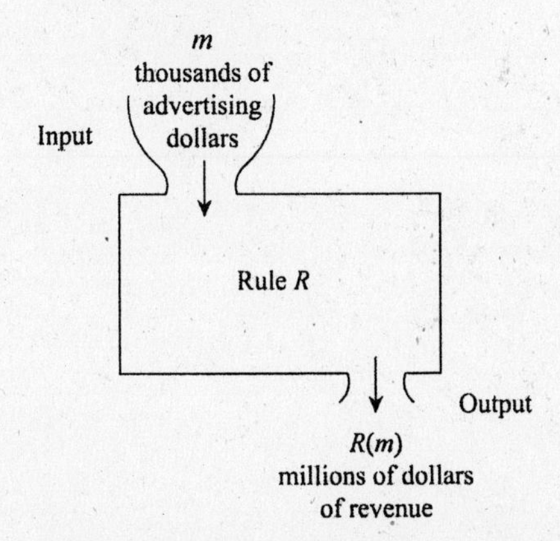

c. When m thousand dollars are being spent on advertising, the annual revenue is $R(m)$ million dollars.

23. a. $\frac{\text{milligrams per liter}}{\text{hour}}$

b.

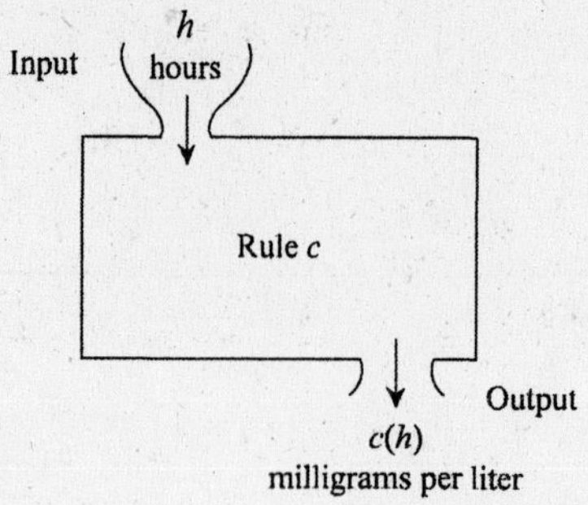

c. The concentration of a drug in the bloodstream is $c(h)$ milligrams per liter h hours after the drug is given.

Section 6.4 The Fundamental Theorem

1. b

3. c

5. Quill Activity

7. Quill Activity

9. $\int 19.436(1.07^x)dx = 19.436\int 1.07^x\,dx$

$$= 19.436\left(\frac{1}{\ln 1.07}\right)1.07^x + C$$

$$= \frac{19.436(1.07^x)}{\ln 1.07} + C$$

Check: $\frac{d}{dx}\left(\frac{19.436(1.07^x)}{\ln 1.07} + C\right) = \frac{19.436(1.07^x)(\ln 1.07)}{\ln 1.07} = 19.436(1.07^x)$

11. $\int\left[6e^x + 4(2^x)\right]dx = 6\int e^x dx + 4\int 2^x dx$

$$= 6e^x + 4\left(\frac{1}{\ln 2}\right)2^x + C$$

$$= 6e^x + \frac{4(2^x)}{\ln 2} + C$$

Check: $\frac{d}{dx}\left(6e^x + \frac{4(2^x)}{\ln 2} + C\right) = 6e^x + \frac{4(2^x)(\ln 2)}{\ln 2} = 6e^x + 4(2^x)$

13. $\int\left(10^x + 4\sqrt{x} + 8\right)dx = \int 10^x dx + 4\int x^{1/2}\,dx + \int 8dx$

$$= \frac{10^x}{\ln 10} + \frac{4}{3/2}x^{3/2} + 8x + C$$

$$= \frac{10^x}{\ln 10} + \frac{8}{3}x^{3/2} + 8x + C$$

Check: $\frac{d}{dx}\left(\frac{10^x}{\ln 10} + \frac{8}{3}x^{3/2} + 8x + C\right) = \frac{10^x(\ln 10)}{\ln 10} + \frac{8}{3}\left(\frac{3}{2}x^{1/2}\right) + 8 = 10^x + 4\sqrt{x} + 8$

15. $S(m) = \int s(m)dm$

$= \int 6250(0.92985^m)dm$

$= 6250 \int 0.92985^m \, dm$

$= 6250\left(\frac{1}{\ln 0.92985}\right)0.92985^m + C$

$= \dfrac{6250(0.92985^m)}{\ln 0.92985} + C$ CDs

m months after the beginning of the year

17. $C(x) = \int c(x)dx$

$= \int\left(\frac{0.7925}{x} + 0.3292(0.009324^x)\right)dx$

$= 0.7925 \int \frac{1}{x} dx + 0.3292 \int 0.009324^x \, dx$

$= 0.7925 \ln|x| + \dfrac{0.3292(0.009324^x)}{\ln 0.009324} + K$ dollars per unit when x units are produced

Note: If the production is presumed positive, the absolute value symbol is not necessary.

19. Find the general antiderivative:

$F(x) = \int f(x)dx$

$= \int (t^2 + 2t)dt$

$= \int t^2 dt + 2\int t \, dt$

$= \frac{t^3}{3} + 2\left(\frac{t^2}{2}\right) + C$

$= \frac{t^3}{3} + t^2 + C$

Solve for C: $F(12) = 700$

$\frac{12^3}{3} + 12^2 + C = 700$

$720 + C = 700$

$C = -20$

The specific antiderivative is $F(t) = \frac{1}{3}t^3 + t^2 - 20$

21. Find the general antiderivative:

$$F(z) = \int f(z)dz$$
$$= \int \left(z^{-2} + e^z\right)dz$$
$$= \frac{z^{-1}}{-1} + e^z + C$$
$$= \frac{-1}{z} + e^z + C$$

Solve for C: $F(2) = 1$

$$\tfrac{-1}{2} + e^2 + C = 1$$
$$C = \tfrac{3}{2} - e^2$$

The specific antiderivative is $F(z) = \frac{-1}{z} + e^z + \left(\tfrac{3}{2} - e^2\right)$

23. Find the general antiderivative:

$$w(t) = \int \frac{7.372}{t}dt = 7.372\ln t + C$$

(Assume t is positive.)
Solve for C. At 9 weeks, $t + 2 = 9$, so $t = 7$ and $w(7)=26$.

$$7.372\ln 7 + C = 26$$
$$C = 26 - 7.372\ln 7 \approx 11.655$$

The specific antiderivative is $w(t) = 7.372\ln t + 26 - 7.372\ln 7$
$\approx 7.372\ln t + 11.655$ grams after $(t + 2)$ weeks

This specific antiderivative is the formula for the accumulation function of W passing through the point (7, 26).

25. a. Find the general antiderivative: $G(t) = \int \left[(1.667\cdot 10^{-4})t^2 - 0.01472t - 0.103\right]dt$

$$= \frac{1.667\cdot 10^{-4}}{3}t^3 - \frac{0.01472}{2}t^2 - 0.103t + C$$

Solve for C using the fact that $G(70) = 94.8$:

$$\frac{1.667\cdot 10^{-4}}{3}(70^3) - \frac{0.01472}{2}(70^2) - 0.103(70) + C = 94.8$$
$$C \approx 119.015$$

The specific antiderivative is $G(t) = \frac{1.667\cdot 10^{-4}}{3}t^3 - \frac{0.01472}{2}t^2 - 0.103t + 119.015$

males per 100 females t years after 1900.

b. This specific antiderivative is the formula for the accumulation function of g passing through (70, 94.8).

27. a. Find the general antiderivative: $P(x) = \int 891.6(1.5^x)dx = \frac{891.6(1.5^x)}{\ln 1.5} + C$

Solve for C using the fact that $P(0)$=2.069: $\frac{891.6(1.5^0)}{\ln 1.5} + C = 2.069$

$C \approx -2196.887$

The specific antiderivative is $P(x) = \frac{891.6(1.5^x)}{\ln 1.5} - 2196.887$ thousand cellular phone subscribers x years after 1988

b. The number of cell phone subscribers grew by about $P(6) - P(2) \approx 22{,}850.6 - 2750.8 \approx$ 20.100 thousand subscribers (or about 20 million) between 1990 and 1994.

29. a. Find the general antiderivative: $v(t) = \int a(t)dt = \int(-32)dt = -32t + C$

Solve for C:

$$v(0) = 0$$
$$-32(0) + C = 0$$
$$C = 0$$

The specific antiderivative is $v(t) = -32t$ ft/sec t seconds after the penny is dropped.

b. Find the general antiderivative:

$$s(t) = \int v(t)dt = \int -32t\,dt = -32\int t\,dt = -32\frac{t^2}{2} + K = -16t^2 + K$$

Solve for C:

$$s(0) = 540$$
$$-16(0)^2 + K = 540$$
$$K = 540$$

The specific antiderivative is $S(t) = -16t^2 + 540$ ft above ground level t seconds after the penny is dropped.

c.

$$s(t) = 0$$
$$-16t^2 + 540 = 0$$
$$t^2 = 33.75$$
$$t \approx \pm 5.809$$

The penny will hit the ground approximately 5.8 seconds after it is dropped.

d. $v(5.809) = -32(5.809) \approx -185.9$ feet per second

$$-185.9 \text{ feet per second} = \left(\frac{-185.9 \text{ feet}}{1 \text{ second}}\right)\left(\frac{3600 \text{ seconds}}{1 \text{ hour}}\right)\left(\frac{1 \text{ mile}}{5280 \text{ feet}}\right)$$

$$= \frac{-126.75 \text{ miles}}{1 \text{ hour}} \text{ or } -126.75 \text{ mph}$$

31. a. $a(t) = -32 \text{ ft/sec}^2$

$v(t) = \int a(t)dt = -32t + C$

Because $v(0) = 0$, $v(t) = -32t$ feet per second after t seconds.

$s(t) = \int v(t)dt = -16t^2 + C$

Because $s(0) = 66$, $s(t) = -16t^2 + 66$ feet after t seconds.

Solving $s(t) = 0$ gives $t \approx 2.031$ sec.

The impact velocity is $v(2.031) \approx -64.99$ ft/sec.

$$\left(\frac{-64.99 \text{ ft}}{1 \text{ sec}}\right)\left(\frac{3600 \text{ sec}}{1 \text{ hour}}\right)\left(\frac{1 \text{ mile}}{5280 \text{ ft}}\right) \approx \frac{-44.31 \text{ miles}}{1 \text{ hour}}$$

The impact velocity is –44.31 mph.

b. Air resistance probably accounts for the difference.

33. a. Find the general antiderivative: $N(x) = \int n(x)dx = \int\left(\frac{593}{x} + 138\right)dx$

$$= 593\int\frac{1}{x}dx + \int 138dx = 593 \ln|x| + 138x + C$$

Solve for C:

$$N(5) = 896$$
$$593\ln 5 + 138(5) + C = 896$$
$$C \approx -748.397$$

The specific antiderivative is $N(x) = 593 \ln x + 138x - 748.397$ employees x years after 1996. *Note:* Because x is positive for the years when the model is valid, the absolute value symbol is not needed.

b. The function in part *a* applies from 1997 ($x = 1$) through 2002 ($x = 6$).

c. There are two ways to estimate the number of employees the company hired. If we consider the function to be continuous with discrete interpretation, then the number can be calculated from the function n by summing the yearly totals:

$n(1) + n(2) + n(3) + n(4) + n(5) + n(6) \approx 2281$

We can also estimate the total number of employees hired between 1997 and 2002 as $\int_1^6 n(x)dx = N(6) - N(1) \approx 1753$. This estimate treats the function as continuous without restriction and is probably less accurate than the first estimate.

If any employees were fired or quit between 1997 and 2002, an estimate of the number of employees hired would not represent the number of employees at the end of 2002.

Section 6.5 The Definite Integral

1. c **3.** c **5.** b **7.** a

9. a.

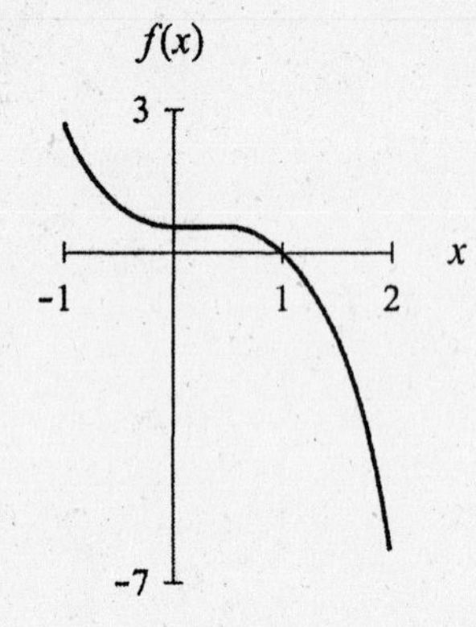

b. Begin by finding the general antiderivative of f:

$$\int f(x)dx = \int\left(-1.3x^3 + 0.93x^2 + 0.49\right)dx$$

$$= \frac{-1.3x^4}{4} + \frac{0.93x^3}{3} + 0.49x + C$$

$$= -0.325x^4 + 0.31x^3 + 0.49x + C$$

By solving $f(x) = 0$, we find that the graph crosses the horizontal axis at $x \approx 1.0544$.

$$\text{Area} = \int_{-1}^{1.0544} f(x)dx - \int_{1.0544}^{2} f(x)dx$$

$$= \left(-0.325x^4 + 0.31x^3 + 0.49x\right)\Big|_{-1}^{0.544} - \left(0.325x^4 + 0.31x^3 + 0.49x\right)\Big|_{1.0544}^{2}$$

$$\approx 0.478 - (-1.125) - [-1.74 - 0.478]$$

$$\approx 3.822$$

Because the graph of f crosses the x-axis between $x = -1$ and $x = 2$, $\int_{-1}^{2} f(x)dx$ is not the area found in part b.

c. $\int_{-1}^{2} f(x)dx = -0.615$

11. a.

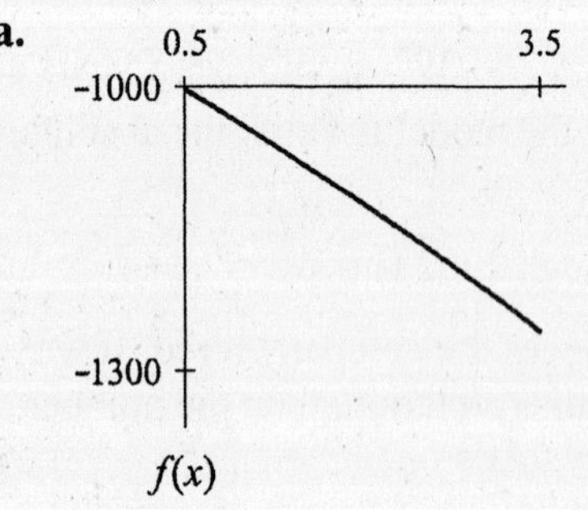

b. $\text{Area} = -\int_{0.5}^{3.5} f(x)dx$

$$= -\int_{0.5}^{3.5} -965.27(1.079^x)dx$$

$$= \int_{0.5}^{3.5} 965.27(1.079^x)dx$$

$$= \frac{965.27(1.079^x)}{\ln 1.079}\Bigg|_{0.5}^{3.5}$$

$$\approx 16{,}565.788 - 13{,}187.054$$

$$\approx 3378.735$$

Because the graph of f lies below the x-axis, $\int_{0.5}^{3.5} f(x)dx$ is the signed area (the negative of the area) calculated in part b.

c. $\int_{0.5}^{3.5} f(x)dx \approx -3378.735$

13. $\int_{5}^{15} P(x)dx \approx 2305.357$ Between 1985 and 1995, the number of international calls billed in the United States increased by 2305.4 million calls.

15. $\int_{0}^{5} r(x)dx = \int_{0}^{5} \left(9.907x^2 - 40.769x + 58.492\right)dx$

$$\approx \left(3.302x^3 - 20.385x^2 + 58.492x\right)\Big|_0^5$$

$$\approx 195.639 - 0$$

$$= 195.639$$

The corporation's revenue increased by \$195.6 million between 1987 and 1992.

17. a. $\int_{0}^{70} s(t)dt = \int_{0}^{70} (0.00241t + 0.02905)dt$

$$= \left(0.001205t^2 + 0.02905t\right)\Big|_0^{70}$$

$$= 7.938 - 0$$

$$= 7.938$$

In the 70 days after April 1, the snow pack increased by 7.938 equivalent cm of water.

b. $\int_{72}^{76} s(t)dt = \int_{72}^{76} \left(1.011t^2 - 147.941t + 5406.578\right)$

$$\approx \left(0.337t^3 - 73.9855t^2 + 5406.578t\right)\Big|_{72}^{76}$$

$$= 131{,}494.592 - 131{,}517.36$$

$$= -22.768$$

Between 72 and 76 days after April 1, the snow pack decreased by 22.768 equivalent centimeters of water.

c. It is not possible to find $\int_{0}^{76} s(t)dt$ because $s(t)$ is not defined between $t = 70$ and $t = 72$.

19. a. Using technology, we find that the graph crosses the horizontal axis at $t \approx 0.8955$.

$$\text{Area} \approx \int_{0}^{0.8955} T(h)dh$$

$$= \int_{0}^{0.8955} \left(9.07h^3 - 24.69h^2 + 14.87h - 0.03\right)dh$$

$$= \left(2.2675h^4 - 8.23h^3 + 7.435h^2 - 0.03h\right)\Big|_0^{0.8955}$$

$$\approx 1.483 - 0 = 1.483$$

From 8:30 a.m. to 8:54 a.m., the temperature increased by 1.48°F.

b. Area $\approx -\int_{0.8955}^{1.75} T(h)dh$

$= -\int_{0.8955}^{1.75} \left(9.07h^3 - 24.69h^2 + 14.87h - 0.03\right)dh$

$= -\left(2.2675h^4 - 8.23h^3 + 7.435h^2 - 0.03h\right)\Big|_{0.8955}^{1.75}$

$\approx -(-0.124 - 1.483) = 1.607$

After rising 1.48°F, the temperature decreased by 1.61°F between 8:54 a.m. to 10:15 a.m.

c. No, the highest temperature reached was 71 + 1.48 = 72.48°F.

21. a. An exponential model for the data is $f(x) = 0.161(1.076186^x)$ trillion cubic feet per year x years after 1900.

b. $\int_{40}^{60} f(x)dx = \int_{40}^{60} 0.161(1.076186^x)dx = \frac{0.161(1.076186^x)}{\ln 1.076186}\Big|_{40}^{60} \approx 179.725 - 41.387 = 138.338$

From 1940 through 1960, 138.3 trillion cubic feet of natural gas was produced.

c. $\int_{40}^{60} f(x)dx$

23. a. A quadratic model for the data is $C'(x) = \left(7.714 \cdot 10^{-5}\right)x^2 - 0.047x + 8.940$ dollars per CD, when x CDs are produced each hour.

b. Find the general antiderivative: $C(x) = \int C'(x)dx$

$= \int \left[\left(7.714 \cdot 10^{-5}\right)x^2 - 0.047x + 8.94\right]dx$

$= \left(2.571 \cdot 10^{-5}\right)x^3 - 0.024x^2 + 8.940x + K$

Use the fact that C(150)=750 to solve for K:

$\left(2.571 \cdot 10^{-5}\right)(150)^3 - 0.024(150)^2 + 8.940(150) + K = 750$

$K \approx -143.893$

The hourly cost model is $C(x) = \left(2.571 \cdot 10^{-3}\right)x^3 - 0.024x^2 + 8.940x - 143.893$ dollars when x CDs are produced each hour.

c. $\int_{200}^{300} C'(x) = C(x)\Big|_{200}^{300} = C(300) - C(200) \approx 1096.82 - 900.68 = \$196.14$

When production is increased from 200 to 300 CDs per hour, cost increases by about $196.14.

25. a, b.

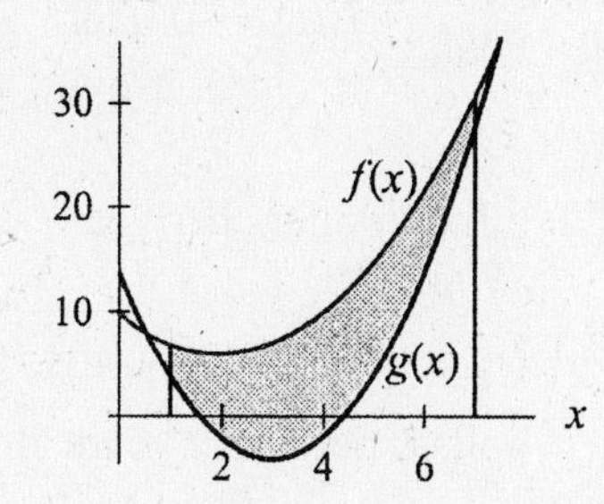

c. Area:

$\int_a^b [f(x) - g(x)]dx$

$= \int_1^7 [(x^2 - 4x + 10) - (2x^2 - 12x + 14)]dx$

$= \int_1^7 (-x^2 + 8x - 4)dx$

$= \left(\frac{-x^3}{3} + 4x^2 - 4x\right)\Big|_1^7$

$= \frac{161}{3} - \left(\frac{-1}{3}\right) = 54$

27. a, c.

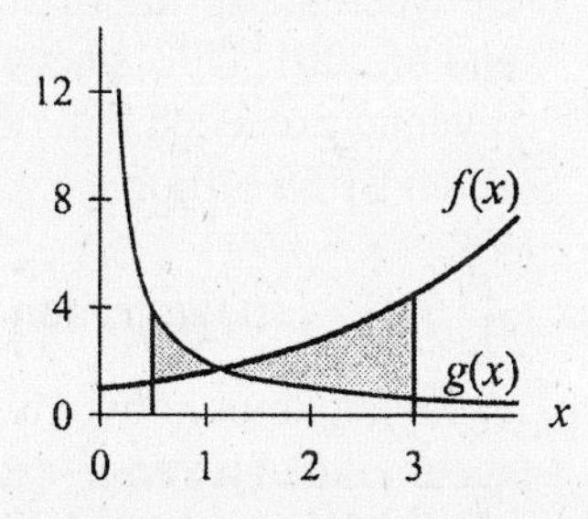

b. Using technology, we find that the curves intersect at $x \approx 1.134$.

d. Difference $= \int_{0.5}^{3} [f(x) - g(x)]dx$

$= \int_{0.5}^{3} \left[e^{0.5x} - \frac{2}{x}\right]dx$

$= \left[\frac{e^{0.5x}}{0.5} - 2\ln|x|\right]\Big|_{0.5}^{3}$

$= (2e^{0.5x} - 2\ln x)\Big|_{0.5}^{3}$

$\approx 6.7662 - 3.9543 \approx 2.812$

e. Area of left region $\approx \int_{0.5}^{1.134} [g(x) - f(x)]dx$

$= (2\ln x - 2e^{0.5x})\Big|_{0.5}^{1.134}$

$\approx -3.274 - (-3.954)$

$= 0.680$

Area of right region $\approx \int_{1.134}^{3} [f(x) - g(x)]dx$

$= (2e^{0.5x} - 2\ln x)\Big|_{0.5}^{3}$

$\approx 6.766 - 3.274$

$= 3.492$

Total area $\approx 0.680 + 3.492 = 4.172$

29. a. When the amount invested in capital increases from \$1500 to \$5500, profit increases by approximately \$13.29 million.

b. Area $= \int_{1.5}^{5.5} [r'(x) - c'(x)]dx = 13.29$

31. a. The population of the country grew by 3690 people in January.

b. The population declined by 9720 people between the beginning of February and the beginning of May.

c. The change in population from the beginning of January through the end of April was
3690 – 9720 = –6030 people

d. Because the graphs intersect, the area of R_1 represents an increase in population and the area of region R_2 represents a decrease in population. The net change is the difference:
Area of R_1 – Area of R_2
The total area is
Area of R_1 + Area of R_2

33. a. Before fitting models to the data, add the point (0, 0), and convert the data from miles per hour to feet per second by multiplying each speed by $\left(\frac{5280 \text{ feet}}{1 \text{ mile}}\right)\left(\frac{1 \text{ hour}}{3600 \text{ seconds}}\right)$.

For each car, the speeds in the revised data set are 0, 44, $58\frac{2}{3}, 73\frac{1}{3}, 88, 102\frac{2}{3}, 117\frac{1}{3}, 132,$ and $146\frac{2}{3}$ feet per second.

The speed of the Supra after t seconds can be modeled as

$s(t) = -0.702t^2 + 20.278t + 2.440$

feet per second.

The speed of the Carrera after t seconds can be modeled as

$c(t) = -0.643t^2 + 18.963t + 5.252$

feet per second.

b. $\int_0^{10} [s(t) - c(t)]dt$

$\approx \int_0^{10} (-0.059t^2 + 1.315t - 2.811)dt$

$\approx (-0.020t^3 + 0.657t^2 - 2.811t)\Big|_0^{10}$

$\approx 17.965 - 0 = 17.965$

The Supra travels about 17.96 feet farther.

c. $\int_5^{10} [s(t) - c(t)]$

$= (-0.020t^3 + 0.657t^2 - 2.811t)\Big|_0^{10}$

$\approx 17.965 - (-0.079) = 18.044$

The Supra travels about 18.04 feet farther.

35. a. FedEx: $F(t) = -0.026t^3 + 0.198t^2 + 0.06t + 0.317$

billion dollars per year t years after 1993

UPS: $U(t) = 0.15t + 0.572$ billion dollars per year t years after 1993

b. Area of region on left $\approx$ \$1.28 billion

Area of region in middle $\approx$ \$1.20 billion

Area of region on right $\approx$ \$1.59 billion

Between the beginning of 1993 and late 1996 ($t \approx 2.8$), UPS's accumulated revenue exceeded that of FedEx by about \$1.28 billion.

Between late 1996 and the spring of 2000 ($t \approx 6.3$), FedEx's accumulated revenue exceeded that of UPS by approximately \$1.2 billion.

Between the spring of 2000 and the end of 2001, UPS's accumulated revenue exceeded that of FedEx by about \$1.59 billion.

c. $\int_0^8 [F(t) - U(t)]dt$ billion. This value is the net amount by which FedEx's accumulated revenue exceeded that of UPS between 1993 and 2001.

37. a. Multiply the output data by 22 to convert the data to total absorption for all 22 hectares of trees. Finding a logistic model for these converted data, we have

$$f(x) = \begin{cases} 0 \text{ tons per year} & \text{when } 0 \le x < 5 \\ \dfrac{557.960}{1 + 91.202e^{-0.318025x}} \text{ tons per year} & \text{when } x \ge 5 \end{cases}$$

x years after 1990.

b,c.

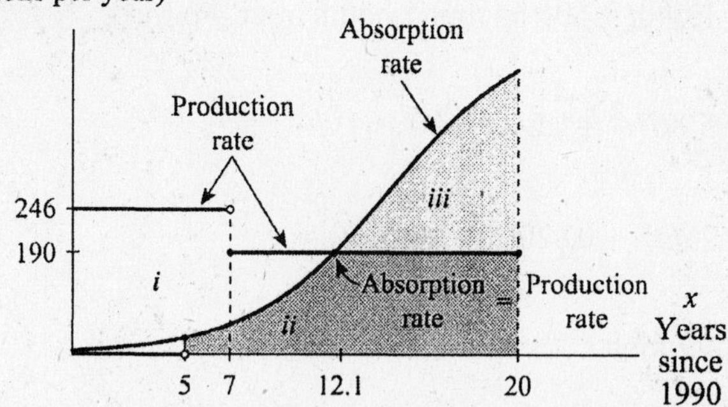

d. i. Between 1990 and 1995, the factory produced (246 tons per year)(5 years) = 1230 tons. Between 1995 and 1997, the amount the factory produced that was not absorbed by the trees is calculated as $\int_5^7 [246 - f(x)]dx \approx 414.2$ tons. To find the time when the trees began absorbing more than the factory produced, we solve $f(x) = 190$ and find $x \approx 12.1$. Between 1997 and early in 2003, the amount the factory produced that was not absorbed by the trees is calculated as $\int_7^{12.1} [190 - f(x)]dx \approx 410.7$ tons.

Summing these three totals, we have the total emissions produced by the factory and not absorbed by the trees: $1230 + 414.2 + 410.7 = 2054.9$ tons

ii. The emissions produced by the factory and absorbed by the trees is found as

$$\int_5^{12.1} f(x)dx + (190)(20 - 12.1) = 638.5 + 1498.6 = 2137.1 \text{ tons}$$

iii. The emissions absorbed by the trees from sources other than the factory is calculated as $\int_{12.1}^{20} [f(x) - 190]dx \approx 1269.0$ tons

e. The factory produces (246 tons per year)(7 years) + (190 tons per year)(13 years) = 4192 tons, and the trees absorb $\int_5^{20} f(x)dx \approx 3406.2$ tons. This does not comply with the federal regulation.

39. Excel Activity

$B(x) = -0.000011x^3 + 0.001x^2 - 0.035x - 0.369$ percent per year (of age) when x is age

$W(x) = -0.000008x^3 + 0.001x^2 - 0.038x + 0.117$ percent per year (of age) when x is age

41. Quill Activity

Section 6.6 Average Value and Average Rates of Change

1. a. $V(t) = -1.664t^3 + 5.867t^2 + 1.640t + 60.164$ mph t hours after 4 p.m.

b.
$$\frac{\int_0^3 V(t)dt}{3-0} \approx \frac{1}{3}\int_0^3 (-1.664t^3 + 5.867t^2 + 1.640t + 60.164)dt$$
$$\approx \frac{1}{3}(-0.416t^4 + 1.956t^3 + 0.820t^2 + 60.164t)\Big|_0^3$$
$$\approx \frac{1}{3}(206.97) \approx 68.99 \text{ mph}$$

c.
$$\frac{\int_1^3 V(t)dt}{3-1} \approx \frac{1}{2}(-0.416t^4 + 1.956t^3 + 0.820t^2 + 60.164t)\Big|_1^3$$
$$\approx \frac{1}{2}(206.97 - 62.52) \approx 72.23\text{mph}$$

3. a.
$$R(y) = \begin{cases} \dfrac{0.719}{1+0.005e^{0.865563y}} + 0.62 \text{ dollars per minute} & \text{when } 2 \le y < 10 \\ -0.045y + 1.092 \text{ dollars per minute} & \text{when } 10 \le y \le 20 \end{cases}$$

y years after 1980

b.
$$\frac{\int_2^{10} R(y)dy}{10-2} = \frac{1}{8}\int_2^{10}\left(\frac{0.719}{1+0.005e^{0.865563y}} + 0.62\right)dy \approx \$0.99 \text{ per minute}$$

c.
$$\frac{\int_2^{20} R(y)dy}{20-2} = \frac{1}{18}\left[\int_2^{10}\left(\frac{0.719}{1+0.005e^{0.865563y}} + 0.62\right) + \int_{10}^{20}(-0.045y + 1.092)dy\right]$$
$$\approx \frac{1}{18}(7.925 + 4.167) \approx \$0.67 \text{ per minute}$$

5. a. Note that the beginning of 1980 is the end of 1979, corresponding to $t = 58$.
Let $P(t) = 7.391(1.02695^t)$.

$$\frac{\int_{79}^{89} P(t)dt}{89-79} = \frac{1}{10}\left(\frac{7.391(1.02695^t)}{\ln 1.02695}\right)\Big|_{79}^{89}$$
$$\approx \frac{1}{10}(2963.60 - 2271.57)$$
$$\approx 69.2 \text{ million people}$$

b. Solving $P(t) \approx 69.2$ gives $t \approx 84.1$ years after the end of 1900. This corresponds to early 1985.

c. $\text{Average rate} = \dfrac{P(89) - P(79)}{89 - 79}$

$\approx \dfrac{78.81 - 60.41}{10} \approx 1.84$ million people per year

7. a. Because the model is linear, the coefficient of x gives the rate of change: -100.6 yearly accidents per year.

b, c.

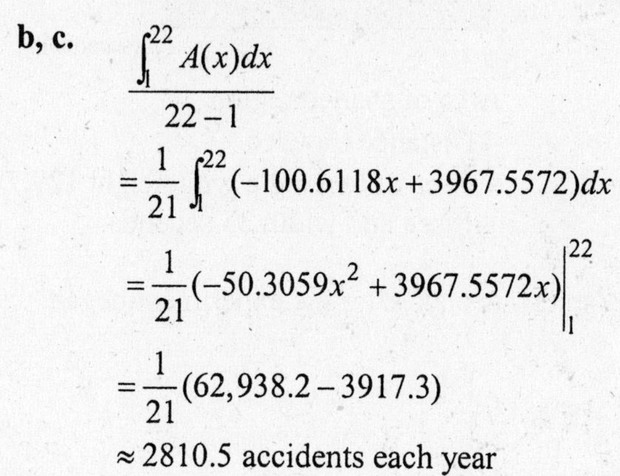

$$\frac{\int_1^{22} A(x)dx}{22-1}$$

$$= \frac{1}{21}\int_1^{22}(-100.6118x + 3967.5572)dx$$

$$= \frac{1}{21}(-50.3059x^2 + 3967.5572x)\Big|_1^{22}$$

$$= \frac{1}{21}(62{,}938.2 - 3917.3)$$

≈ 2810.5 accidents each year

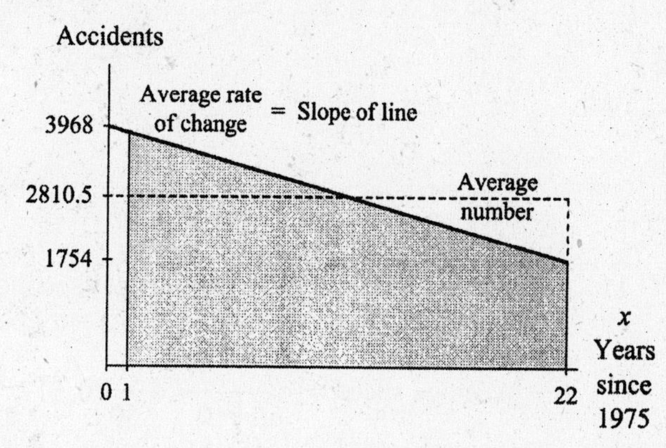

9. a. $\text{Average acceleration} = \dfrac{\int_0^{35} a(t)dt}{35-0} = \dfrac{1}{35}\int_0^{35}(0.024t^2 - 1.72t + 22.58)dt$

$$= \frac{1}{35}(0.008t^3 - 0.86t^2 + 22.58t)$$

$$= \frac{1}{35}(79.8 - 0) = 2.28 \text{ ft/sec}^2$$

b. $v(t) = \int_0^t a(x)dx = 0.008t^3 - 0.86t^2 + 22.58t$ feet per second after t seconds

$\text{Average velocity} = \dfrac{\int_0^{35} v(t)dt}{35} = \dfrac{1}{35}\int_0^{35}(0.008t^3 - 0.86t^2 + 22.58t)dt$

$$\approx \frac{1}{35}(0.002t^4 - 0.287t^3 + 11.29t^2)\Big|_0^{35}$$

$$\approx \frac{1}{35}(4540.67 - 0) \approx 129.7 \text{ feet/second}$$

c. $s(35) = \int_0^{35} v(t)dt$

This is the same integral that was evaluated in part *b*, so the distance is about 4540.7 ft.

d. It would have traveled (129.7 ft/sec)(35 sec) $\approx$ 4540.7 ft (using unrounded values).

e.

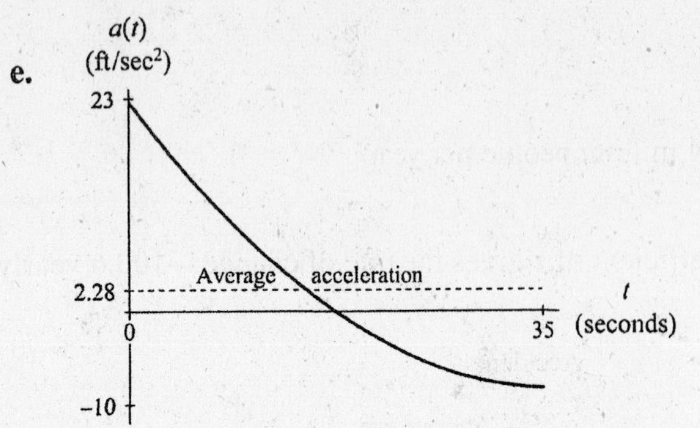

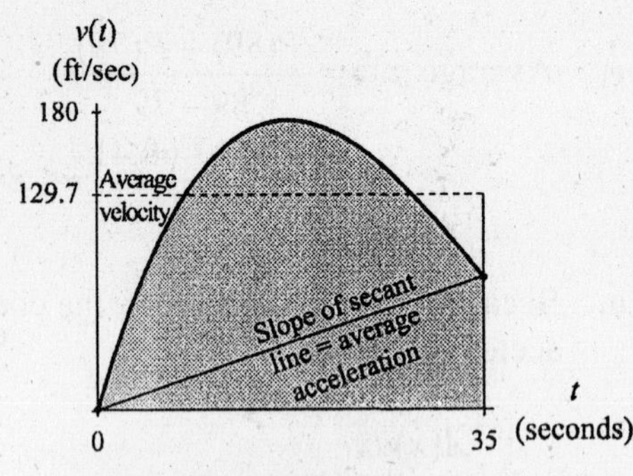

Area of shaded region
= Distance traveled
= Area of rectangle with height 129.7 ft/sec and width 35 seconds

11. a. $V(t) = 1.033t + 138.413$ meters per second t microseconds after the experiment began.

b. $$\text{Average speed} = \frac{\int_{10}^{60} V(t)dt}{60-10} \approx \frac{1}{50}(0.517t^2 + 138.43t)\Big|_{10}^{60}$$
$$\approx \frac{1}{50}(10{,}164.97 - 1435.80) \approx 174.58 \text{ meters per second}$$

13. a. $B(t) = 0.030t^2 - 0.718t + 3.067$ mm Hg per hour t hours after 8 a.m.

b. Note that $B(t)$ is the rate of change, not the actual blood pressure.

$$\text{Average rate} = \frac{\int_0^{12} B(t)dt}{12-0} \approx \frac{1}{12}(0.010t^3 - 0.359t^2 + 3.067t)\Big|_0^{12}$$
$$\approx \frac{1}{12}(2.54 - 0) \approx 0.21 \text{ mm Hg per hour}$$

c. Let $D(t)$ represent the diastolic blood pressure t hours after 8 a.m. Then $D(t) = \int B(t)dt \approx 0.010t^3 - 0.359t^2 + 3.067t + C$. Solving $D(4) = 95$ mm Hg gives $C \approx 87.831$, so $D(t) \approx 0.010t^3 - 0.359t^2 + 3.067t + 87.831$.

$$\text{Average blood pressure} = \frac{\int_0^{12} D(t)}{12}$$
$$\approx \frac{1}{12}\int_0^{12}(0.010t^3 - 0.359t^2 + 3.067t + 87.831)$$
$$\approx \frac{1}{12}\left(0.0025t^4 - 0.120t^3 + 1.534t^2 + 87.831t\right)\Big|_0^{12}$$
$$\approx \frac{1}{12}(1120.3) \approx 93.4 \text{ mm Hg}$$

15. $$\text{Average rate} = \frac{\int_{56}^{100} P(t)dt}{100-56} = \frac{1}{44}\int_{56}^{100}[0.0106t - 1.148]dt \approx -0.32 \text{ seconds per year.}$$

17. a. The average of the data is highest between 10 a.m. and 6 p.m.

b. $$C(x) = \begin{cases} 4.75x + 2.833 \text{ ppm} & \text{when } 0 \le x < 4 \\ 0.536x^2 - 7.871x + 45.200 \text{ ppm} & \text{when } 4 \le x \le 12 \\ -5.5x + 93.667 \text{ ppm} & \text{when } 12 < x \le 16 \end{cases}$$

x hours after 6 a.m.

The average concentration between 10 a.m. and 6 p.m. is

$\frac{1}{12-4}\int_4^{12} C(x)dx = \frac{1}{8}\int_4^{12}(0.536x^2 - 7.871x + 45.200)dx \approx 19.4$ ppm.

c. Average concentration

$= \frac{1}{16-0}\int_0^{16} C(x)dx$

$= \frac{1}{16}\left[\int_0^4 (4.75x + 2.833)dx + \int_4^{12}(0.536x^2 - 7.871x + 45.200)dx + \int_{12}^{16}(-5.5x + 93.667)dx\right]$

$\approx \frac{1}{16}(49.332 + 155.157 + 66.668)$

$= \frac{1}{16}(271.157) \approx 16.94$ ppm

Severe pollution warning.

19. Consider the two graphs of a function f shown below, where A is the average value of f from a to b and k is an arbitrary constant.

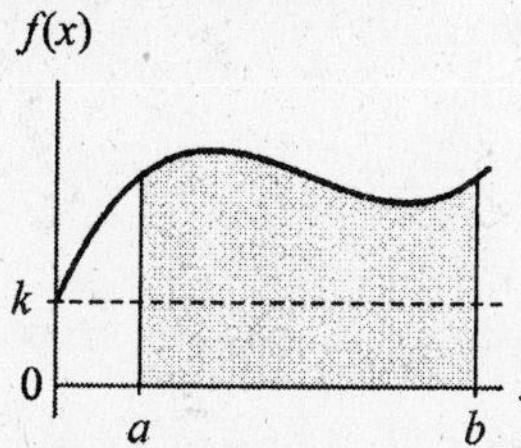

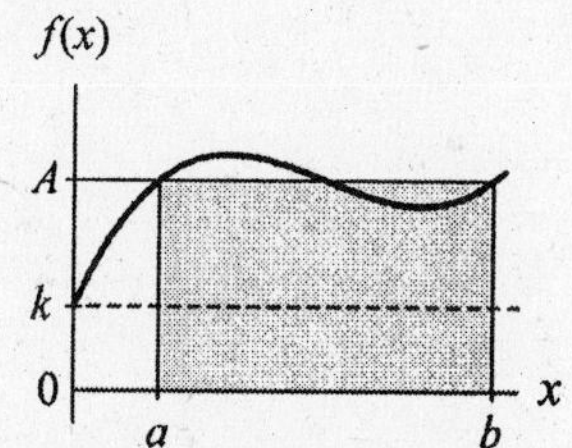

We know that the areas of the two shaded regions are equal. If we remove from each graph the rectangular region with height k and width $b - a$, then the areas of the resulting regions are still equal, because we have removed the same area from each.

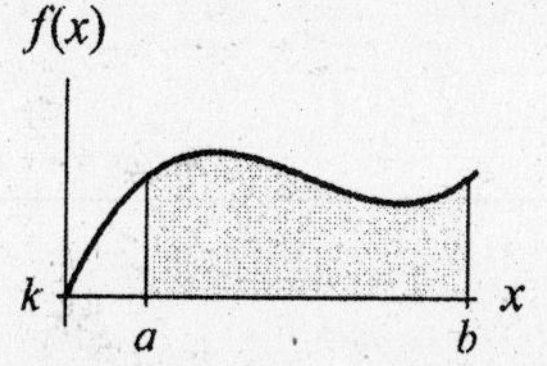

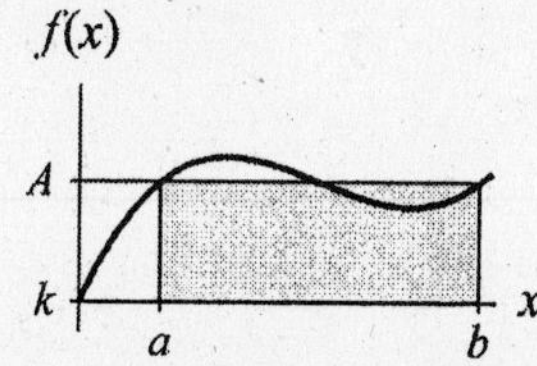

It is true for the graphs shown in this section with vertical axis shown from k rather than from zero that the area of the region between the function and $y = k$ from a to b is the same as the area of the rectangle with height equal to the average value minus k and width equal to $b - a$.

21. Excel Activity

a. $N(x) = -0.049x^3 + 1.678x^2 - 2.166x + 82.151$ inmates per 100,000 residents x years after 1977

b. $\frac{N(12) - N(0)}{12 - 0} \approx 10.9$ inmates per 100,000 per year

$$\frac{N(24)-N(12)}{24-12} \approx 8.4 \text{ inmates per 100,000 per year}$$

c. $$\frac{\int_{0}^{24} N(x)dx}{24-0} \approx 207.6 \text{ inmates per 100,000}$$

Section 6.7 Antiderivative Limitations

1. $\int 2e^{2x}dx = e^{2x} + C$

3. Not possible using the techniques discussed in this text.

5. $\int (1+e^x)^2 dx = \int (1+2e^x + e^{2x})dx$

$$= x + 2e^x + \frac{e^{2x}}{2} + C$$

7. Not possible using the techniques discussed in this text.

9. Using technology, we estimate the value of the integral as approximately 2.5452.

11. $\int_2^5 \frac{\ln x}{x}dx = \frac{(\ln x)^2}{2}\bigg|_2^5 = \frac{(\ln 5)^2}{2} - \frac{(\ln 2)^2}{2}$ (exact)

13. Using technology, we estimate the value of the integral as approximately 3.6609.

15. $\int_3^4 \frac{2x}{x^2+1}dx = \ln(x^2+1)\Big|_3^4 = \ln 17 - \ln 10$ (exact)

17. Using technology, we estimate the value of the integral as approximately 8.7595.

19. $\int_3^4 \frac{x^2+1}{x^2}dx = \int_3^4 \left(1+\frac{1}{x^2}\right)dx$ (exact)

$$= \left(x - \frac{1}{x}\right)\bigg|_3^4$$

$$= (4-\tfrac{1}{4}) - (3-\tfrac{1}{3}) = 1\tfrac{1}{12}$$

Chapter 6 Review Test

1. a.

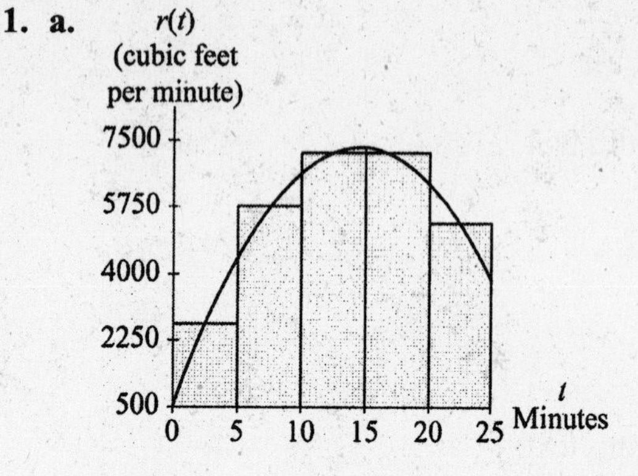

b. The width of each rectangle is $\frac{25-0}{5} = 5$.

Interval	Midpoint t minutes	Rectangle height $r(t)$ cubic feet per minute	Rectangle area (5 minutes)[$r(t)$ ft³ per minute]→ cubic feet
0 to 5	2.5	2639.5	13,197.5
5 to 10	7.5	5704.5	28,522.5
10 to 15	12.5	7169.5	35,847.5
15 to 20	17.5	7034.5	35,172.5
20 to 25	22.5	5299.5	26,497.5
			Midpoint rectangle area = 139,237.5 cubic feet

c. In the first 25 minutes that oil was flowing into the tank, approximately 139,238 cubic feet of oil flowed in.

2. a. A quadratic model for the data is
$S(t) = -1.643t^2 + 16.157t + 0.2$ miles per hour t hours after midnight

b.

Number of rectangles	Approximation of area
5	127.131
10	126.869
20	126.803
40	126.786
80	126.782
160	126.781
	Trend ≈ 126.8

The hurricane traveled about 126.8 miles.

3. a. The area beneath the horizontal axis represents the amount of weight that the person lost during the diet.

b. The area above the axis represents the amount of weight that the person regained between weeks 20 and 30.

c. Because –26.7 + 15.4 = –11.3, the person's weight was 11.3 pounds less at 30 weeks than it was at 0 weeks.

d.

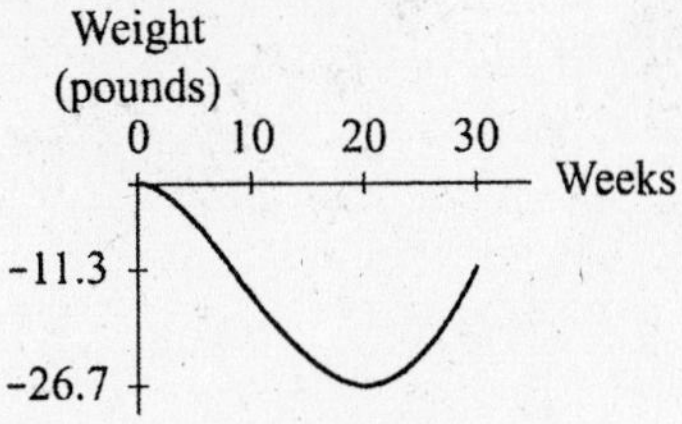

Weight (pounds)
0 10 20 30 Weeks
-11.3
-26.7

e. The graph in part *d* represents the change in the person's weight as a function of the number of weeks since the person began the diet.

4. a. Find the general antiderivative: $R(t) = \int r(t)dt = 10\left(\frac{-3.2}{3}t^3 + \frac{93.3}{2}t^2 + 50.7t\right) + C$

$$= \frac{-32}{3}t^3 + 466.5t^2 + 507t + C$$

Solve for C:

$$R(0) = 5000$$

$$\frac{-32}{3}(0)^3 + 466.5(0)^2 + 507(0) + C = 5000$$

$$C = 5000$$

The specific antiderivative is $R(t) = \frac{-32}{3}t^3 + 466.5t^2 + 507t + 5000$ cubic feet after t minutes

b. $R(10) - R(0) \approx 46{,}053.3 - 5000 = 41{,}053.3$ ft^3

c. Use technology to solve $R(t) = 150{,}000$, which gives $t \approx 27.55$ or $t \approx 31.73$. The tank will be full after about 27.6 minutes.

5. a. $\int_0^{2.75} a(x)dx = \int_0^{2.75} 840(1.08763^x)dx$

$$= \frac{840(1.08763^x)}{\ln 1.0763}\Bigg|_0^{2.75}$$

$$\approx 12{,}598.475 - 9999.879 \approx \$2598.60$$

b. At the end of the third quarter of the third year, the \$10,000 had increased by \$2598.60 so that the total value of the investment was \$12,598.60.

6. $\int_{79}^{88} [m(t) - w(t)]\, dt \approx \$123$ thousand

Between the beginning of 1980 and the end of 1988, a man earning the average full-time wage would have earned approximately \$123,000 more than a woman earning the average full-time wage.

Chapter 7

Section 7.1 Perpetual Accumulation and Improper Integrals

1. $\int_0^\infty 3e^{-0.2t}\,dt = \lim_{N\to\infty}\int_0^N 3e^{-0.2t}\,dt$

$= \lim_{N\to\infty}\left(\frac{3}{-0.2}e^{-0.2t}\right)\Big|_0^N$

$= \lim_{N\to\infty}\left(\frac{3}{-0.2}e^{-0.2N} - \frac{3}{-0.2}e^{-0.2(0)}\right)$

$= \lim_{N\to\infty}\left(\frac{3}{-0.2}e^{-0.2N} + 15\right) = 0 + 15 = 15$

3. $\int_{10}^\infty 3x^{-2}\,dx = \lim_{N\to\infty}\int_{10}^N 3x^{-2}\,dx$

$= \lim_{N\to\infty}\left(-3x^{-1}\Big|_{10}^N\right)$

$= \lim_{N\to\infty}\left(\frac{-3}{N} - \frac{-3}{10}\right)$

$= 0 + \frac{3}{10} = \frac{3}{10} = 0.3$

5. $\int_{-\infty}^{-10} 4x^{-3}\,dx = \lim_{N\to-\infty}\int_N^{-10} 4x^{-3}\,dx$

$= \lim_{N\to-\infty}\left(-2x^{-2}\Big|_N^{-10}\right)$

$= \lim_{N\to-\infty}\frac{-2}{(-10)^2} - \frac{-2}{N^2}$

$= -0.2 - 0 = -0.2$

7. $\int_{0.36}^\infty 9.6x^{-0.432}\,dx$

$= \lim_{N\to\infty}\int_{0.36}^N 9.6x^{-0.432}\,dx$

$= \lim_{N\to\infty}\left(\frac{9.6}{0.568}x^{0.568}\Big|_{0.36}^N\right)$

$= \lim_{N\to\infty}\left(\frac{9.6}{0.568}N^{0.568} - \frac{9.6}{0.568}\left(0.36^{0.568}\right)\right) \to \infty$

This integral diverges.

9. $\int_2^{\infty} \frac{2x}{x^2+1}dx$

$= \lim_{N\to\infty} \int_2^N \frac{2x}{x^2+1}dx$

$= \lim_{N\to\infty} \left[\ln(x^2+1)\Big|_2^N \right]$

$= \lim_{N\to\infty} \left[\ln(N^2+1) - \ln 5 \right] \to \infty$

This integral diverges.

11. $\int_a^{\infty} [f(x)+k]dx = \int_a^{\infty} f(x)dx + \int_a^{\infty} kdx$

The second integral can be thought of as the area of a rectangle with height k and infinite width. The second integral diverges, so the original integral diverges.

13. a. Amount after 100 years $= \int_0^{100} -1.55(0.9999999845^t)\cdot 10^{-6}dt$

$= \frac{-1.55(0.9999999845^t)\cdot 10^{-6}}{\ln 0.9999999845}\Big|_0^{100}$

$= \frac{-1.55(0.9999999845^{100})\cdot 10^{-6}}{\ln 0.9999999845} - \frac{-1.55(0.9999999845^{0})\cdot 10^{-6}}{\ln 0.9999999845}$

≈ -0.0002 milligram

After 100 years, only 0.0002 mg of ^{238}U will have decayed.

Amount after 1000 years

$= \int_0^{1000} -1.55(0.9999999845^t)\cdot 10^{-6}dt = \frac{-1.55(0.9999999845^t)\cdot 10^{-6}}{\ln 0.9999999845}\Big|_0^{1000}$

$= \frac{-1.55(0.9999999845^{1000})\cdot 10^{-6}}{\ln 0.9999999845} - \frac{-1.55(0.9999999845^{0})\cdot 10^{-6}}{\ln 0.9999999845}$

≈ -0.0015 milligram

After 1000 years, only 0.0015 mg of ^{238}U will have decayed.

b. $\int_0^{\infty} -1.55(0.9999999845^t)\cdot 10^{-6}dt = \lim_{N\to\infty} \int_0^N -1.55(0.9999999845^t)\cdot 10^{-6}dt$

$= \lim_{N\to\infty} \left(\frac{-1.55(0.9999999845^t)\cdot 10^{-6}}{\ln 0.9999999845}\Big|_0^N \right)$

$= \lim_{N\to\infty} \left[\frac{-1.55(0.9999999845^N)\cdot 10^{-6}}{\ln 0.9999999845} - \frac{-1.55(0.9999999845^0)\cdot 10^{-6}}{\ln 0.9999999845} \right]$

$= 0 - \frac{-1.55(0.9999999845^0)\cdot 10^{-6}}{\ln 0.9999999845} = -99.99999923 \approx -100$ milligrams

Eventually, all of the ^{238}U will decay.

15. a. Solving $150 = 499.589(0.958^p)$ yields $p_0 \approx \$28.04$.

b. (Note that we use the unrounded value of p_0 in calculations)

$$C = q_0 p_0 + \int_{p_0}^{\infty} D(p)dp \approx 150(28.04) + \int_{28.04}^{\infty} 499.589(0.958^p)dp$$

$$= 4206 + \lim_{N\to\infty} \int_{28.04}^{N} 499.589(0.958^p)dp = 4206 + \lim_{N\to\infty}\left[\left.\frac{499.589(0.958^p)}{\ln 0.958}\right|_{28.04}^{N}\right]$$

$$= 4206 + \lim_{N\to\infty}\left[\frac{499.589(0.958^N)}{\ln 0.958} - \frac{499.589(0.958^{28.04})}{\ln 0.958}\right]$$

$$= 4206 + 0 - \frac{499.589(0.958^{28.04})}{\ln 0.958} \approx \$7702 \text{ thousand}$$

Consumers are willing and able to spend \$7.7 million for 150,000 books.

17. $$\int_0^{\infty} 0.1e^{-0.1x}dx = \lim_{N\to\infty}\int_0^N 0.1e^{-0.1x}dx = \lim_{N\to\infty}\left(\left.-e^{-0.1x}\right|_0^N\right)$$

$$= \lim_{N\to\infty}\left[-e^{-0.1N} - (-e^0)\right]$$

$$= \lim_{N\to\infty} -e^{-0.1N} + \lim_{N\to\infty}(e^0) = 0 + 1 = 1$$

Section 7.2 Streams in Business and Biology

1. a. i. $R(m) = 0.2\left(\frac{\$47{,}000}{12}\right) = \$783.33$ per month

ii. $R(m) = 0.2\left(\frac{47{,}000}{12} + 100m\right)$

$= 783.33 + 20m$ dollars/month

after m months

iii. $R(m) = 0.2\left(\frac{47{,}000}{12}(1.005^m)\right)$

$= 7833.33(1.005^m)$

dollars per month after m months

b. i. $\sum_{m=0}^{59} \frac{9400}{12}\left(1.05^{60-m}\right) = \$53{,}493.40$

ii. $\sum_{m=0}^{59} \left(\frac{9400}{12} + 20m\right)\left(1 + \frac{0.05}{12}\right)^{60-m}$

$= \$92{,}082.72$

iii. $\sum_{m=0}^{59} \frac{9400}{12}(1.005^m)\left(1 + \frac{0.05}{12}\right)^{60-m}$

$= \$61{,}818.49$

The first option is the only one that will not result in the amount needed for the down payment.

3. $\int_0^4 (0.125)(17.628)(1.05^t)e^{0.07(4-t)}\,dt = \$11.2$ billion

5. a. $\int_0^7 (177.26)(0.03)e^{0.088(7-x)}\,dx \approx \$51.5$ billion

b. $P = \frac{51.5}{e^{0.088(7)}} \approx \$27.8$ billion

7. a. Because \$500 per month is equivalent to \$6000 per year, use $R(t) = 6000$ dollars per year.

Future value $= \int_0^T R(t)e^{r(T-t)}\,dt = \int_0^6 6000e^{0.0634(6-t)}\,dt = \int_0^6 6000e^{0.0634(6)}e^{-0.0634t}\,dt$

$= \left.\frac{6000e^{0.0634(6)}e^{-0.0634t}}{-0.0634}\right|_0^6 \approx 43{,}804.70$

The investments will be worth about \$43,804.70.

b. $\sum_{m=0}^{71} 500\left(1+\frac{0.0634}{12}\right)^{72-m} \approx 43{,}896.84$

The investments will be worth about $43,896.84.

c. Answers may vary.

9 a. $r(q) = 82.1(1.05^q)(0.15)$ billion dollars per quarter q quarters after the third quarter of 2002.

b. $R(q) = 82.1(1.05^q)(0.15)\left(1.09^{16-q}\right)$ million dollars per quarter for money invested q quarters after the third quarter of 2002.

c. If the investment begins with the 4th quarter 2002 profits, then the initial investment is based on a profit of $(82.1)(1.05) = \$86.205$ billion. Thus we calculate

$$\sum_{q=0}^{15} 86.205(1.05^q)(0.15)\left(1.09^{16-q}\right) \approx 629.8 \text{ billion dollars}$$

(Note that this value does not include the contribution made at the end of year 2006.)

11. a. The income is earned at a rate of

$$R(t) = \left(\frac{36{,}400 \text{ liters}}{3 \text{ years}}\right)\left(\frac{\$0.80}{1 \text{ liter}}\right)$$
$$= \$9706.67 \text{ per year}$$

Future value

$$= \int_0^T R(t)e^{r(T-t)}\,dt$$
$$= \int_0^7 (9706.67)e^{0.045(7-t)}\,dt$$
$$= \int_0^7 (9706.67)e^{0.045(7)}e^{-0.045t}\,dt$$
$$= \left.\frac{(9706.67)e^{0.045(7)}e^{-0.045t}}{-0.045}\right|_0^7$$
$$= \left.\frac{(9706.67)e^{0.045(7-t)}}{-0.045}\right|_0^7$$
$$\approx -215{,}703.70 - (-295{,}570.01)$$
$$= 79{,}866.31 \text{ dollars}$$

Pepsi will make about $79,866.

b. Let P = present value. Since $Pe^{rt} = Pe^{0.045(7)} \approx 79{,}866.31$, the present value is $P \approx \frac{79{,}866.31}{e^{0.045(7)}} \approx 58{,}285$ dollars. Pepsi would have had to invest about $58,285.

13. a. Let P = present value, and use the result of Activity 7a.

$$Pe^{rt} = \text{future value}$$
$$Pe^{0.062(6)} \approx 43{,}804.70$$
$$P \approx \frac{43{,}804.70}{e^{0.062(6)}} \approx 29{,}944.36$$

You would need to invest about $29,944.

Note: This amount can also be determined by calculating $\int_0^6 6000e^{-0.0634t}\,dt$.

b. Let P = present value, and use the result of Activity 7b.

$$P\left(1+\frac{0.0634}{12}\right)^{12(6)} \approx 43{,}804.70$$

You would need to invest about $30,037.41.

c. Answers may vary.

15. a. Present value = $\int_0^T R(t)e^{-rt}\,dt$

$$= \int_0^{20} 850e^{-0.15t}\,dt = \left.\frac{850e^{-0.15t}}{-0.15}\right|_0^{20}$$
$$\approx -282.2 - (-5666.7)$$
$$= \$5384.5 \text{ million}$$

The 20-year present value is about $5384.5 million, or $5.4 billion.

b. Present value = $\int_0^T R(t)e^{-rt}\,dt$

$$= \int_0^{20} 850e^{-0.13t}\,dt = \left.\frac{850e^{-0.13t}}{-0.13}\right|_0^{20}$$
$$\approx -485.63 - (-6538.46)$$
$$\approx \$6052.8 \text{ million}$$

The 20-year present value is about $6052.8 million, or $6.1 billion.

c. Use $R(t) = 850(1.1^t)$ million dollars per year.

Present value = $\int_0^T R(t)e^{-rt}\,dt$

$$= \int_0^{20} 850(1.1^t)e^{-0.14t}\,dt$$
$$= \int_0^{20} 8.5(1.1e^{-0.14})^t\,dt$$
$$= \left.\frac{8.5(1.1e^{-0.14})^t}{\ln(1.1e^{-0.14})}\right|_0^{20}$$
$$\approx -7781.1 - (-19{,}020.0)$$
$$= \$11{,}238.9 \text{ million}$$

The 20-year present value is about $11,238.9 million, or $11.2 billion.

Note: We have assumed that the annual returns given refer to APRs, compounded continuously. If they are interpreted as APYs, the answers to parts *a*, *b*, and *c* are \$5410.2 million, \$6351.3 million, and \$12,148.5 million, respectively.

17. a. Use $R(t) = 2(0.95^t)$ billion dollars per year.

$$\text{Present value} = \int_0^T R(t)e^{-rt}\,dt$$

$$= \int_0^{10} 2(0.95^t)e^{-0.2t}\,dt = \int_0^{10} 2(0.95e^{-0.2})^t\,dt = \left.\frac{2(0.95e^{-0.2})^t}{\ln(0.95e^{-0.2})}\right|_0^{10}$$

$$\approx -0.645 - (-7.959) = \$7.314 \text{ billion}$$

The 10-year present value is about \$7.3 billion. This is less than the \$8.1 billion offered by CSX.

b. Use $R(t) = 1.2(1.02^t)$ billion dollars per year.

$$\text{Present value} = \int_0^T R(t)e^{-rt}\,dt$$

$$= \int_0^{10} 1.2(1.02^t)e^{-0.2t}\,dt$$

$$= \int_0^{10} 1.2(1.02e^{-0.2})^t\,dt$$

$$= \left.\frac{1.2(1.02e^{-0.2})^t}{\ln(1.2e^{-0.2})}\right|_0^{10}$$

$$\approx -1.0986 - (-6.6594)$$

$$\approx \$5.561 \text{ billion}$$

The 10-year present value is about \$5.6 billion.

c. Answers will vary.

Note: We have assumed that the annual returns given refer to APRs, compounded continuously. If they are interpreted as APYs, the answers to part *a* and *b* are \$7.7 billion (less than CSX offered), and \$5.930 billion, respectively.

19. Use $R(t) = 1.2(1.06^t)$ million dollars per year.

Present value $= \int_0^T R(t)e^{-rt}\,dt$

$$= \int_0^5 1.2(1.06^t)e^{-0.12t}\,dt$$

$$= \int_0^5 1.2(1.06e^{-0.12})^t\,dt$$

$$= \left.\frac{1.2(1.06e^{-0.12})^t}{\ln(1.06e^{-0.12})}\right|_0^5$$

$$\approx -14.28 - (-19.44)$$

$$\approx \$5.16 \text{ million}$$

The capital value, or 5-year present value, is about $5.2 million.

21. Answers will vary, depending on the current year. The given solution is based on the end of 2001.

a. $(12 \text{ million})(0.83^{22}) \approx 0.20$ million terns

b. Use $r(t) = 2.04$ million terns per year, and $s = 0.83$. The desired expression is $r(t)s^{22-t} = 2.04(0.83)^{22-t}$ million terns hatched t years after 1979.

c. Future value $= Ps^b + \int_0^b r(t)s^{b-t}\,dt$

$$= 12(0.83^{22}) + \int_0^{22} 2.04(0.83^{22-t})dt$$

$$\approx 0.20 + \int_0^{22} 2.04(0.83^{22})(0.83^{-1})^t\,dt$$

$$= \left.\frac{2.04(0.83^{22})(0.83^{-1})^t}{\ln(0.83^{-1})}\right|_0^{22} + 0.20$$

$$= \left.\frac{2.04(0.83^{22-t})}{-\ln 0.83}\right|_0^{22} + 0.20$$

$$\approx 10.95 - 0.18 + 0.20$$

$$= 10.97 \text{ million terns}$$

There are about 10.97 million sooty terns at the end of 2001.

23. a. $(200 \text{ thousand})(0.67^{50}) \approx 4 \cdot 10^{-7}$ thousand ≈ 0

None of the current population will still be alive.

b. Use $r(t) = 60 - 0.5t$ thousand seals per year, and $s = 0.67$. The desired expression is $r(t)s^{50-t} = (60 - 0.5t)(0.67^{50-t})$ thousand seals born t years from now.

c. Use technology to evaluate the definite integral.

Future value $= Ps^b + \int_0^b r(t)s^{b-t}\,dt \approx 0 + \int_0^{50} (60 - 0.5t)(0.67)^{50-t}\,dt \approx 90.5$ thousand seals

There will be about 90.5 thousand seals.

Section 7.3 Integrals in Economics

1. **a.** The demand function **b.** The supply function

   c. The producers' surplus **d.** The consumers' surplus

3. **a.** To find the price P above which consumers will purchase none of the goods or services, either find the smallest positive value for which the demand function is zero, $D(p) = 0$, or, if $D(p)$ is never exactly zero but approaches zero as p increases without bound, then let $P \to \infty$.

   b. The supply function S is a piecewise continuous function with the first piece being the 0 function. The value p at which $S(p)$ is no longer 0 is the shutdown price. The shutdown point is $(p_1, S(p_1))$.

   c. The market equilibrium price p_0 can be found as the solution to $S(p) = D(p)$. That is, it is the price at which demand is equal to supply. The equilibrium point is the point $(p_0, D(p_0)) = (p_0, S(p_0))$.

5. **a.** $q_0 \approx 27.5$ thousand units

   b. and c. See figures in Answer Key page A-52 in Text.

7. **a.** See figures in Answer Key page A-52 in Text. $p^* \approx 2.5$ dollars per unit, $q^* \approx 2.5$ million units, $p_1 \approx 0$ dollars per unit, $P \approx 0$ dollars per unit

   b. and c. See figures in Answer Key page A-52 in Text.

   d. Total Social Gain = $\int_0^{2.5} S(p)dp + \int_{2.5}^{5} D(p)dp$

9. **a.** D is an exponential demand function and so does not have a finite value p at which $D(p) = 0$. Thus the model does not indicate a price above which consumers will purchase none of the goods or services.

   b. Solving $D(p) = 18$ gives $p_0 \approx 90.98$, so the demand is $q_0 = 18$ thousand ceiling fans when the price is $p_0 \approx 90.98$ dollars. The maximum amount consumers are willing and able to spend is

$$p_0 q_0 + \int_{p_0}^{P} D(p)dp \approx 18(90.98) + \int_{90.98}^{\infty} 25.92(0.996^p)dp$$

$$= 1637.61 + \lim_{N\to\infty} \int_{90.98}^{N} 25.92(0.996^p)dp$$

$$= 1637.61 + \lim_{N\to\infty} \left. \frac{25.92(0.996^p)}{\ln 0.996} \right|_{90.98}^{N} \approx 1637.61 + 0 - (-4490.99)$$

$= 6128.60$

The units of this quantity are (thousands of fans)(dollars per fan) = thousands of dollars. Consumers are willing and able to spend about \$6128.6 thousand for 18 thousand fans.

c. $D(100) \approx 17.4$ thousand fans

d. Consumers' surplus $= \int_{100}^{\infty} D(p)dp = \lim_{N\to\infty} \int_{100}^{N} 25.92(0.996^p)\,dp$

$$= \lim_{N\to\infty} \left.\frac{25.92(0.996^p)}{\ln 0.996}\right|_{100}^{N} \approx 0 - (-4331.5)$$

$$= 4331.5$$

As in part *b*, the units are thousands of dollars. The consumers' surplus is about \$4331.5 thousand.

11. a. $D(p) = 0.025p^2 - 1.421p + 19.983$ lanterns when the market price is \$$p$ per lantern

b. Note that $q_0 = D(p_0) = D(12.34) \approx 6.263$. Also, $D(p) = 0$ when $p \approx 25.701$ or $p \approx 31.075$, so $P \approx 25.701$. The amount consumers are willing and able to spend is

$$p_0 q_0 + \int_{p_0}^{P} D(p)dp \approx 6.263(12.34) + \int_{12.34}^{25.701} (0.025p^2 - 1.421p + 19.983)dp$$

$$\approx 77.288 + \left.(0.008p^3 - 0.710p^2 + 19.983p)\right|_{12.34}^{25.701}$$

$$\approx 77.288 - 0 + 186.000 - 154.106 = 109.18$$

The units of this quantity are (lanterns)(dollars per lantern) = dollars. Consumers are willing and able to spend \$109.18 each day.

c. The consumers' surplus is the second integral shown in the solution to part *b*; that is,

$\int_{p_0}^{P} D(p)dp \approx 186.000 - 154.106 = 31.894$ dollars or \$31.89.

13. a. $S(40) = 0.024(40)^2 - 2(40) + 60 = 18.4$ thousand answering machines

$S(150) = 0.024(150)^2 - 2(150) + 60 = 300$ thousand answering machines

Producers will supply 18,400 answering machines at \$40, and 300,000 answering machines at \$150.

b. Because $S(99.95) = 99.86006$ thousand answering machines, the producers' revenue is (\$99.95 per answering machine)(99.86006 thousand answering machines) $\approx$ \$9981 thousand, or \$9,981,000.

Producers' surplus $= \int_{p_1}^{p_0} S(p)dp$

$$= \int_{20}^{99.95} (0.024p^2 - 2p + 60)dp$$

$$= \left.0.008p^3 - p^2 + 60p\right|_{20}^{99.95} \approx 3995.0 - 864.0 = 3131$$

The units of this quantity are

(dollars per answering machine)(thousands of answering machines) = thousands of dollars, so the producers' surplus is about \$3131 thousand, or \$3,131,000.

15. a. $S(p) = \begin{cases} 0 \text{ hundred prints} & \text{when } p < 5 \\ 0.300p^2 - 3.126p + 10.143 \text{ hundred prints} & \text{when } p \geq 5 \end{cases}$

where p hundred dollars is the price of a print.

b. Solving $S(p) = 5$ gives $p \approx 8.3712$ hundred dollars. Producers will supply 500 prints at a price of \$837.12.

c. Because $S(6.3) \approx 2.358$ hundred prints, the producers' revenue is (\$630 per print)(235.8 prints) $\approx$ \$148,500.

$$\text{Producers' surplus} = \int_{p_1}^{p_0} S(p)dp \approx \int_5^{6.3} (0.300p^2 - 3.126p + 10.143)dp$$

$$= 0.100p^3 - 1.563p^2 + 10.143p\Big|_5^{6.3} \approx 26.875 - 24.123 = 2.732$$

The units of this quantity are
(hundreds of dollars per print)(hundreds of prints) = tens of thousands of dollars, so the producers' surplus is about \$27,300.

17. a. Because the solutions to $D(p) = 0$ are $p \approx -41.20$ and $p \approx 20.577$, we use $P \approx 20.577$. Solving $D(p) = 20$ gives $p \approx 20.252$, so we use $p_0 \approx 20.252$ and $q_0 = 20$.
The amount consumers are willing and able to spend is

$$p_0 q_0 + \int_{p_0}^{P} D(p)dp \approx 20(20.252) + \int_{20.252}^{20.577} (-1.003p^2 - 20.689p + 850.375)dp$$

$$= 405.04 + \left(\frac{-1.003p^3}{3} - 10.3445p^2 + 850.375p\right)\Big|_{20.252}^{20.577}$$

$$\approx 405.04 + 10{,}205.27 - 10{,}202.02$$
$$= 408.30$$

The units of this quantity are
(hundreds of dollars per sculpture)(sculptures) = hundreds of dollars, so consumers are willing and able to spend about \$408.3 hundred or \$40,830.

b. Because $D(5) = 850.375$ and $S(5) = 297.157$, the quantity supplied at \$500 per sculpture is 297 sculptures, and supply will not exceed demand.

c. Solving $D(p) = S(p)$ gives $p \approx 13.21$ hundred dollars, so the equilibrium price is about \$1321 per sculpture. The positive solution to $D(p) = 0$ is $p \approx 20.58$ hundred dollars, so consumers will not purchase when the price is over \$2058 and we use $P \approx 20.58$.
The total social gain is

$$\int_{p_1}^{p^*} S(p)dp + \int_{p^*}^{P} D(p)dp$$

$$\approx \int_{4.5}^{13.21} (0.256p^2 + 8.132p + 250.097)dp + \int_{13.21}^{20.58} (-1.003p^2 - 20.689p + 850.375)dp$$

$$= \left(\frac{0.256p^3}{3} + 4.066p^2 + 250.097p\right)\Big|_{4.5}^{13.21} + \left(\frac{-1.003p^3}{3} - 10.3445p^2 + 850.375p\right)\Big|_{13.21}^{20.58}$$

$\approx (4209.07 - 1215.55) + (10{,}205.27 - 8656.63)$
≈ 4542.15 (using unrounded values)

The units of this quantity are hundreds of dollars, so the total social gain is about $4542.15 hundred or $454,215.

19. a. $D(p) = 499.589(0.958086^p)$ thousand books when the market price is $\$p$ per book.

b. $S(p) = \begin{cases} 0 \text{ thousand books} & \text{when } p < 18.97 \\ 0.532p^2 - 20.060p + 309.025 \text{ thousand books} & \text{when } p \geq 18.97 \end{cases}$

where $\$p$ is the price of a book.

c. Solving $D(p) = S(p)$ gives an equilibrium price of $p^* \approx \$27.15$ per book. Since $D(p^*) = S(p^*) \approx 156.2$ thousand books, approximately 156.2 thousand books will be supplied and demanded.

d. Total social gain $= \int_{p_1}^{p^*} S(p)dp + \int_{p^*}^{P} D(p)dp$

$$\approx \int_{18.97}^{27.15} (0.532p^2 - 20.060p + 309.025)d + \int_{27.15}^{\infty} 499.589(0.958086^p)dp$$

$$\approx (0.177p^3 - 10.030p^2 + 309.025p)\Big|_{18.97}^{27.15} + \lim_{N\to\infty}\left(\frac{499.589(0.958086^p)}{\ln 0.958086}\right)\Bigg|_{27.15}^{N}$$

$\approx 4542.71 - 3462.26 + [0 - (-3648.16)]$
≈ 4728.62 (using unrounded values)

The units of this quantity are (dollars per book)(thousands of books) = thousands of dollars, so the total social gain is about $4728.6 thousand.

Section 7.4 Probability Distributions and Density Functions

1. a. There is a 46% chance that any telephone call made on a computer software technical support line will be 5 minutes or more.

b. The likelihood that any two cars on a certain two-lane road are less than 7 feet apart is approximately 25%.

c. New Orleans will receive between 2 and 4 inches of rain during the month of March 15% of the time.

3. a. $f(x) \geq 0$ for all x.

$$\int_{-\infty}^{\infty} f(x)dx = \int_0^1 1.5\left(1 - x^2\right)dx = 1.5\left(x - \tfrac{1}{3}x^3\right)\Big|_0^1 = 1.5\left(1 - \tfrac{1}{3}\right) = 1$$

Yes, *f* is a probability density function.

b. $h(x) \geq 0$ for all x.

$$\int_{-\infty}^{\infty} h(x)dx = \int_0^1 6\left(x - x^2\right)dx = 6\left(\tfrac{1}{2}x^2 - \tfrac{1}{3}x^3\right)\Big|_0^1 = 6\left(\tfrac{1}{2} - \tfrac{1}{3}\right) = 1$$

Yes, *h* is a probability density function.

c. $r(t) \geq 0$ for all t.

The area between the graph of r and the t-axis is $\frac{1}{2}(0.5)(1.2) + \frac{1}{2}(1)(1.2) = 0.3 + 0.6 \neq 1$.

No, *r* is not a probability density function.

d. Because $s(c)$ is negative for some values of c, s is not a probability density function.

5. a. $P(x < 1) = \int_0^1 y(x)dx = \int_0^1 0.32x\,dx = (0.16x^2)\Big|_0^1 = 0.16 - 0 = 0.16$

b. $\mu = \int_{-\infty}^{\infty} xy(x)dx = \int_0^{2.5} 0.32x^2 dx = \left(\frac{0.32}{3}x^3\right)\Big|_0^{2.5} = \frac{0.32}{3}(2.5^3 - 0) \approx 1.67$

The mean is about 167 gallons.

c.

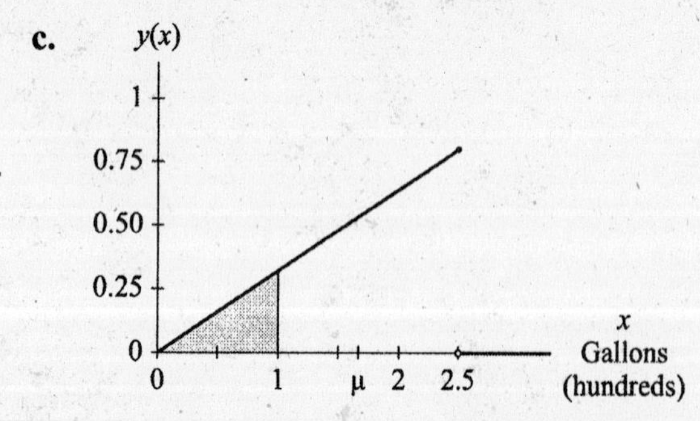

7. Because the values of $f(x)$ are all non-negative and $\int_{-\infty}^{\infty} f(x)dx = 1$, the probability (which is the area between the graph of *f* and the input axis) must always be between 0 and 1. Another explanation is that the probability of some occurrence is the proportion of times it is expected to happen, and all proportions are fractions between 0 and 1.

9. a. $P(20<t<30)=\int_{20}^{30}0.2e^{-0.2t}\,dt=-e^{-0.2t}\Big|_{20}^{30}=-e^{-6}+e^{-4}\approx 0.016$

The probability that successive arrivals are between 20 and 30 seconds apart is approximately 1.6%.

b. $P(t\le 10)=\int_{-\infty}^{10}e(t)dt=\int_{0}^{10}0.2e^{-0.2t}\,dt=\int_{0}^{10}0.2e^{-0.2t}\,dt=\left(-e^{-0.2t}\right)\Big|_{0}^{10}=-e^{-2}+e^{0}\approx 0.865$

The probability that successive arrivals are 10 seconds or less apart is approximately 86.5%.

c. $P(t>15)=\int_{15}^{\infty}e(t)dt=\lim\limits_{N\to\infty}\int_{15}^{N}0.2e^{-0.2t}\,dt=\lim\limits_{N\to\infty}\left(-e^{-0.2t}\Big|_{15}^{N}\right)$

$=\lim\limits_{N\to\infty}\left(-e^{-0.2N}+e^{-3}\right)=0+e^{-3}\approx 0.050$

The probability that successive arrivals are more than 15 seconds apart is about 5%.

11. a. $\mu=\int_{-\infty}^{\infty}tP(t)dt=\int_{0}^{4}\frac{3}{32}(4t^2-t^3)dt=\frac{3}{32}\left(\frac{4}{3}t^3-\frac{1}{4}t^4\right)\Big|_{0}^{4}=\frac{3}{32}\left(\frac{256}{3}-64\right)=2$

The mean time is 2 minutes.

b. $\sigma^2=\int_{-\infty}^{\infty}(t-2)^2P(t)dt=\int_{0}^{4}\frac{3}{32}(t^2-4t+4)(4t-t^2)dt=\frac{3}{32}\int_{0}^{4}(-t^4+8t^4-20t^2+16t)dt$

$=\frac{3}{32}\left(-\frac{1}{5}t^5+2t^4-\frac{20}{3}t^3+8t^2\right)\Big|_{0}^{4}=\frac{3}{32}\left(-\frac{1024}{5}+512-\frac{1280}{3}+128\right)$

$=\frac{3}{32}\left(\frac{128}{15}\right)=0.8\sigma=\sqrt{0.8}\approx 0.894$

The standard deviation is approximately 0.894 minute.

c. $P(0\le t\le 1.5)=\int_{0}^{1.5}P(t)dt=\int_{0}^{1.5}\frac{3}{32}(4t-t^2)dt=\frac{3}{32}\left(\frac{4}{2}t^2-\frac{1}{3}t^3\right)\Big|_{0}^{4}=\frac{3}{32}(4.5-1.125)\approx 0.316$

The likelihood that any child between the ages of 8 and 10 learns the rules of the board game in 1.5 minutes or less is about 31.6%

d. $P(t\ge 3)=\int_{3}^{\infty}P(t)dt=\int_{3}^{4}\frac{3}{32}(4t-t^2)dt=\frac{3}{32}\left(\frac{4}{2}t^2-\frac{1}{3}t^3\right)\Big|_{3}^{4}=\frac{3}{32}\left(32-\frac{64}{3}-18+9\right)\approx 0.156$

The is about a 15.6% chance that any child between the ages of 8 and 10 takes between 2 and 4 minutes to learn the rules of the board game.

13. a. The number of customers who require daily ATM service is increasing the fastest at one standard deviation less than the mean, 167 – 30 = 137 customers.

b. Answers will vary.

c. Use technology to approximate the values of the definite integrals.

i. $P(150\le x\le 200)=\int_{150}^{200}\frac{1}{30\sqrt{2\pi}}e^{\frac{-(x-167)^2}{2(30^2)}}dx$

≈ 0.579

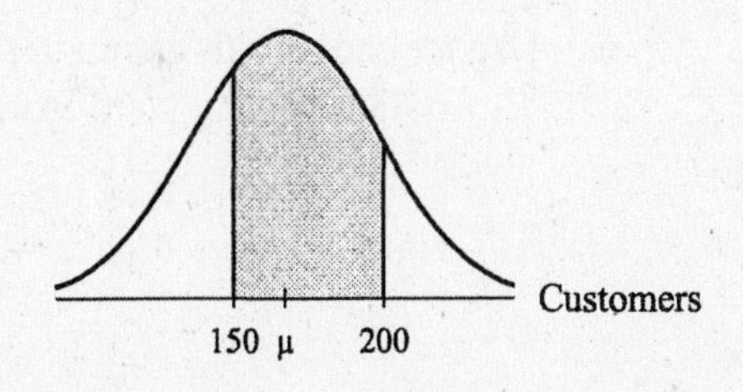

ii. $P(x < 220) = P(x \le \mu) + P(\mu < x < 220)$

$$= 0.5 + \int_{167}^{200} \frac{1}{30\sqrt{2\pi}} e^{\frac{-(x-167)^2}{2(30^2)}} dx$$

$$\approx 0.5 + 0.461 = 0.961$$

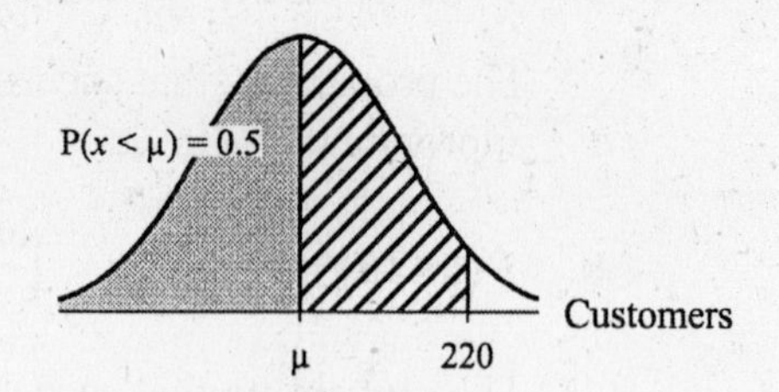

iii. $P(x > 235) = P(x \ge \mu) - P(\mu < x < 235)$

$$= 0.5 - \int_{167}^{235} \frac{1}{30\sqrt{2\pi}} e^{\frac{-(x-167)^2}{2(30^2)}} dx$$

$$\approx 0.5 - 0.49 = 0.01$$

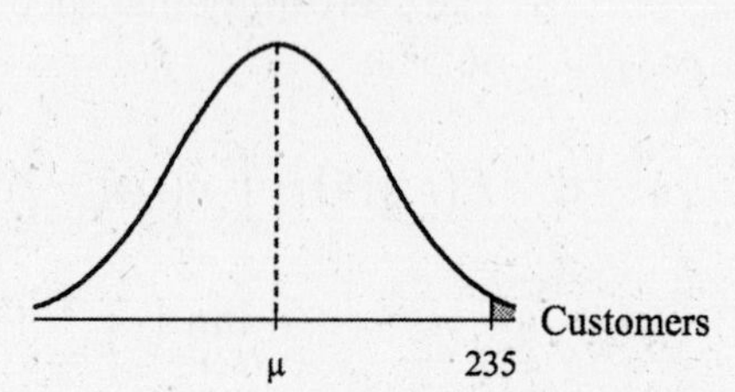

15. a. $6\sigma = 800 - 200 = 600$, so $\sigma = 100$.

b. The realigned mean score is more for each distribution, because the recentering puts the mean of each distribution at 500.

c. 50% of students were expected to make a math score of at least 475 under the original score system because 475 was the mean before the recentering.

d. No, there is not enough information because we do not know the standard deviation.

e. Yes, the shape of the distribution changes in order to make comparison between the math and verbal scores. Note that there is no shift since the overall scale remains from 200 to 800.

f. The scale was recentered for interpretation purposes. The recentering does not reflect any change in student performance. Entrance requirements and other comparisons will now be made on the new scale.

17. a. Use technology to estimate the value of the integral.

$$P(60 \le x \le 80) = \int_{60}^{80} \frac{1}{28.65\sqrt{2\pi}} e^{\frac{-(x-72.3)^2}{2(28.65^2)}} dx \approx 0.272$$

Approximately 27.2% of the students are likely to make a score between 60 and 80.

b. Use technology to estimate the value of the integral.

$$P(x > 90) = 0.5 - \int_{72.3}^{90} \frac{1}{28.65\sqrt{2\pi}} e^{\frac{-(x-72.3)^2}{2(28.65^2)}} dx \approx 0.5 - 0.232 = 0.268$$

Approximately 26.8% of the students are likely to make a score of at least 90.

c. Use technology to estimate the value of the integral.
$P(x < 43.65 \text{ or } x > 100.95) = 1 - P(43.65 \le x \le 100.95)$

$$= 1 - \int_{43.65}^{100.95} \frac{1}{28.65\sqrt{2\pi}} e^{\frac{-(x-72.3)^2}{2(28.65^2)}} dx \approx 1 - 0.683 = 0.317$$

Approximately 31.7% of the students made a score that was more than one standard deviation away from the mean.

d. The rate of change is a maximum at one standard deviation less than the mean, $72.3 - 28.65 = 43.65$.

19. a. $\mu = \int_{-\infty}^{\infty} xu(x)dx = \frac{1}{b-a}\int_a^b xdx = \frac{1}{b-a}\left(\frac{x^2}{2}\Big|_a^b\right) = \frac{1}{b-a}\left(\frac{b^2-a^2}{2}\right) = \frac{b+a}{2}$

b. $\sigma^2 = \frac{1}{b-a}\int_a^b (x-\mu)^2 dx = \frac{1}{b-a}\left[\frac{(x-\mu)^3}{3}\Big|_a^b\right] = \frac{1}{3(b-a)}\left[(b-\mu)^3 - (a-\mu)^3\right]$

$= \frac{1}{3(b-a)}\left[\left(b - \frac{b+a}{2}\right)^3 - \left(a - \frac{b+a}{2}\right)^3\right] = \frac{1}{3(b-a)}\left[\left(\frac{b-a}{2}\right)^3 - \left(\frac{a-b}{2}\right)^3\right]$

$= \frac{1}{24(b-a)}\left[(b-a)^3 - (a-b)^3\right] = \frac{1}{24(b-a)}\left[(b-a)^3 + (b-a)^3\right]$

$= \frac{2(b-a)^3}{24(b-a)} = \frac{(b-a)^2}{12}$

So $\sigma = \frac{b-a}{\sqrt{12}}$

c. When $x < a$, $F(x) = \int_{-\infty}^{x} f(t)dt = \int_{-\infty}^{x} 0dt = 0$

When $a \le x \le b$, $F(x) = \int_{-\infty}^{x} f(t)dt = \int_a^x \frac{1}{b-a}dt = \frac{1}{b-a}\left(t\Big|_a^x\right) = \frac{1}{b-a}(x-a) = \frac{x-a}{b-a}$

When $x > b$, $F(x) = \int_{-\infty}^{x} f(t)dt = \int_a^b \frac{1}{b-a}dt = \frac{1}{b-a}\left(t\Big|_a^b\right) = \frac{1}{b-a}(b-a) = 1$

21. a. See figure in Answer Key in back of Text.
b. See figure in Answer Key in back of Text.

23. a. When $x < 0$, $F(x) = 0$. When $0 \le x < 1$, $F(x) = \int_{-\infty}^{x} f(t)dt = \int_0^x 2tdt = t^2\Big|_0^x = x^2$

When $x \ge 1$, $F(x) = 1$.

Thus $F(x) = \begin{cases} 0 & \text{when } x < 0 \\ x^2 & \text{when } 0 \le x < 15 \\ 1 & \text{when } x \ge 1 \end{cases}$

b. $P(x < 0.67) = \int_{-\infty}^{0.67} f(x)dx = \int_0^{0.67} 2xdx = x^2\Big|_0^{0.67} = 0.67^2 = 0.4489$

$P(x < 0.67) = F(0.67) = 0.67^2 = 0.4489$

c. $P(x > 0.25) = 1 - P(x \le 0.25) = 1 - F(0.25) = 1 - 0.25^2 = 0.9375$

d.

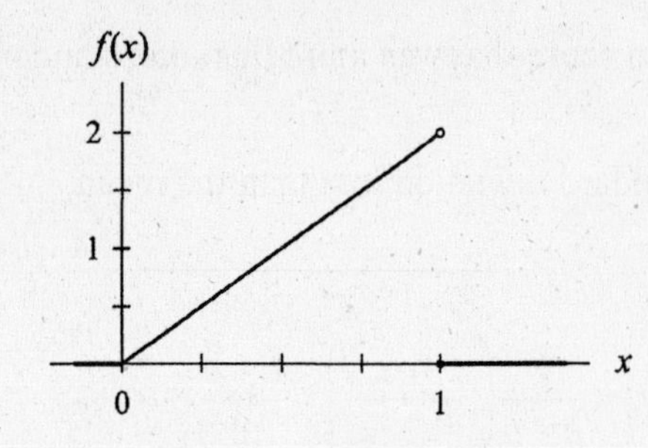

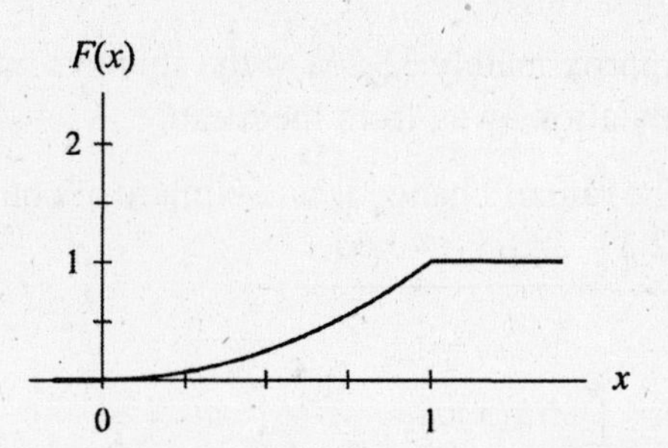

25. a.

x	5	10	15	25	35	50	75	100
$F(x)$	8.2	17.8	27.6	46.4	62.2	78.8	92.1	96.4

x	2.5	7.5	12.5	20	30	42.5	62.5	87.5
$f(x)$	1.64	1.92	1.96	1.88	1.58	1.107	0.532	0.172

c. Yes, there appears to be an inflection point near $x = 40$, and the limiting value as x increases appears to be 0.

d. The fit appears to be good.

e. Use technology to estimate the value of the definite integral.

$$\int_0^{100} f(x)dx = \int_0^{100} 0.85403395x^{0.56}e^{-0.043798079x}dx \approx 96.41\% \approx F(100)$$

f. Note that we evaluate $\int_0^{100} \frac{xf(x)}{100}dx$ to find the mean. Because f is a function that gives a percent, we divide by 100 to convert it to a probability density function. Use technology to approximate the value of the integral.

$$\mu = \int_0^{100} \frac{xf(x)}{100}dx = \int_0^{100} 0.85403395x^{1.56}e^{-0.043798079x}dx \approx 31.117$$

The integral is in thousands of dollars. Thus the mean is approximately $31,000.

27. a. The distribution on the top (the bell-shaped curve) is characteristic of a normal breeding population. The graph on the bottom (with three peaks) is what would be expected from a controlled population in which compys were introduced in three batches.

b. The dinosaurs were supposed to be all female and, therefore, not to reproduce, so Malcolm saw the normal curve.

c. Neither graph is a probability density function because the area under each curve is greater than 1.

Chapter 7 Review Test

1. **a.** $R(t) = 0.10(3000(12) + 500t) = 0.10(36{,}000 + 500t) = 3600 + 50t$ dollars per year after t years

   b. Use technology to estimate the value of the integral:

   Six-year future value $= \int_0^T R(t)e^{r(T-t)}dt = \int_0^6 (3600 + 50t)e^{0.083(6-t)}dt = \$29{,}064$

   c. Present value:

$$Pe^{rt} = \$29{,}064$$

$$Pe^{(0.083)(6)} = 29{,}064$$

$$P = \frac{29{,}064}{e^{(0.083)(6)}} \approx \$17{,}664$$

2. **a.** The monthly rate of investment is $R(t) = 0.07[3100(1.0004^t)] = 217(1.0004^t)$ dollars per month after t months. There are 8(12)=96 compounding periods.

   96-month future value $= \sum_{t=0}^{95} 217(1.0004^t)\left(1 + \frac{0.082}{12}\right)^{96-t} = \$35{,}143.80$

   b. Present value:

$$P\left(1 + \frac{r}{n}\right)^{nt} = \$35{,}143.80$$

$$P\left(1 + \frac{0.082}{12}\right)^{12(8)} = 35{,}143.80$$

$$P = \frac{35{,}143.80}{\left(1 + \frac{0.082}{12}\right)^{96}} = \$18{,}277.65$$

3. **a.** $10{,}000(0.63^{20}) \approx 1$ fox

   b. $500\left(0.63^{20-t}\right)$ foxes born t years after 1990 that will still be alive in 2010

   c. $10{,}000(0.63^{20}) + \int_0^{20} 500\left(0.63^{20-t}\right)dt \approx 1083$ foxes

4. **a.** By solving $D(p) = 0$ we find that the demand curve crosses the p-axis at $p \approx 21.42$ hundred dollars. By solving $D(p) = 30$ we find that the price associated with a demand of 30 fountains is $p \approx 20.94$ hundred dollars. Consumers' willingness and ability to spend is calculated as

   (30 fountains)(20.94 hundred dollars per fountain) + $\int_{20.94}^{21.42} D(p)dp$

   $\approx \$62{,}827 + \$716 = \$63{,}543$

   b. At \$1000 each, suppliers will supply $S(10) = 0.3(100) + 8.1(10) + 300 = 411$ fountains. At \$1000 each, the demand will be $D(10) = -100 - 20.6(10) + 900 = 594$ fountains. Supply will not exceed demand at this price.

c. We find the equilibrium point by finding where $S(p) = D(p)$. The input of the equilibrium point is $p \approx 13.11$ hundred dollars. The total social gain is

$$\int_2^{13.11} S(p)dp + \int_{13.11}^{21.42} D(p)dp$$

$$= 4239.66 + 1996.35 \approx 6236.0 \text{ hundred dollars} = \$623{,}600$$

5. a. $P(x < 3.8) = \int_0^{3.8} 0.125x\,dx = 0.0625x^2\Big|_0^{3.8} = 0.9025$. The value of *x* will be smaller than 3.8 about 90% of the time. This event is likely to occur.

b.

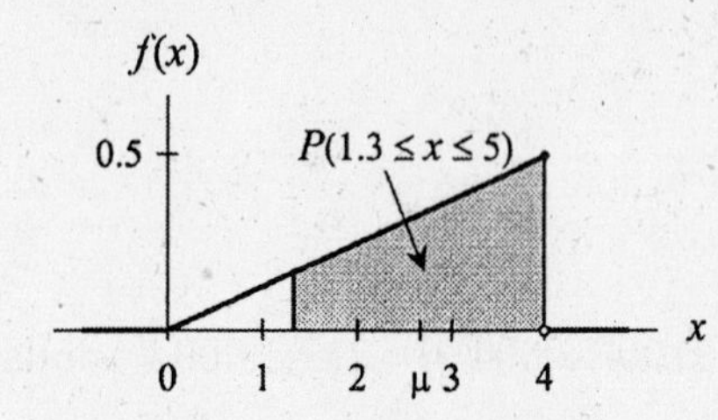

$$P(1.3 \le x \le 5) = \int_{1.3}^{4} 0.125x\,dx$$

$$= 0.0625x^2\Big|_{1.3}^{4}$$

$$= 0.894375$$

c. $\mu = \int_0^4 x(0.125x)dx = \frac{0.125}{3}x^3\Big|_0^4 \approx 2.6667$

d.

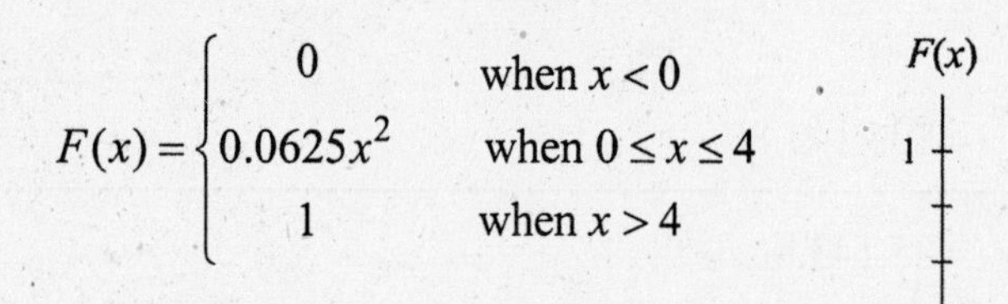

$$F(x) = \begin{cases} 0 & \text{when } x < 0 \\ 0.0625x^2 & \text{when } 0 \le x \le 4 \\ 1 & \text{when } x > 4 \end{cases}$$

e. $P(1.3 < x < 5) = F(5) - F(1.3) = 1 - 0.0635(1.3^2) = 1 - 0.105625 = 0.894375$

Chapter 8

Section 8.1 Functions of Angles: Sine and Cosine

1.

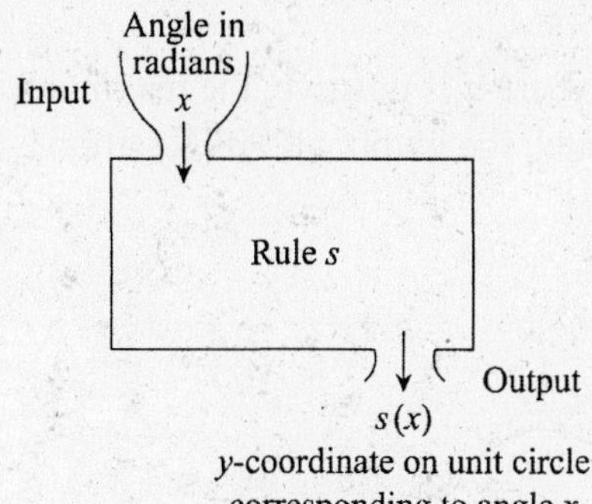

3. a. The maximum value is 1 and occurs at $x = \frac{\pi}{2}$.

b. The absolute maximum value is 1 and occurs at $x = \frac{\pi}{2} + 2k\pi$, where k is an integer.

c. The minimum value is -1 and occurs at $\frac{3\pi}{2}$.

d. The absolute minimum value is -1 and occurs at $x = \frac{3\pi}{2} + 2k\pi$, where k is an integer.

5. a.

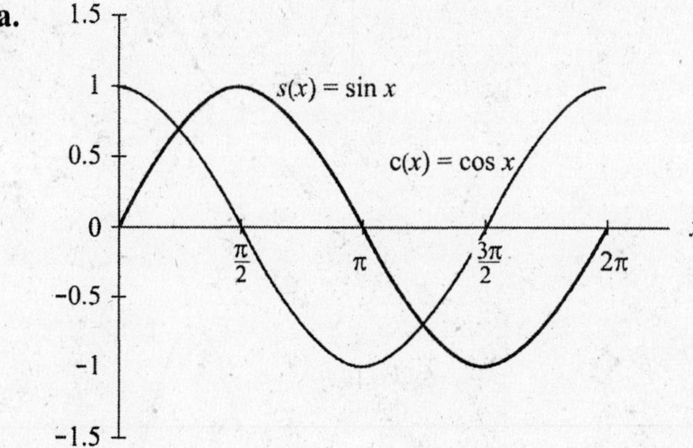

b. $c(x) = s(x) \approx 0.707$ when $x \approx 0.785$, and $c(x) = s(x) \approx -0.707$ when $x \approx 3.927$.

c. $x \approx 0.785 + 2k\pi$, where k is an integer, and $x \approx 3.927 + 2k\pi$, where k is an integer.

7. Amplitude: 1

Frequency: 1

Period: $\frac{2\pi}{1} = 2\pi$

Vertical shift: none

Horizontal shift: right $\frac{\pi}{1} = \pi$

Horizontal axis reflection: yes

At $x = \frac{-h}{b} = \pi$: decreasing

9. Amplitude: 235

Frequency: 300

Period: $\frac{2\pi}{300} \approx 0.02$

Vertical shift: down 65

Horizontal shift: right $\frac{100}{300} = \frac{1}{3}$

Horizontal axis reflection: yes

At $x = \frac{-h}{b} = \frac{1}{3}$: decreasing

11. Observe that for a graph of the form $y = a \sin x + d$, where a is positive, the maximum is $d + a$ and the minimum is $d - a$. Use this information to match the graph with its equation.

a.	iii	**b.**	vi	**c.**	i
d.	v	**e.**	ii	**f.**	iv

13. Amplitude: 1

Frequency: 2

Period: $\frac{2\pi}{2} = \pi$

Vertical shift: up 3

Horizontal shift: none

Horizontal axis reflection: no

At $x = \frac{-h}{b} = 0$: increasing

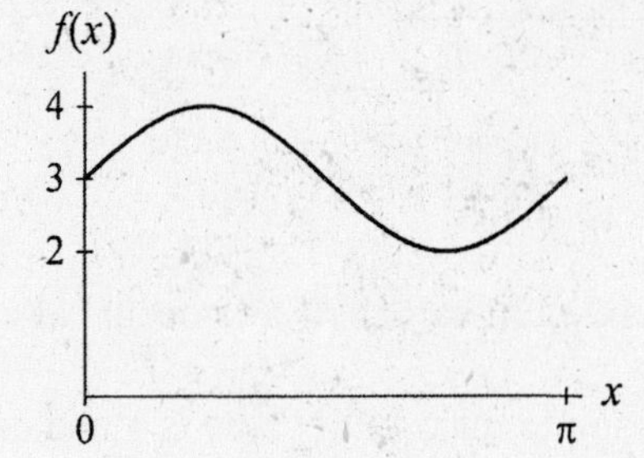

15. Amplitude: 2

Frequency: 1

Period: $\frac{2\pi}{1} = 2\pi$

Vertical shift: none

Horizontal shift: left $\frac{3}{1} = 3$

Horizontal axis reflection: no

At $x = \frac{-h}{b} = -3$: increasing

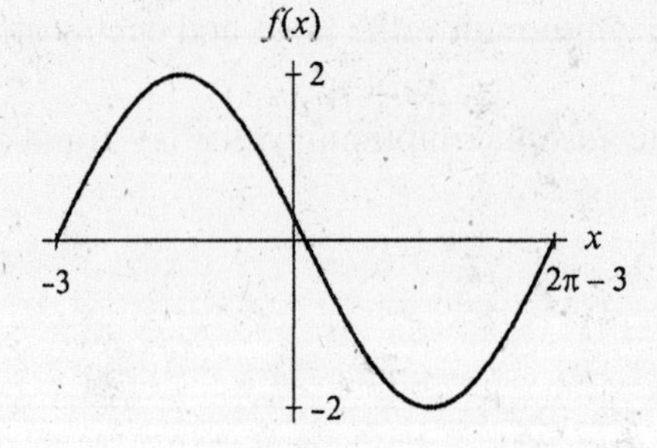

17. Amplitude: 2

Frequency: 2

Period: $\frac{2\pi}{2} = \pi$

Vertical shift: none

Horizontal shift: right $\frac{3}{2}$

Horizontal axis reflection: yes

At $x = \frac{3}{2}$: decreasing

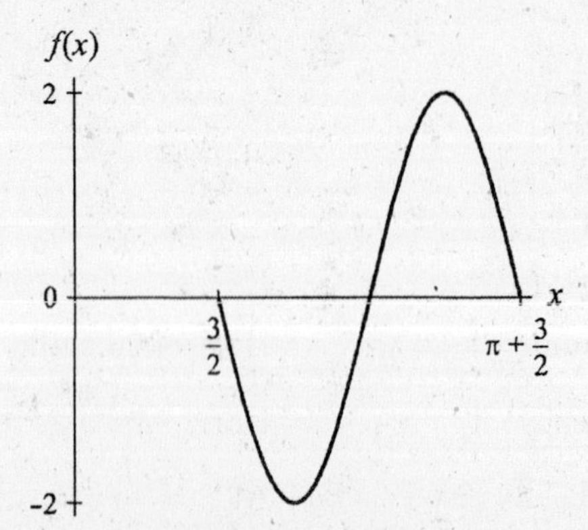

19. The members of each pair of graphs are identical. Generalization: Begin with a sine graph that is not vertically shifted, and reflect it across the horizontal axis. Begin with the same sine graph and shift it right or left by an integer multiple of half the period of the original graph. The new graph will be identical to the reflected graph.

21. a. See figure on Answer Key page A-57 in Text. The graph appears to have rotated about the horizontal axis as a result of taking the negative of the input values of the function.

b. See figure on Answer Key page A-57 in Text. The graph appears not to have rotated at all. In actuality the graph has been rotated about the vertical axis.

c. See figure on Answer Key page A-57 in Text.

23. Quill Activity

Section 8.2 Sine Functions as Models

1. **a.** The data are cyclic because the residential gas usage is determined by the daily temperature, which usually repeats itself over a year's time.

   b. Amplitude $\approx \frac{3.3 - 0.3}{2} = 1.5$ therms per day

   c. Period $\approx$ 12 months
   horizontal shift $\approx$ 0.57 months to the left

   d. $G(x) = 1.5\sin(0.52x + 0.57) + 1.8$ therms per day of gas used daily in month x

3. **a.**

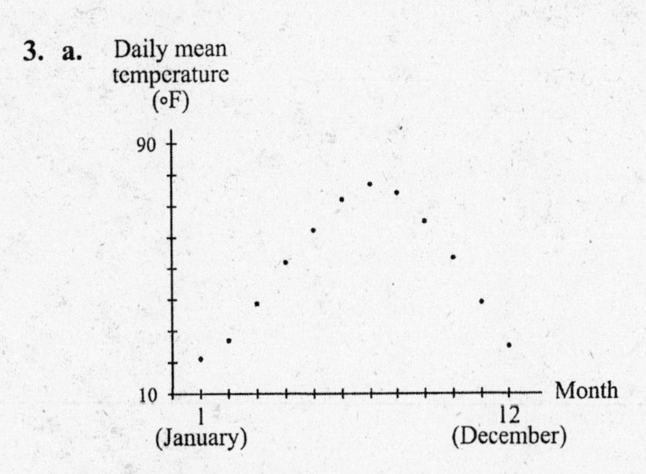

The data are cyclic because temperatures vary as a result of seasons, which are cyclic. The average daily temperature will repeat itself every year.

   b. Amplitude $= \frac{76.9 - 21.1}{2} = 27.9$ °F; Vertical shift = 21.1 + 27.9 = 49°F

   c. Period $\approx$ 12 months

   To estimate the horizontal shift, we find the value half way between the highest and lowest temperatures. This value is 51.9°F, occurring in April. Thus the data are shifted right by 4 months: Horizontal shift $\approx$ 4 months to the right

   d. On the basis of the answers to parts b and c, we have $a = 27.9$, $b = \frac{2\pi}{12} = \frac{\pi}{6} \approx 0.52$,

   $h = 4\left(\frac{\pi}{6}\right) = \frac{2\pi}{3} \approx 2.09$, and $k = 49$.

   $m(x) = 27.9\sin\left(\frac{\pi}{6}x - \frac{2\pi}{3}\right) + 49 \approx 27.9\sin(0.52x - 2.09) + 49$ °F

   during the xth month of the year

e.

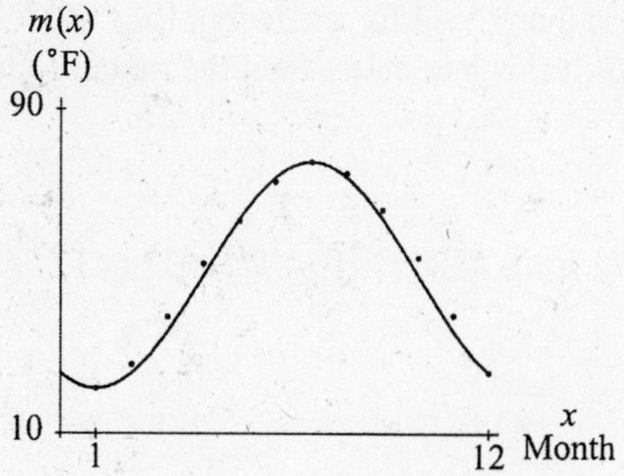

$m(x)$
(°F)
90

10
1
12
x
Month

The fit appears to be reasonable.

f. $x = 7$ in July and $f(7) = 27.9$

$\sin(0.52 \cdot 7 - 2.09) + 49 \approx 76.9$.

The mean daily temperature in July of this year is about 76.9 °F. This result agrees with the value in the table. Although the function is an excellent description of the normal average daily temperature for the years 1961-1990, it cannot predict with any certainty the mean daily temperatures in Omaha for a particular July.

5. a. The low temperatures should have the same period and approximately the same amplitude and horizontal shift as the average temperatures, but they should have a smaller vertical shift.

b. Amplitude $\approx \dfrac{65.9 - 10.9}{2} = 27.5$ °F; Vertical shift $\approx 10.9 + 27.5 = 38.4$°F

Horizontal shift $\approx$ 4 months to the right; Period $\approx$ 12 months

As expected, the period and horizontal shift are the same. The amplitude is slightly lower than for the mean temperature data, and the vertical shift is less.

c. We calculate b using the equation $12 = \frac{2\pi}{b}$, so $b = \frac{\pi}{6}$. We calculate h using the equation $\frac{|h|}{b} = 4$ and the information that the horizontal shift is to the right, so h is negative: $h = -4b = \frac{-2\pi}{3}$. Using $a = 27.5$ and $k = 38.4$, we have $f(x) = 27.5 \sin(\frac{\pi}{6}x - \frac{2\pi}{3}) + 38.4$ °F during the xth month of the year.

7. a. A possible model is $g(x) = 1.610 \sin(0.534x - 0.211) + 1.591$ therms per day where x is the number of months after November 2000.

Amplitude = 1.610 therms per day

Period $\approx \dfrac{2\pi}{0.534} \approx 11.8$ months

Frequency $\approx$ 0.534 periods

Horizontal shift $= \dfrac{0.211}{0.534} \approx 0.395$ month to the right

Vertical shift $\approx$ up 1.591 therms per day

b. The expected value is approximately 1.6 therms per day.

c. The average gas bill is approximately
(1.591 therms per day)(30 days)($0.56532 per therm) $\approx$ $26.89.

9. Using technology, a model is $g(x) = 28.5328 \sin(0.4789x - 1.8013) + 47.9847$ °F during the xth month of the year. The b and k values in Activity 2 are slightly more than those for the technology model, while a and h are slightly less than those for the technology model. The function obtained by using technology seems to fit the data better than the model found in Activity 3.

11. a. Using technology, a model is $g(x) = 27.3572 \sin(0.4886x - 1.8777) + 37.6413$ °F during the xth month of the year.

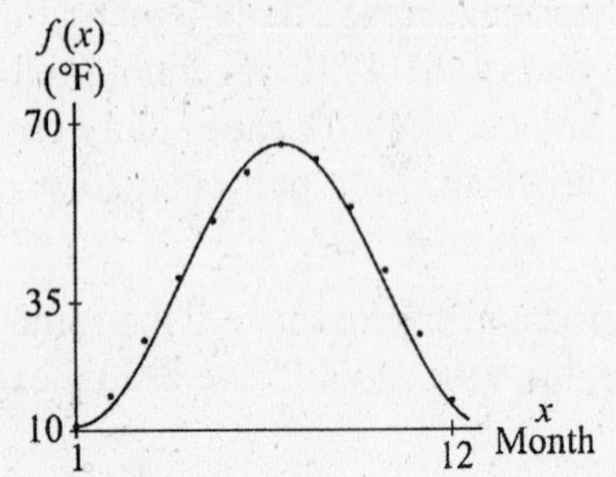

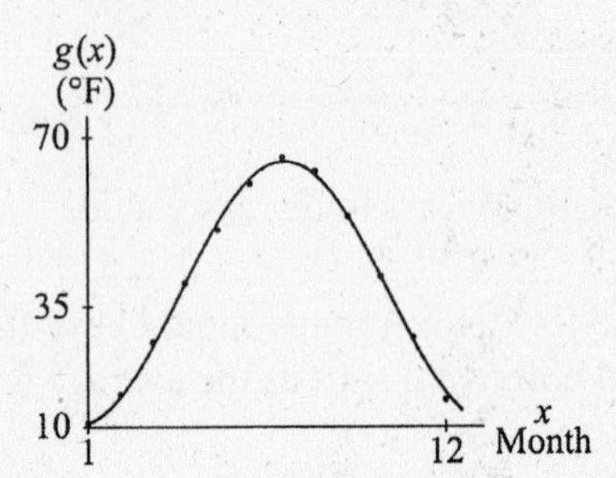

Both models appear to fit the data well.

b. Answers will vary.

13. a. $M(x) = 0.932\sin(0.467x - 2.940) + 8.737$ billion trips x years after 1992

b. $T(x) = -0.0052x^3 + 0.127x^2 - 0.573x + 8.56$ billion trips x years after 1992

c. The cubic function fits the data fairly well, but the sine function better follows the curvature of the data and provides a very good fit.

d. $M(14) \approx 8.32$ billion trips

$T(14) \approx 11.24$ billion trips

Discussion will vary.

15. a. The highest mean daily temperature is $a + k = 37 + 25 = 62$ °F. The lowest mean daily temperature is $-a + k = -37 + 25 = -12$ °F.

b. The average of the highest and lowest mean daily temperatures is $\dfrac{62 + (-12)}{2} = 25$°F, which is k, the vertical shift of the graph of f.

17. a. Let x be the number of years since 1949. The data appear concave up between 1949 and 1955, concave down between 1956 and 1960, and concave up between 1961 and 1963. A cubic model is not a good fit because more than one inflection point are indicated by the data. Possibly a piecewise continuous model that is quadratic between 1949 and 1955 and cubic between 1956 and 1963 would fit the data. However, the sine model seems most appropriate.

b. Using technology, a sine model is $f(x) = 41.5473 \sin(0.5284x - 2.9132) + 194.4930$ aircraft x years after 1949.

c. 1964 is 15 years since 1949. $f(15) \approx 155$ aircraft

d. Answers will vary.

19. a.

Hours	°F	Hours	°F

5	37	65	60
11	44.5	71	47
17	52	77	34
23	47	83	45
29	42	89	56
35	49	95	46.5
41	56	101	37
47	49	107	45.5
53	42	113	54
59	51		

b. Using technology, a sine model is $f(x) = 8.7094 \sin(0.2608x - 2.8404) + 47.0895$ °F x hours after midnight on Wednesday. The period of this model is $\frac{2\pi}{0.2608} \approx 24$ hours.

c. High temperature $= k + a \approx 56$°F; low temperature $= k - a \approx 38$°F

d. The greatest discrepancies between the model and the data occur on Friday afternoon and Saturday morning. The hydroelectric plant's energy may also be needed Wednesday afternoon and Sunday morning.

21. Answers will vary.

23. Without using technology:

Vertical shift $= k = \frac{\text{max} + \text{min}}{2} = \frac{18.5 + 4.5}{2} = 11.5$

Amplitude $= a = \frac{18.5 - 4.5}{2} = 7$

Period $= 365 = \frac{2\pi}{b}$, so $b = \frac{2\pi}{365}$

Horizontal shift $= \frac{181 + 361}{2} = 271$ to the right

$\frac{|h|}{b} = 271$, so $|h| = 271b = \frac{271(2\pi)}{365}$ Because the shift is to the right, h is negative.

The model is $D(t) = 7 \sin\left(\frac{2\pi}{365}t - \frac{542\pi}{365}\right) + 11.5$ hours of daylight t days after December 31 of the previous year.

Using technology, the model is $d(t) = 6.6767 \sin(0.0164t - 1.9083) + 11.7299$ hours of daylight t days after December 31 of the previous year.

25. Excel Activity

Use technology to find sine models for each odor. For each model, you may need to give an approximate period to avoid a singular matrix in creating a sine model. To estimate the period, determine when the firing rate first returns to its beginning value. For example, for cherry odors the starting value is 14 and the value is 15 at 50 m/s, so we use 50 as an approximate period. Models may vary.

Cherry odor (approximate a period of 50 ms):
$c(x) = 6.9302 \sin(0.1129x + 2.0378) + 8.8556$ firings/ms with period $\approx$ 55.7 ms

Apple odor (approximate a period of 40 ms):
$a(x) = 7.0726 \sin(0.1327x + 1.6205) + 10.5255$ firings/ms with period $\approx$ 47.3 ms

Cherry/apple odor (approximate a period of 50 ms):
$t(x) = 3.2213 \sin(0.1203x + 1.9576) + 5.6179$ firings/ms with period $\approx$ 52.2 ms

The periods appear to be close, although we do not have enough information to determine if the difference between them is significant or not.

27. Excel Activity

a. Amplitude $\approx \dfrac{33-0}{2} = 16.5$ thousand lizards; period $\approx$ 12 months

vertical shift $\approx$ 16.5 thousand lizards

b. $L(m) = 16.298\sin(0.537m - 1.801) + 11.206$ thousand lizards in month m where m is the number of months since the beginning of year 1. The amplitude of this model is approximately 16.3 thousand lizards, only 0.2 thousand lizards less than the estimated amplitude in part a. The period is $\dfrac{2\pi}{0.537} \approx 11.7$ months, compared with the 12 months estimated in part a. The vertical shift is significantly less than the one estimated in part a.

c.

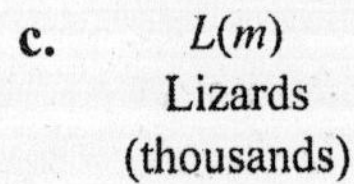

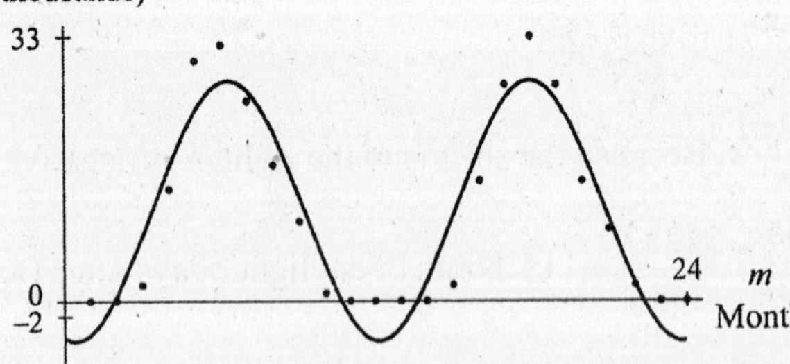

The graph appears to match the period of the data well, but the graph falls below the horizontal axis, and doesn't reach the highest values.

d. $l(x) = 15.388\sin(0.787x - 1.667) + 16.346$ thousand lizards in month x where $x = 1$ through $x = 8$ correspond to March through October of year 1 and $x = 9$ through $x = 16$ correspond to March through October of year 2. The amplitude of this function is 15.388, about 1 less than the amplitude of the equation in part b. The period is approximately 8 months instead of the 11.9 months in part b. This smaller period is the result of deleting the data for 4 months of each year. The vertical shift of 16.3 is greater than the vertical shift of the equation in part b. This vertical shift better matches the shift calculated in part a. This equation fits the modified data well.

Section 8.3 Rates of Change and Derivatives

1. $f'(x) = \frac{d}{dx}(\sin 3x) + 0 = 3\cos 3x$

3. $\frac{dt}{dr} = 5.2\frac{d}{dr}\cos(0.45r + \pi) + 80 - 0 = 5.2(-0.45)\sin(0.45r + \pi) + 80$

5. $h'(x) = 2.08\frac{d}{dx}\sin(-0.16x + 12.3) + 3.58\frac{d}{dx}\cos(2.7x + 8.1) - 0$

$$= 2.08(-0.16)\cos(-0.16x + 12.3) + 3.58(2.7)[-\sin(2.7x + 8.1)]$$

$$= -0.3328\cos(-0.16x + 12.3) - 9.666\sin(2.7x + 8.1)$$

$$h''(x) = -0.3328\frac{d}{dx}\cos(-0.16x + 12.3) - 9.666\frac{d}{dx}\sin(2.7x + 8.1)$$

$$= -0.3328(-0.16)[-\sin(-0.16x + 12.3)] - 9.666(2.7)\cos(2.7x + 8.1)$$

$$= -0.053248\sin(-0.16x + 12.3) - 26.0982\cos(2.7x + 8.1)$$

7. **a.** It takes the satellite 138 – 18 = 120 minutes = 2 hours to complete one orbit.

b. The greatest distance south of the equator is 3050 kilometers at 48 minutes after launching.

c.

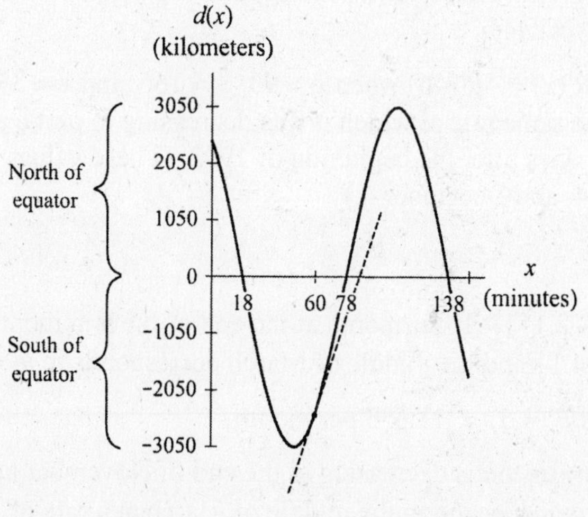

d(x) (kilometers)

3050, 2050, 1050, 0, -1050, -2050, -3050

North of equator

South of equator

18, 60, 78, 138

x (minutes)

d. The rate of change is approximately 93.9 kilometers per minute.

e. No, the answer to part *d* is the rate of change of the distance from the equator with respect to time. It is not speed, which is the rate of change of the total distance traveled with respect to the traveling time.

9. **a.** Rate of change in 1992 ≈ -0.4 billion trips per year
Rate of change in 1996 ≈ 0.2 billion trips per year
Rate of change in 2000 ≈ 0.3 billion trips per year

b. Approximately 9.7 billion trips

11. a. Estimates will vary. About $9 - 17 = -8$ mm/day

b. Percentage change $= \frac{9-17}{17} 100\% \approx 47.1\%$ decline

Average rate of change $= \frac{9-17}{11-6} \approx -1.6$ mm/day/month

c. Estimates will vary. About 2.4 mm per day per month. Extraterrestrial radiation in Amarillo is increasing by 2.4 millimeters of equivalent water evaporation per day per month in March.

d. Estimates will vary. Percentage rate of change $= \frac{2.3}{12.5} 100\% \approx 18.4\%$ per month

13. a. $R(x) = 0.7c(x) + 0.2p(x)$

$= 1.32727 \sin(0.0186x + 1.1801) + 1.46052 \cos(0.0197x - 3.7526) + 7.90076$

million dollars, where x is the day of the year

b. $R'(x) = 1.32727(0.0186)\cos(0.0186x + 1.1801) -$

$1.46052(0.0197)\sin(0.0197x - 3.7526)$ million dollars per day

where x is the day of the year

c. $R'(46) \approx -0.003$ million dollars per day
Combined sales were decreasing by about \$3000 per day on February 15, 1992.

d. Using technology, we find that $R'(x) = R'(46)$ when $x \approx 220$ and $x \approx 293$. Thus, the revenue was decreasing at the same rate as that found in part c at approximately 220 and 293 days after the beginning of 1992. (These values correspond to August 7 and October 19. Recall that 1992 was a leap year.)

e. Using technology, we find that $R'(x) = -R'(46)$ when $x \approx 90$, $x \approx 166$, and $x \approx 339$. Thus the revenue was increasing at the same rate at which it was decreasing in part c at approximately 90, 166, and 339 days after the beginning of 1992. (These values correspond to March 30, June 14, and December 4.)

15. $B'(m) = 22.926(0.510)\cos(0.510m - 2.151)$ °F per month at the end of the mth month. The end of November corresponds to $m = 11$ and the middle of March corresponds to $m = 4.5$.

$B'(11) \approx -11.1$°F per month and $B'(4.5) \approx 11.6$°F per month

The normal daily high temperature in Boston is decreasing at the end of November at a rate of approximately 11.11°F per day and is increasing in the middle of March at a rate of approximately 11.57°F per day.

17. a. From 6 m/sec to 10 m/sec: change $= P(10) - P(6) \approx -17.42$ kW
From 10 m/sec to 14 m/sec: change $= P(14) - P(10) \approx -17.26$ kW
From 14 m/sec to 18 m/sec: change $= P(18) - P(14) \approx -0.29$ kW

b. From 6 m/sec to 10 m/sec: $\text{Percentage change} = \frac{P(10)-P(6)}{P(6)}100\% = 38.4\%$ decline

$\text{Average rate of change} = \frac{P(10)-P(6)}{10-6} \approx -4.36 \text{ kW/m/sec}$

From 10 m/sec to 14 m/sec: $\text{Percentage change} = \frac{P(14)-P(10)}{P(10)}100\% = 61.7\%$ decline

$\text{Average rate of change} = \frac{P(14)-P(10)}{14-10} \approx -4.31 \text{ kW/m/sec}$

From 18 m/sec to 14 m/sec: $\text{Percentage change} = \frac{P(18)-P(14)}{P(14)}100\% = 2.7\%$ decline

$\text{Average rate of change} = \frac{P(18)-P(14)}{18-14} \approx -0.07 \text{ kW/m/sec}$

c. $P'(s) = 20.204(0.258)\cos(0.258s + 0.570)$ kW/m/sec when the wind speed is s m/sec

d. $P'(6) \approx -2.71$ kW/m/sec
The power output of the engine is decreasing at the rate of 2.7 kilowatts per meter per second at a wind speed of 6 m/s.

$P'(10) \approx -5.21$ kW/m/sec
When the wind speed is 10 m/s and increases to 11 m/s, the power output of the engine decreases by approximately 5.2 kilowatts

$P'(14) \approx -2.64$ kW/m/sec
When the wind speed is 14 m/s and increases to 15 m/s, the power output of the engine decreases at by approximately 2.6.

$P'(18) \approx 2.51$ kW/m/sec
At a wind speed of 18 m/s, the power output of the engine is increasing at the rate of 2.5 kilowatts per meter per second

e. At 6 meters per second: $\text{Percentage rate of change} = \frac{P'(6)}{P(6)}100\% \approx -6.0\%$ per m/sec

At 10 meters per second: $\text{Percentage rate of change} = \frac{P'(10)}{P(10)}100\% \approx -18.6\%$ per m/sec

At 14 meters per second: $\text{Percentage rate of change} = \frac{P'(14)}{P(14)}100\% \approx -24.7\%$ per m/sec

At 18 meters per second: $\text{Percentage rate of change} = \frac{P'(18)}{P(18)}100\% \approx -24.1\%$ per m/sec

19. a. Using technology, a sine model is $d(t) = 60.7407 \sin(0.1206t + 0.3134) + 99.9194$ deaths per 100,000 people per week, where t is the number of weeks since January 1, 1923. (Answers will vary.)

b. $d'(t) = 60.7407(0.1206)\cos(0.1206t + 0.3134)$ deaths per 100,000 people per week per week, where t is the number of weeks since January 1, 1923
The middle of 1924 corresponds to $t = 78$.
$d'(78) \approx -7.01$ deaths per 100,000 people per week per week

The end of 1924 corresponds to $t = 104$.
$d'(104) \approx 7.02$ deaths per 100,000 people per week per week

21. a. $D(d) = 23.677\sin(0.017d - 1.312) - 0.292$ degrees on the dth day of the year
The amplitude of the model is 23.7 degrees. This is the greatest angle of declination the sun reaches. The period is approximately 374 days (calculated with the unrounded b value). We expect the sun's declination to complete a cycle every year. The period in the model is slightly long.

b. The equinoxes occur at $d \approx 78.7$ and $d \approx 264.4$.

$D'(78.7) \approx 0.4$ degrees per day; $D'(264.4) \approx -0.4$ degree per day

At the equinoxes, the declination of the sun is changing by about 0.4 degree per day.

c. The summer and winter solstices occur when the declination is greatest in both north and south directions. On the graph, the summer solstice corresponds to the maximum and the winter solstice corresponds to the minimum. At these points, the rate of change is zero.

23. Inside: $u(x) = 2.4^x$ Outside: $y(u) = \sin u$

Derivative of inside: $u'(x) = (\ln 2.4)2.4^x$ Derivative of outside: $y'(u) = \cos u$

$$y'(x) = u'(x)y'(u(x)) = (\ln 2.4)2.4^x \cos u = (\ln 2.4)2.4^x \cos(2.4^x)$$

25. Inside: $u(x) = \sin x - 7$ Outside: $y(u) = 4u^2 + 8u + 13$
Derivative of inside: $u'(x) = \cos x$ Derivative of outside: $y'(u) = 8u + 8$

$$y'(x) = u'(x)y'(u(x)) = (\cos x)(8u + 8) = (\cos x)[8(\sin x - 7) + 8]$$
$$= 8\cos x(\sin x - 7 + 1) = 8\cos x(\sin x - 6)$$

27. Inside: $u(x) = \sin x$ Derivative of inside: $\cos x$

Outside: $y(u) = \ln u$ Derivative of outside: $y'(u) = \frac{1}{u}$

$$y'(x) = u'(x)y'(u(x)) = (\cos x)\frac{1}{u} = (\cos x)\frac{1}{\sin x} = \frac{\cos x}{\sin x}$$

29. Excel Activity

a. Using technology, a sine model is $f(x) = 59.6582 \sin(0.5250x - 1.9621) + 105.2150$ thousands of dollars in sales, where x is the number of months since December of the year before Year 1. The fit is fairly close, except near the maxima and minima for the later years.

b. The average of the high and low sales for Year 1 is $\frac{167+50}{2} = \$108.5$ thousand

The average of the high and low sales for Year 2 is $\frac{174+52}{2} = \$113$ thousand

The average of the high and low sales for Year 3 is $\frac{175+55}{2} = \$115$ thousand

The average of the high and low sales for Year 4 is $\frac{180+58}{2} = \$119$ thousand

c. Enter the following data: (5, 108.5), (17, 113), (29, 115), (41, 119). Using technology, a linear function is $y = 0.2792x + 107.4542$ thousand dollars (average of high and low sales), where x is the number of months since December of the year before Year 1.

d. The new model is $f(x) = 59.6582 \sin(0.5250x - 1.9621) + 0.2792x + 107.4542$ thousands of dollars in sales, where x is the number of months since December of the year before Year 1. The fit is better than that of the function in part *a*.

e. $f'(x) = 59.6582(0.5250)\cos(0.5250x - 1.9621) + 0.2792$ thousands of dollars in sales per month, where x is the number of months since December of the year before Year 1.

September of the third year corresponds to $x = 33$.
$f'(33) \approx -\$29.5$ thousand dollars in sales per month
In September of the third year, the sales of ice cream were decreasing at a rate of approximately \$29,500 per month.

January of the fourth year corresponds to $x = 37$.
$f'(37) \approx 6.1$ thousand dollars in sales per month
In January of the third year, the sales of ice cream were increasing at a rate of approximately \$5700 per month.

Section 8.4 Extrema and Points of Inflection

1. a. Between $x = 0$ and $x \approx 13.45$, the maximum value of 9.67 occurs at $x \approx 9.66$. The minimum value of 7.81 occurs at $x \approx 2.93$.

b. The greatest number of yearly mass transit trips between 1992 and 2000 was approximately 9.66 billion (in 2002). The least number of yearly mass transit trips between 1992 and 2000 was approximately 7.81 billion (in 1995).

3. a. The maximum daily mean temperature is approximately 76.9°F (in July).
The minimum daily mean temperature is approximately 21.1°F (in January).

b. Using technology, a sine model is $t(x) = 28.5328 \sin(0.4789x - 1.8013) + 47.9847$ °F in the xth month of the year.
The derivative is $t'(x) = 28.5328(0.4789) \cos(0.4789x - 1.8013)$ °F per month
For $0 < x \le 12$, $t'(x) = 0$ when $x \approx 0.481$ and $x \approx 7.041$, respectively. The maximum point on the model is (7.041, 76.517) and the minimum point on the model is (0.481, 19.452). Thus, according to the model, the maximum normal daily mean temperature in Omaha between 1961 and 1990 was about 76.5°F and occurred at the beginning of August. The minimum normal daily mean temperature was about 19.5°F and occurred near the middle of January.

c. The maximum and minimum temperatures obtained with the model are slightly lower than those found using the data. Because the data are based on a monthly average over a 30-year period, actual high and low temperatures in any given year are likely to vary from those found in parts *a* and *b*.

5. a. Answers may vary since the intervals for 1 year are small and difficult to read from the graph. Some of the peaks of the sine model that appear to be within 1 year of a peak in the wheat price index include the peak between 1762 and 1770, the peak between 1778 and 1786, and the peak between 1834 and 1842.

b. Answers may vary since the intervals for 1 year are small and difficult to read from the graph. Some of the valleys of the sine model that appear to be within 1 year of a valley in the wheat price index include the valley near 1770, the valley near 1778, and the valley near 1858.

c. The next maximum and minimum for the model $y = 100.2 + 4.4 \sin(\frac{\pi}{4}t + \frac{62}{45})$ for $t > 113$ occurs when $t \approx 116.25$ (a relative minimum of about 104.6) and when $t \approx 120.25$ (a relative maximum of about 95.8). The next maximum would occur in the middle of 1882 and the next minimum would occur in the middle of 1878.

d. Answers will vary.

7. a. The blue graph with the summer solstice labeled is the number of hours of daylight. The black graph is the number of hours of darkness.

b. Period $\approx$ 12 months
From the graph, the highest point is about 20 hours and the lowest point is about 5 hours.
Thus amplitude $\approx \dfrac{20-5}{2} = 7.5$ hours and vertical shift $\approx \dfrac{20+5}{2} = \approx 12.5$ hours. (Answers may vary.)

c. Estimates may vary.

Month	Hours	Month	Hours
Jan	5	July	19
Feb	8	Aug	17
Mar	10.5	Sept	14.5
Apr	13.5	Oct	11.5
May	16.5	Nov	8.5
June	19	Dec	6

d. Using technology, a sine model is $H(m) = 7.021 \sin(0.4803m - 1.6189) + 11.7855$ hours at the beginning of the mth month.

Period $= \dfrac{2\pi}{0.4803} \approx 13$ months, amplitude ≈ 7.0 hours, vertical shift ≈ 11.8 hours

9. Answers will vary.

11. a. $a'(m) = 27.1(0.485)\cos(0.485x - 1.707)$ watts per centimeter squared per month when $m = 1$ in January, $m = 2$ in February, and so on.

b. Between $m = 0$ and $m = 12$, $a'(m) = 0$ when $m \approx 0.2808$ and $m \approx 6.7853$.
$a(0.2808) \approx 5.8$ and $c(6.7853) \approx 60$. We compare these outputs with those at the endpoints: $a(0) \approx 6.1$ and $c(61) \approx 10.5$.
Radiation is at its highest level in July and at its lowest level in January.

c. The highest level of radiation is 60 watts per centimeter squared. The lowest level of radiation is 5.8 watts per centimeter squared.

13. a. $T'(x) = 0.565\cos(0.469x - 2.293)$ percent per year x years after 1988.

b. For one period of T, between $x = 0$ and $x \approx 13.4$, solve the equation $T'(x) = 0$ to find that $x \approx 1.54$ (input for relative minimum) and $x \approx 8.24$ (input for relative maximum). Compare the outputs at these inputs to $T(0)$ and $T(13.4)$ to find the absolute extrema. The smallest after-tax profit rate on investment, about 5.18%, occurred in 1990 and the largest rate, about 7.5%, occurred in 1996.

$$T''(x) = 0.565\left(\frac{d}{du}\cos(u)\right)\left(\frac{d}{dx}(0.469x - 2.293)\right) \quad \text{where } u = 0.469x - 2.293$$
$$= 0.565(-\sin(0.469x - 2.293))(0.469)$$
$$= 0.565(0.469)(-\sin(0.469x - 2.293))$$

Solve $T''(x) = 0$ to find that the inputs at the inflection points are $x \approx 4.89$ and $x \approx 11.59$. The most rapid increase in the after-tax profit rate was 0.56 percent per year (in 1993), and the most rapid decrease in the rate was 0.55 percent per year (in 2000).

c. See figure in Answer Key page A-62 in Text. Discussion will vary.

15. a. period $= \dfrac{2\pi}{0.01345} \approx 467.2$ milliseconds

b. $p'(s) = 40.5(0.01345)\cos(0.01345s - 1.5708)$ counts per second per millisecond
$p'(s) = 0$ when $s \approx 4.2$, $s \approx 233.6$, $s \approx 467.2$, and $s \approx 700.8$
The pulse speed was at its highest value, 227 counts per second, after about 233.6

milliseconds and 700.7 milliseconds. The pulse speed was at its lowest value, 146 counts per second, after about 4.2 milliseconds and 467.2 milliseconds.

c. $p''(s) = -40.5(0.01345)(0.01345)\sin(0.01345s - 1.5708)$ counts/second/millisecond^2
$p''(s) = 0$ when $s \approx 116.8$ and $s \approx 350.4$
$p'(116.8) \approx 0.545$ and $p(116.8) \approx 186.5$; $p'(350.4) \approx -0.545$ and $p(350.4) \approx 186.5$
The speed of the pulses emitted from the star was increasing the fastest at approximately 116.8 milliseconds. The speed at that time was about 186.5 counts per second, the speed of the pulses emitted from the star was decreasing the fastest after approximately 350.4 milliseconds when the speed was also 186.5 counts per second.

17. a. Using technology, a model is $d(t) = 60.7407\sin(0.1206t + 0.3134) + 99.9194$ deaths per 100,000 people per week, where t is the number of weeks since January 1, 1923.

b. $d'(t) = 60.7407(0.1206)\cos(0.1206t + 0.3134)$ weekly deaths/100,000 people per week
$d'(t) = 0$ and $d(t)$ is a maximum first beyond the data given when $t \approx 166.8$ weeks after January 1, 1923. Thus the next peak beyond the data given is approximately 166.8 weeks after January 1, 1923.
Answers will vary.

c. $d''(t) = -60.7407(0.1206)(0.1206)\sin(0.1206t + 0.3134)$ weekly deaths/100,000 people per week per week
$d''(t) = 0$ first beyond the data given when $t \approx 153.70$ weeks after January 1, 1923. The first time after 150 weeks past January 1, 1923 that the weekly pneumonia death rate was increasing the fastest was about 153.7 weeks. The weekly death rate at that time was approximately $d(153.7) \approx 100$ deaths per 100,000 people per week.

19. a. $a''(m) = -27.1(0.485)(0.485)\sin(0.485x - 1.707)$ watts per centimeter squared per month per month when $m = 1$ in January, $m = 2$ in February, and so on.

b. Between $x = 60$ and $x = 61$, $a''(m) = 0$ and $a'(m) > 0$ when $m \approx 3.5$ and $a''(m) = 0$ and $a'(m) < 0$ when $m \approx 10.0$.

Radiation is increasing most rapidly in mid-April and decreasing most rapidly at the beginning of October.

c. $a(3.5196) \approx 32.9$ and $a(9.9971) \approx 32.9$

The radiation received at the times in part *b* is 32.9 watts per centimeter squared.

21. a. $R(x) = 0.7c(x) + 0.2p(x)$

$= 1.32727\sin(0.0186x + 1.1801) + 1.46052\cos(0.0197x - 3.7526) + 7.90076$
million dollars, where x is the day of the year

b. $R'(x) = 1.32727(0.0186)\cos(0.0186x + 1.1801) -$
$1.46052(0.0197)\sin(0.0197x - 3.7526)$
$= 0.0247\cos(0.0186x + 1.1801) - 0.0288\sin(0.0197x - 3.7526)$
million dollars per day, where x is the day of the year

c. For x between 0 and 365, $R'(x) = 0$ and $R(x)$ is maximum when $x \approx 192.95$.
$R(192.95) \approx 8.03$; The highest revenue is about \$8.03 million on July 12.

d. For x between 0 and 365, $R'(x) = 0$ and $R(x)$ is minimum when $x \approx 66$ and $x \approx 318$. $R(66.27226) \approx 7.6625$, $R(317.6030) \approx 7.6832$

The lowest revenue is about \$7.66 million on March 7. There is also a local minimum with a revenue of about 7.68 million on November 14.

Section 8.5 Accumulation in Cycles

1. 

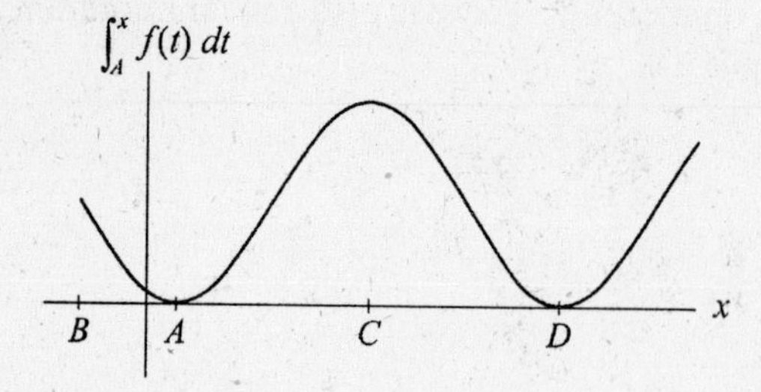

2.

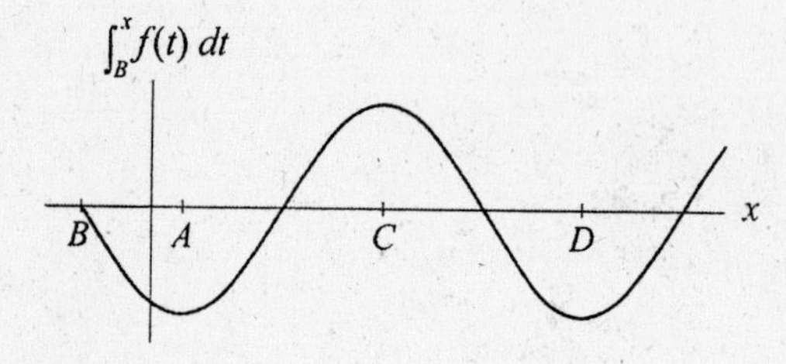

3. $\int_0^a f(x)dx = 1$

4. $\int_{-a}^{c} f(x)dx = 0$

5. $\int_{-a}^{a} [f(x)+1]dx = \int_{-a}^{a} f(x)dx + \int_{-a}^{a} dx = 2 + 2a$

6. $\int_{-c}^{a} 3|f(x)|dx = 3\int_{-c}^{a} |f(x)|dx = 3(4) = 12$

7. $\int (7.3\sin x + 12)dx = -7.3\cos x + 12x + C$

8. $\int 7.3\sin(x+12)dx = -7.3\cos(x+12) + C$

9. $\int \sin(7.3x+12)dx = \frac{-1}{7.3}\cos(7.3x+12) + C$

10. $\int [7.3\sin(7.3x+12)+12]dx = -\frac{7.3}{7.3}\cos(7.3x+12) + 12x + C = -\cos(7.3x+12) + 12x + C$

11. $\int [4.67\sin(0.024x+3.211)+14.63]dx = \frac{-4.67}{0.024}\cos(0.024x+3.211) + 14.63x + C$

$\approx -194.583\cos(0.024x+3.211) + 14.63x + C$

12. $\int [49.88\cos(3.54t-4.86)+7.02]dt = \frac{49.88}{3.54}\sin(3.54t-4.86) + 7.02t + C$

$\approx 14.090\sin(3.54t-4.86) + 7.02t + C$

13. a. Height units are dollars per year, and width units are years. Area units are dollars.

b. The t-intercepts of w between $t = 7$ and $t = 50$ are $t \approx 21.74$ and $t \approx 43.41$. The area between $t = 7$ and $t \approx 21.74$ is \$1.71, the area between $t \approx 21.74$ and $t \approx 43.41$ is \$2.23, and the area between $t \approx 43.41$ and $t = 50$ is \$0.47. The total area is approximately \$4.41.

c. $\int_7^{50} w(t)dt \approx -0.04$ dollars, which is not the same as the result of part b because w crosses the t-axis. The federal minimum wage, expressed in constant 2000 dollars, decreased by approximately 4 cents between 1957 and 2000.

d. See figure in Answer Key page A-63 in Text.

e. The federal minimum wage, expressed in constant 2000 dollars, decreased by approximately 2.16 dollars between 1970 and 1995.

14. a.
$$\int 23.944\cos(0.987t+1.276)dt = \frac{23.944}{0.987}\sin(0.987t+1.276)+C$$
$$\approx 24.259\sin(0.987t+1.276)+C$$

We use the fact that $S(1) = 54$ and solve for C to get $C \approx 35.324$.

$S(t) = 24.259\sin(0.987t + 1.276) + 35.324$ thousand dollars, where $t = 1$ in January, $t = 2$ in February, and so on.

b. Using technology, we find that $S'(t) = 0$ between $t = 0$ and $t = 12$ when $t \approx 0.2987$, $t \approx 3.4817$, $t \approx 6.6646$, and $t \approx 9.8476$.

We compare $S(t)$ values for each of these t-values and conclude that peak sales occur on January 10 ($t \approx 0.2987$) and July 21 ($t \approx 6.6646$) and sales are lowest on April 15 ($t \approx 3.4817$) and October 27 ($t \approx 9.8476$).

c.
$$\int_1^6 23.944\cos(0.987t+1.276)dt = \left(\frac{23.944}{0.987}\sin(0.987t+1.276)\right)\Bigg|_1^6 \approx \$0.548 \text{ thousand}$$

June sales were about \$548 more than January sales.

d. Between $t = 1$ and $t = 6$, $S'(t)$ is below the t-axis for $t < 3.4817$.

$$\int_1^6 |S'(t)|dt = -\int_1^{3.4817} 23.944\cos(0.987t+1.276)dt + \int_{3.4817}^6 23.944\cos(0.987t+1.276)dt$$
$$= -\left(\frac{23.944}{0.987}\sin(0.987t+1.276)\right)\Bigg|_1^{3.4817} + \left(\frac{23.944}{0.987}\sin(0.987t+1.276)\right)\Bigg|_{3.4817}^6$$
$$\approx 42.93524 + 43.483701 \approx \$86.411 \text{ thousand}$$

e.
$$\frac{\int_1^6 S(x)dx}{6-1} = \frac{\int_1^6 [24.259\sin(0.987t+1.276)+35.324]dx}{5}$$
$$= \frac{1}{5}\left(\frac{24.259}{0.987}\sin(0.987t+1.276)+35.324x\right)\Bigg|_1^6 \approx \$29.2 \text{ thousand}$$

Sales average \$29.2 thousand between January and June.

15. a. $T(x) = -1.205\cos(0.469x - 0.722) + 6.404$ percent x years after 1988

b. The after-tax profit rate was lowest in 1990 ($x \approx 1.54$) at approximately 5.2% and was highest in 1996 ($x \approx 8.24$) at approximately 7.6%.

c. Approximately 6.4%

d. The values are close, but not identical, because a model (and not the actual data) was used to determine the average value in part *c*.

16. a. $$\int 12.059\cos(0.524x - 2.27)dx = \frac{12.059}{0.524}\sin(0.524x - 2.27) + C$$
$$\approx 23.013\sin(0.524x - 2.27) + C$$

$T(7) = 73.5$, so solving for C, we get $C \approx 50.830$.
$T(x) = 23.013\sin(0.524x - 2.27) + 50.830$ °F where $x = 1$ in January, $x = 2$ in February, and so on.

b. $T(12) \approx 33.1$°F

c. $$\int_2^8 12.059\cos(0.524x - 2.27)dx = \left(\frac{12.059}{0.524}\sin(0.524x - 2.27)\right)\Bigg|_2^8 \approx 43.2$$

From February to August, the change in temperature is about 43.2°F.

d. Between $x = 2$ and $x = 8$, $T'(x)$ is below the x-axis for $x > 7.330$.

$$\int_2^8 |T'(x)|dx = \int_2^{7.330} 42.059\cos(0.524x - 2.27)dx - \int_{7.330}^8 42.059\cos(0.524x - 2.27)dx$$
$$= \left(\frac{12.059}{0.524}\sin(0.524x - 2.27)\right)\Bigg|_2^{7.330} - \left(\frac{12.059}{0.524}\sin(0.524x - 2.27)\right)\Bigg|_{7.330}^8 \approx 46.0$$

17. a. $$\int_1^{16}[15.388\sin(0.787x - 1.667) + 16.346]dx = \left[\frac{-15.388}{0.787}\cos(0.787x - 1.667) + 16.346x\right]\Bigg|_1^{16}$$
$$\approx 262.915 - 3.888 \approx 259 \text{ thousand lizards}$$

b. $\sum_{x=1}^{16} L(x) \approx 261$ thousand lizards. This value is 2 thousand more than the answer to part *a*.

c. Answers will vary.

18. If it is not a leap year, March 1 is the $31 + 28 + 1 = 60$th day of the year.

$$\int_0^{60} 0.398\sin(0.0168d + 0.259)dd = \left[\frac{-0.398}{0.0168}\cos(0.0168d + 0.259)\right]\Bigg|_0^{60}$$
$$= -7.087 - (-22.900) \approx 15.8 \text{ degrees}$$

Technically, we have calculated the change in the declination of the sun from the beginning of January 1 through the end of March 1. It is equally valid to use ends of each day (the interval 1 to 60, which gives a change of about 15.7 degrees) or the beginnings of each day (the interval 0 to 59, which gives a change of about 15.4 degrees).

19. $\int_0^{40} r(t)dt \approx -4.2$

From 1900 to 1940, the ratio in the number of males per 100 females decreased by about 4.2 males per 100 females.

$\int_{50}^{100} r(t)dt \approx -2.33$

From 1950 to 2000, the ratio in the number of males per 100 females decreased by about 2.33 males per 100 females.

$\int_0^{100} r(t)dt \approx -8.8$

From 1900 to 2000, the ratio in the number of males per 100 females decreased by about 8.8 males per 100 females.

20. a. The graph of $d'(m)$ lies above the m-axis for $0.0879 < m < 6.2000$.

$$\int_{0.0879}^{6.2000} 1.18\cos(0.514t - 1.61)dt = \left(\frac{1.18}{0.514}\sin(0.514t - 1.616)\right)\Bigg|_{0.0879}^{6.2000} \approx 4.59$$

The total number of daylight hours gained over the year is about 4.6.

b. The graph of $d'(m)$ lies below the m-axis for $0 \le m < 0.0879$, $6.2000.< m \le 12$.

$$\int_0^{0.0879} 1.18\cos(0.514t - 1.61)dt + \int_{6.200}^{12} 1.18\cos(0.514t - 1.61)dt$$

$$= \left(\frac{1.18}{0.514}\sin(0.514t - 1.616)\right)\Bigg|_0^{0.0879} + \left(\frac{1.18}{0.514}\sin(0.514t - 1.616)\right)\Bigg|_{6.2}^{12} \approx -4.56$$

The total number of daylight hours lost over the year is about 4.6.

21. a. The height units are counts per second. The width units are milliseconds. The units for the area are (counts per second)(millisecond).

b. $p(s) = 0.0405 \sin(0.01345s - 1.5708) + 0.1865$ counts per millisecond after s milliseconds

c. $$\int_0^{467.151324} [0.0405\sin(0.0135s - 1.5708) + 0.1865]ds$$

$$= \left(\frac{-0.0405}{0.0135}\cos(0.0135s - 1.5708) + 0.1865s\right)\Bigg|_0^{467.151324} \approx 87$$

Approximately 87 pulses are emitted by the star over one period (about 467 milliseconds or about 0.5 second).

22. Quill Activity

Chapter 8 Review Test

1. **a.** Both vertical shift and amplitude are based on the highest and lowest output values. In the table, the highest value is 1500, and the lowest value is 75.

   $$\text{Vertical shift} \approx \frac{1500+75}{2} = 787.5$$

   $$\text{Amplitude} \approx \frac{1500-75}{2} = 712.5$$

   Period $\approx$ 12 months

   To determine the horizontal shift, we find the first output value closest to the vertical shift. In this case, the March value of 800 lawn mowers fits this description. Thus we conclude that

   Horizontal shift $\approx$ 3 months to the right.

   b. Using technology, a sine model is $l(x) = 713.2507 \sin(0.5250x - 1.5569) + 777.8826$ lawn mowers ordered x months since December of the previous year. The amplitude is very close, whereas the vertical shift of the model is about 10 lawn mowers less than that found using the data. The period of the model is approximately 12 months. The horizontal shift and period are very close to those found using the data.

   c. $l(16) \approx 1157$ lawnmowers. For this estimate to be valid, the cyclic pattern shown by the data should continue for the following year.

   d. Using technology, we find that the model appears to have greatest number of orders when $x = 6$ (June).

   e. Find the value of x for which $l''(x) = 0$ and $l'(x) > 0$. Using technology, $x \approx 3$ (March).

2. **a.** Average rate of change $\approx \dfrac{56-103}{30} \approx -1.6$ units per year

   b. Estimate the slope to be about units per year at 1860.

3. **a.** Average rate of change $= \dfrac{S(72)-S(42)}{30} \approx -1.32$ units per year

   b. $S'(t) = 11.2(0.0654)\cos(0.0654t + 5.9690) + 13.8(0.1309)\cos(0.1309t + 0.9599)$
   $+ 5.3(0.3272)\cos(0.3272t + 0.9076)$ per year t years after 1818
   $S'(42) \approx 0.37$ unit per year
   According to the model, the Sauerbeck index was increasing at the rate of approximately 0.37 unit per year in 1860.

   c. $$\frac{1}{72-42}\int_{42}^{72} S(t)dt = \frac{1}{30}\left(88.6t - \frac{11.2}{0.0654}\cos(0.0654t + 5.9690) - \frac{13.8}{0.1309}\cos(0.1309t + 0.09599) - \frac{5.3}{0.3272}\cos(0.3272t + 0.9076\right)\Bigg|_{72}^{42} \approx 90.8$$

4. **a.** $r'(t) = -0.1439(0.0197)\cos(0.0197t - 3.7526)$ million pints per day per day t years after January 1, 1992.

Using technology, $r'(t) = 0$ and $r''(t) < 0$ when $t \approx 111$ days, so the rate of change was greatest on March 20 (1992 was a leap year).

b. $R(t) = \int -0.1439\sin(0.0197t - 3.7526)dt = \dfrac{0.1439}{0.0197}\cos(0.0197t - 3.7526) + C$

Because $R(366) = 14.4$, we solve for C to get $C \approx 20.383$.

$R(t) = 7.3046\cos(0.0197t - 3.7526) + 20.383$ million pints t days after January 1, 1992.

c. The answer to part *a* is the input corresponding to the inflection point for the model in part *b*. This where the slope of the graph of the equation in part *b* is the steepest.

5. $\int_{32}^{61} -0.1439\sin(0.0197t - 3.7526)dt = \dfrac{0.1439}{0.0197}\cos(0.0197t - 3.7526)\Big|_{32}^{61} \approx 1.24$ million pints

At the end of February 1992, Campbell's was selling approximately 1.24 million pints more than it was selling at the beginning of February.

Chapter 9

Section 9.1 Multivariable Functions and Contour Graphs

1. **a.** $P(1.2, s)$ is the profit in dollars from the sale of a yard of fabric as a function of s, the selling price per yard, when the production cost is \$1.20 per yard.

   b. $P(c, 4.5)$ is the profit in dollars from the sale of a yard of fabric as a function of c, the production cost per yard, when the selling price is \$4.50 per yard.

   c. When the production cost is \$1.20 per yard and the selling price is \$4.50 per yard, the profit is \$3.00 for each yard sold.

3. **a.** $P(100{,}000, m)$ is the probability of the senator voting in favor of a the bill as a function of the amount m, in millions of dollars, invested by the tobacco industry lobbying against the bill, when the senator receives 100,000 letters supporting the bill.

   b. $P(l, 53)$ is the probability of the senator voting in favor of the bill as a function of l, the number of letters supporting the bill received by the senator, when the tobacco industry spends \$53 million lobbying against the bill.

5.

l
feet
w
feet
h
feet
Rule V
V(l,w,h)
cubic
feet

7.

Air temperature (°F) \ Relative humidity (%)	40	45	50	55	60	65	70	75	80	85	90	95	100
110	135												
108	130	137											
106	124	130	137										
104	119	124	130	137									
102	114	119	124	130	137								
100	109	113	118	123	129	136							
98	105	108	113	117	122	128	134						
96	101	104	107	111	116	121	126	132					
94	97	100	103	106	110	114	119	124	129	135			
92	94	96	98	101	104	108	112	116	121	126	131		
90	91	92	94	97	99	102	106	109	113	117	122	126	131
88	88	89	91	93	95	97	100	103	106	109	113	117	121
86	85	86	88	89	91	93	95	97	99	102	105	108	111
84	83	84	85	86	87	89	90	92	94	96	98	100	102
82	81	82	83	83	84	85	86	87	88	90	91	93	94
80	80	80	81	81	82	82	83	83	84	85	85	86	87

130

105

90

9. a. There will be 11.97 hours of daylight.

b. There will be 11.97 hours of daylight.

c. Answers will vary.

d.

Latitude North South	Month Jan Jul	Feb Aug	Mar Sep	Apr Oct	May Nov	Jun Dec	Jul Jan	Aug Feb	Sep Mar	Oct Apr	Nov May	Dec Jun
0	12.12	12.12	12.12	12.12	12.12	12.12	12.12	12.10	12.11	12.12	12.12	12.12
5	11.87	11.96	12.08	12.22	12.35	12.41	12.38	12.28	12.16	12.02	11.90	11.83
10	11.61	11.81	12.06	12.35	12.57	12.70	12.64	12.45	12.17	11.91	11.67	11.55
15	11.34	11.66	12.04	12.47	12.82	13.00	12.92	12.62	12.22	11.81	11.44	11.25
20	11.07	11.50	12.01	12.60	13.07	13.32	13.22	12.81	12.26	11.70	11.20	10.94
25	10.78	11.33	11.97	12.74	13.34	13.66	13.53	13.02	12.31	11.58	10.94	10.62
30	10.45	11.14	11.97	12.88	13.65	14.05	13.88	13.23	12.35	11.47	10.67	10.26
35	10.09	10.95	11.95	13.06	13.98	14.47	14.27	13.47	12.42	11.33	10.36	9.86
40	9.68	10.71	11.91	13.25	14.36	14.96	14.71	13.76	12.48	11.18	10.00	9.39
45	9.19	10.45	11.87	13.48	14.82	15.55	15.25	14.09	12.55	11.01	9.60	8.85
50	8.61	10.13	11.84	13.78	15.38	16.29	15.91	14.48	12.66	10.80	9.07	8.17
55	7.83	9.73	11.79	14.10	16.14	17.28	16.78	14.99	12.76	10.55	8.45	7.28
60	6.79	9.21	11.74	14.62	17.10	18.70	18.01	15.67	12.92	10.22	7.60	6.04

12 11 10 9

18 17 16 15 14 13

11.

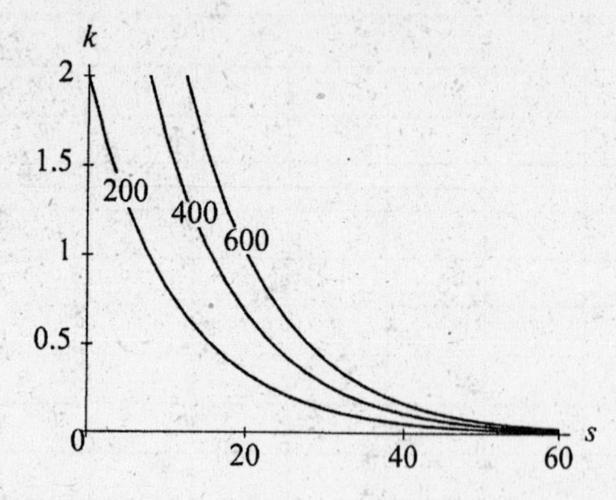

13.

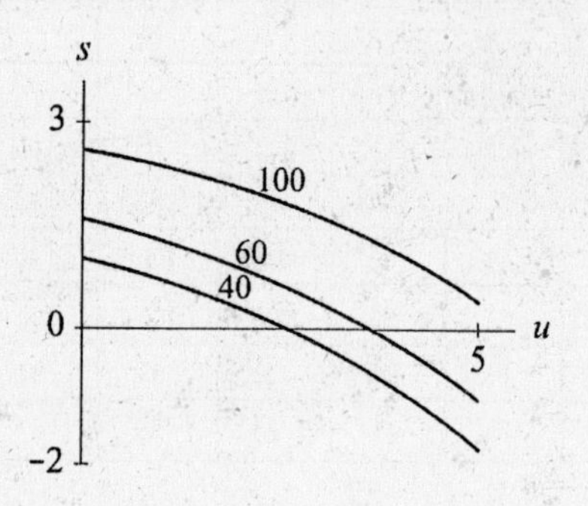

15. a. $P(c, s) = P(c, 100, 10, s) = 0.175c + 0.027s^2 - 0.730s + 108.958$ for a supermarket with s thousand square feet of sales space and a customer base with a per capita income of $\$c$ thousand

b.

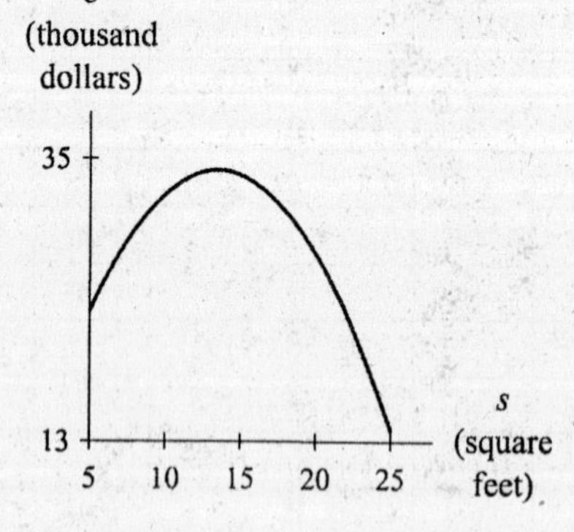

17. a. Answers will vary.

b.

Weight (pounds)	Height (inches)						
	60	62	64	66	68	70	72
90	17.6	16.5	15.4	14.5	13.7	12.9	12.2
100	19.5	18.3	17.2	16.1	15.2	14.3	13.6
110	21.5	20.1	18.9	17.8	16.7	15.8	14.9
120	23.4	21.9	20.6	19.4	18.2	17.2	16.3
130	25.4	23.8	22.3	21.0	19.8	18.7	17.6
140	27.3	25.6	24.0	22.6	21.3	20.1	19.0
150	29.3	27.4	25.7	24.2	22.8	21.5	20.3
160	31.2	29.3	27.5	25.8	24.3	23.0	21.7
170	33.2	31.1	29.2	27.4	25.8	24.4	23.1
180	35.2	32.9	30.9	29.1	27.4	25.8	24.4
190	37.1	34.8	32.6	30.7	28.9	27.3	25.8
200	39.1	36.6	34.3	32.3	30.4	28.7	27.1

15 **20** **25** **35** **30**

c. Replace the output with a constant K and solve for one variable in terms of the other. We choose to solve for w.

$$K = \frac{0.4536w}{0.00064516h^2}$$

$$w = \frac{K(0.00064516)h^2}{0.4536}$$ pounds, where h is the height in inches and K is the BMI

d.

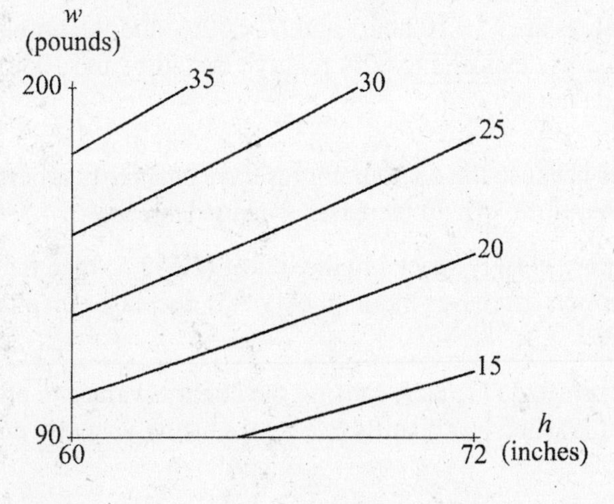

This graph is the upside-down mirror reflection of the graph in part *b*. It is also more accurately drawn.

19. a. Solve for one of the variables in $10.65 + 1.13w + 1.04s - 5.83ws = K$ (we show both).

$$s(1.04 - 5.83w) = K - 10.65 - 1.13w$$
$$s = \frac{K - 10.65 - 1.13w}{1.04 - 5.83w}$$

or

$$w(1.13 - 5.83s) = K - 10.65 - 1.04s$$
$$w = \frac{K - 10.65 - 1.04s}{1.13 - 5.83s}$$

b.

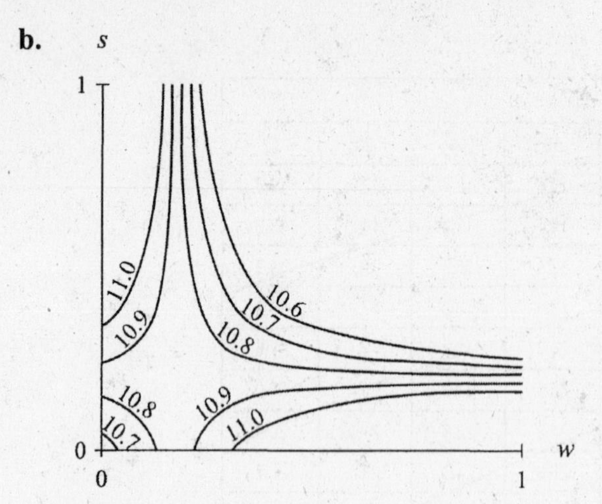

21. a. The center of the 184,000-contour curve corresponds to approximately 10,000 tons of 40% fat cheese and 55,000 tons of regular cheese since is about (10, 55). This point corresponds to a maximum because the contour curves decrease in every direction away from the point.

b. The contour graph shows that the maximum revenue is greater than 184,000,000 guilders but less than 214,000,000 guilders (or else we would see the 214,000-contour curve). From the three-dimensional graph, it appears that the maximum revenue is near 200,000,000 guilders. One possible approximation is 190,000,000 guilders.

23. a. Estimates will vary. The input values are $P \approx 10$ hours and $H \approx 70\%$, and the output value is about 12.5 days. When *C. grandis* is exposed to 70% relative humidity and 10 hours of light, it takes about 12.5 days to develop.

b. Estimates will vary. The input values are $P \approx 10$ hours and $H \approx 60\%$, and the output value is about 11.5 days. When *C. grandis* is exposed to 60% relative humidity and 10 hours of light, it takes about 11.5 days to develop.

25. a. $f(x, y)$ decreases when y increases because the contour curves have smaller numbers when y increases from 2 and x is 1.5; therefore, $f(x, y)$ increases when y decreases.

b. Because the contour curves are more closely spaced to the left of (2.5, 2.5) than they are directly below that point, the function decreases more quickly as x decreases than it does as y decreases.

c. The change is greater when (2, 2) shifts to (1, 2.5), causing the contour values to change from about 21 to about 14, than it is when (1, 0) shifts to (4, 1), causing virtually no change in contour values.

27. a. The three dimensional graph has two peaks the approximately the same height—one around (0.7, 0, 0.25) and one around (–0.7, 0. 0.25)—separated by a valley.

b. The point (0.4, 0.4) lies between the 0 and –0.05 contours. The point (0. 0.3) lies near the –0.1 contour. Thus the descent is greater from (0.7, 0.1) to (0, 0.3).

c. Moving up from (0, 0.1), we encounter increasingly negative contours. Moving right from (0, 0.1) we encounter increasingly positive contours. Thus the function output increases as x increases from (0, 0.1).

d. The point (–0.2, –0.3) lies near the –0.05 contour. Thus we are looking for a point lying on the –0.05 + 0.15 = 0.1 contour. There are infinitely many such points. Two possibilities are (–0.7, 0.38) and (0.94, 0).

29. a.

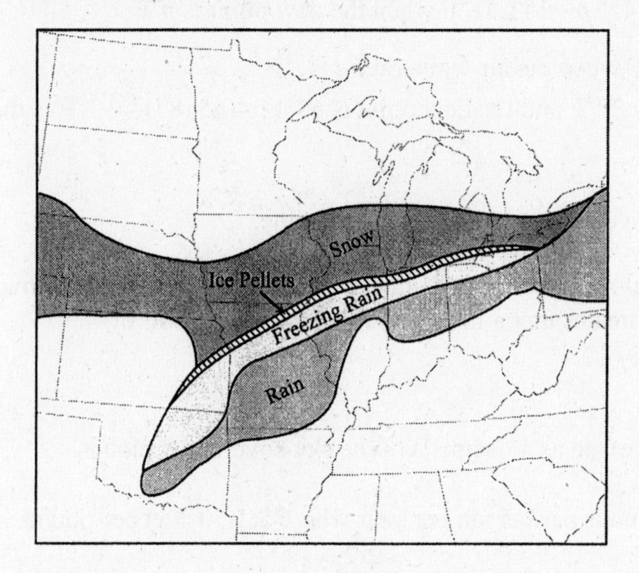

b. Answers may vary. Possible optimal placements are shown on the graph below.

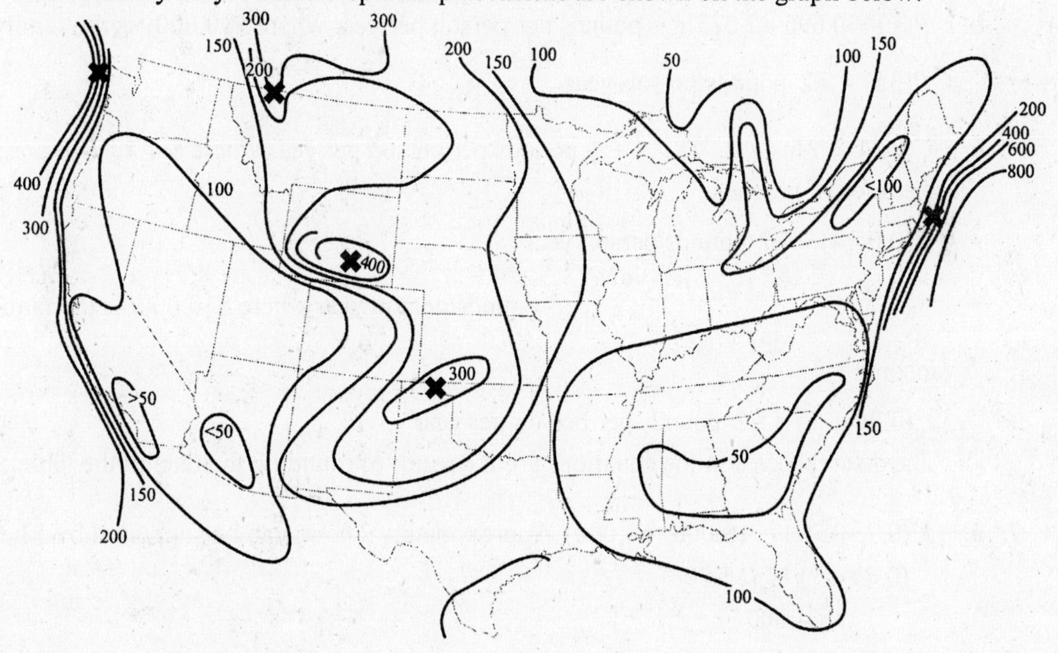

b. We estimate the average available wind power in mid-Texas to be 125 watts per square meter and that in western Nebraska to be 275 watts per square meter. Thus, the difference is approximately 150 watts per square meter.

31. Quill Activity

Section 9.2 Cross-sectional Models and Rates of Change

1. a. $A(p, 95) = 0.0238p^2 - 2.3455p + 151.31$ °F when the dew point is p°F.

b. $A(65, t) = 1.147t - 9.114$ °F when the air temperature is t°F.
When the air temperature is 87°F and the dew point is 65°F, $A(65, 87) \approx 91$°F is the apparent temperature.

3. a. At least $\frac{8}{10}$ of the sky is covered by clouds 60% of the time.

b. A plot of the frequency of cloud cover over Minneapolis in January against the fraction of sky covered at 9:00 a.m. indicates that a linear model would be a good fit.
$$C(9, f) = -1.651f^3 + 2.686f^2 - 1.597f + 1.019$$
where f is the fraction (expressed as a decimal) of the sky covered by clouds.

5. a. $P(x) = -0.857x + 7.781$ pounds per person per year where $\$(1.50 + x)$ per pound is the price of peaches

$P(0.05) \approx 7.7$ pounds/person/year

b. $P(y) = 4.696 + 2.025 \ln y$ pounds per person per year where \$10,000y is the family income

$P(3.5) \approx 7.2$ pounds/person/year

c. $C(p, 4) = 2\ln 4 + 2.7183^{-p} + 4$ pounds per person per year where $\$(1.50 + p)$ per pound is the price of peaches

$C(0.05, 4) \approx 7.7$ pounds/person/year

$$C(0.30, i) = 2\ln i + 2.7183^{-0.30} + 4 \approx 2\ln i + 4.741$$
pounds/person/year where \$10,000i is the family income

$C(0.30, 3.5) \approx 7.2$ pounds per person per year
The discrepancies in the equations are the result of rounding the data in the table.

7. a. $$K(0.7, 0.7, 17, 15{,}000, 64{,}000) = \frac{(0.7)(0.7)(17)(15{,}000)}{64{,}000} \approx 1.95 \text{ cows per hectare}$$
Approximately 2 cows can be supported by 1 hectare.

Using the fact 1 hectare ≈ 2.471 acre, we have
$$\left(\frac{1.95 \text{ cows}}{\text{hectare}}\right)\left(\frac{1 \text{ hectare}}{2.471 \text{ acres}}\right) \approx 0.8 \text{ cow per acre.}$$

b. $K(0.8,\ 0.85,\ G,\ P,\ A)$

$$= \frac{(0.8)(0.85)GP}{A}$$

$$= \frac{0.68GP}{A} \text{ animals per hectare}$$

with variables as defined in the activity statement.

c. $K(0.7,\ 0.8,\ 17,\ P,\ A)$

$$= \frac{(0.7)(0.8)(17)P}{A}$$

$$= \frac{9.52P}{A} \text{ animals per hectare}$$

with variables as defined in the activity statement.

$K(0.7, 0.8, 17, P, A)$ is the carrying capacity of a crop as a function of net crop production and the energy requirement of the animal when 70% of the crop is consumed, 80% of the crop is digested as useful nutrients, and the gross energy content of the crop is 17 megajoules per kilogram.

9. a. $A(0.1,\ 360,\ 532.98)$

$$= \frac{12(532.98)}{0.1}\left[1 - \left(1 + \frac{0.1}{12}\right)^{-360}\right]$$

$$\approx \$60{,}733.51$$

b. $A(0.075,\ n,\ m)$

$$= \frac{12m}{0.075}\left[1 - \left(1 + \frac{0.075}{12}\right)^{-n}\right]$$

$$\approx 160m\left(1 - 1.00625^{-n}\right) \text{ dollars}$$

for a period of n months and a monthly payment of m dollars.

c. $A(r, 36, m)$ is the output of the cross-sectional function for the loan amount (in dollars) when 36 monthly payments are made.

d. $$A(0.06,\ 36,\ m) = \frac{12m}{0.06}\left[1 - \left(1 + \frac{0.06}{12}\right)^{-36}\right] \approx 200m\left(1 - 1.005^{-36}\right)$$

$\approx 32.9m$ dollars when the monthly payment is m dollars.

11. a. $H(15, -15) = (10.45 + 10\sqrt{15} - 15)(33 + 15)$

≈ 1640.63 kilogram calories per square meter of body surface area per hour

b. We solve $H(10, t) = 2000$ for t:

$$(10.45 + 10\sqrt{10} - 10)(33 - t) = 2000$$

$$33 - t = \frac{2000}{10.45 + 10\sqrt{10} - 10}$$

$$t = 33 - \frac{2000}{10.45 + 10\sqrt{10} - 10} \quad t \approx -24.9^\circ\text{C}$$

13. a. $A(85, 90) = -42.379 + 2.049(90) + 10.1433(85) - 0.2248(85)(90) - (6.8378 \cdot 10^{-3})(90^2) - (5.4817 \cdot 10^{-2})(85^2) + (1.2287 \cdot 10^{-3})(90^2)(85) + (8.5282 \cdot 10^{-4})(90)(85^2) - (1.99 \cdot 10^{-6})(90^2)(85^2) \approx 117°\text{F}$

b. Solving for h in the equation

$A(h, 90) = -42.379 + 2.049(90) + 10.1433h - 0.2248h(90) - (6.8378 \cdot 10^{-3})(90^2) - (5.4817 \cdot 10^{-2})h^2 + (1.2287 \cdot 10^{-3})(90^2)h + (8.5282 \cdot 10^{-4})(90)h^2 - (1.99 \cdot 10^{-6})(90^2)h^2 = 100$

yields $h \approx 42\%$.

c. $A(h, 96) = -42.379 + 2.049(96) + 10.1433h - 0.2248h(96) - (6.8378 \cdot 10^{-3})(96^2) - (5.4817 \cdot 10^{-2})h^2 + (1.2287 \cdot 10^{-3})(96^2)h + (8.5282 \cdot 10^{-4})(96)h^2 - (1.99 \cdot 10^{-6})(96^2)h^2$

which simplifies to

$A(h, 96) = 0.0087h^2 - 0.1138h + 91.3078$ °F when the relative humidity is h%

d. $A(70, t) = -42.379 + 2.049t + 10.1433(70) - 0.2248(70)t - (6.8378 \cdot 10^{-3})t^2 - (5.4817 \cdot 10^{-2})(70^2) + (1.2287 \cdot 10^{-3})t^2(70) + (8.5282 \cdot 10^{-4})t(70^2) - (1.99 \cdot 10^{-6})t^2(70^2)$

which simplifies to

$A(70, t) = 399.0487 - 9.50818t + 0.06942t^2$ °F for an air temperature of t°F

15. a. A plot of the apparent air temperature as a function of relative humidity when the air temperature is 96°F indicates that a quadratic model would be a good fit.

$T(h, 96) = 0.010h^2 - 0.260h + 95.298$ °F, when the humidity is h%

When $h = 50\%$, $\frac{dT(h, 96)}{dh} = 0.010(2 \cdot 50) - 0.260$

≈ 0.74°F per percentage point of humidity

b. A plot of the apparent air temperature as a function of air temperature when the relative humidity is 40% indicates that a quadratic model would be a good fit.

$T(50,\ t) = 0.051t^2 - 7.299t + 339.681$°F when the temperature is t °F

When $t = 96$°F, $\dfrac{dT(50,\ t)}{dt} = 0.051(2 \cdot 96) - 7.299 \approx 2.5$°F per °F of air temperature

c. When $h = 50\%$, $\dfrac{dT(h,\ 96)}{dh} \approx \dfrac{111^\circ\text{F} - 104^\circ\text{F}}{55\% - 45\%} = 0.7$ °F per percentage point of humidity

When $t = 96$°F, $\dfrac{dT(50,\ t)}{dt} \approx \dfrac{113^\circ\text{F} - 103^\circ\text{F}}{98^\circ\text{F} - 94^\circ\text{F}} = 2.5$ °F per °F of air temperature

17. a. $A(14{,}000,\ r) = 14{,}000r^2 + 28{,}000r + 14{,}000$ dollars
where $100\%r$ is the annual percentage yield (i.e., r is in decimals).

b. $\dfrac{dA(14{,}000, r)}{dr} = 28{,}000r + 28{,}000$ dollars per 100 percentage points

$\dfrac{dA(14{,}000, 0.127)}{dr} = 28{,}000(0.127) + 28{,}000 = \$31{,}556\ /\ 100$ percentage points

c. $A(14{,}000,\ r) = 1.4r^2 + 280r + 1400$ dollars, where r% is the annual percentage yield

The magnitudes of the coefficients are reduced in this model, the r^2 coefficient by a factor of 10,000, the r coefficient by a factor of 100, and the constant by a factor of 10.

d. $\dfrac{dA(14{,}000, r)}{dr} = 2.8r + 280$ dollars per percentage point

$\dfrac{dA(14{,}000, 12.7)}{dr} = 2.8(12.7) + 280 = \$315.56$ per percentage point

The derivative tells us approximately how much the output will change when the input increases by one unit. If the input is a percentage expressed as a decimal, then an increase in one unit corresponds to 100 percentage points. For example, if $r = 0.127$ and is increased by 1 to $r = 1.127$ the corresponding percentages are 12.7% and 112.7%.

This answer is equivalent to the one found in part *b*.

19. a. $W(t{,}30) = 1.3582t - 25.8649$ °F where t°F is the air temperature;

$$\left.\frac{dW(t{,}30)}{dt}\right|_{t=10} \approx 1.36 \text{ °F per °F of air temperature}$$

The apparent temperature rises by approximately 1.36°F for each °F rise in air temperature when the air temperature is -10°F and the wind speed is 30 mph.

b. $W(20, v) = 48.17 - 27.2v^{0.16}$ °F where v mph is the wind speed;

$$\left.\frac{dW(20, v)}{dv}\right|_{v=25} \approx -0.29 \text{ °F per mph of wind speed}$$

When the air temperature is 20°F and the wind speed is 25 mph, the apparent temperature drops by about 0.29°F for each mph increase in wind speed.

21. $\frac{dP(10,r)}{dr} = -19.4661 + 1.6632r$ percent per proportion of parts of peanut oil to one part sugar where r is the ratio of parts of peanut oil to one part sugar and the processing time is 10 hours

$\left.\frac{dP(10,r)}{dr}\right|_{r=14} = -19.4661 + 1.6632(14) \approx 3.82$ % per proportion of parts of peanut oil to one part sugar when the processing time is 10 hours and 14 parts of peanut oil are used for each part of sugar.

23. Quill Activity

Section 9.3 Partial Rates of Change

1. $\frac{\partial W}{\partial h}$ pounds per inch

3. $\left.\frac{\partial T}{\partial t}\right|_{g=23}$ °F per degree of latitude

5. $\left.\frac{\partial R}{\partial c}\right|_{b=2}$ dollars per cow for $c = 100$

6. $\left.\frac{\partial G}{\partial s}\right|_{h=3.5}$ grade point average per point on SAT for $s = 1048$

7. a. $\left.\frac{\partial P}{\partial m}\right|_{l=100,000}$ is the rate of change of the probability that the senator will vote for the bill with respect to

the amount spent by the tobacco industry on lobbying when the senator receives 100,000 letters in opposition to the bill. We expect this rate of change to be negative because if the number of letters is constant but lobbying funding against the bill increases, the probability that the senator votes for the bill is likely to decline.

b. $\left.\frac{\partial P}{\partial l}\right|_{m=53}$ is the rate of change of the probability that the senator votes for the bill with respect to the number of letters received when \$53 million is spent on lobbying efforts. We expect this rate of chance to be positive because if the number of letters increases (while lobbying funding remains constant) the probability of the senator voting in favor of the bill is likely to increase.

8. a. $\left.\frac{\partial N}{\partial s}\right|_{p=25}$ is the rate of change of the number of skiers on a Saturday at a ski resort in Utah with respect to the number of inches of fresh snow received since the previous Saturday when the price of an all-day lift ticket is \$25. We expect this rate of change to be positive because if the number the number of inches of fresh snow received increases, the number of skiers is likely to increase.

b. $\left.\frac{\partial N}{\partial p}\right|_{s=6}$ is the rate of change of the number of skiers on a Saturday at a ski resort in Utah with respect to the price of an all-day lift ticket when the number of inches of fresh snow received since the previous Saturday is 6 inches. We expect this rate of change to be negative because if the price of an all-day lift ticket increases, the number of skiers is likely to decrease.

9. a. $$\frac{\partial f}{\partial x} = 3(0.529)x^2 + 2(0.375y^3)x + 8.971y + 14.390 + 0$$
$$= 1.587x^2 + 0.75xy^3 + 8.971y + 14.390$$

b. $$\frac{\partial f}{\partial y} = 0 + 3(0.375x^2)y^2 + 8.971x + 0 + 0 = 1.125x^2y^2 + 8.971x$$

c. $$\left.\frac{\partial f}{\partial x}\right|_{y=2} = 1.587x^2 + 0.75x(2^3) + 8.971(2) + 14.390 = 1.587x^2 + 6x + 32.332$$

11. a. $$M_t = s\left(\frac{1}{t}\right) + 0 + 0 = \frac{s}{t}$$

b. $$M_s = \ln t + 3.75$$

c. $$\left.M_s\right|_{t=3} = \ln 3 + 3.75$$

13. a. $$\frac{\partial h}{\partial s} = \frac{1}{t} - 2(st - tr)^1(t - 0) = \frac{1}{t} - 2t(st - tr)$$

b. $$\frac{\partial h}{\partial t} = -1(s)t^{-2} + \frac{1}{r} - 2(st - tr)^1(s - r) = \frac{-s}{t^2} + \frac{1}{r} - 2(st - tr)(s - r)$$

c. $$\frac{\partial h}{\partial r} = 0 + -1(t)r^{-2} - 2(st - tr)^1(0 - t) = \frac{-t}{r^2} + 2t(st - tr)$$

d. $$\left.\frac{\partial h}{\partial r}\right|_{(s,t,r)=(1,2,-1)} = \frac{-2}{(-1)^2} + 2(2)[1(2) - 2(-1)] = -2 + 4(4) = 14$$

15. a. $A(14{,}000, r) = 14{,}000(1+r)^2$ dollars. This function gives the value of an investment after 2 years when the APY is $100r$%.

b. $$\left.\frac{\partial A}{\partial r}\right|_{P=14{,}000} = \frac{dA(14{,}000, r)}{dr} = 28{,}000(1+r) \text{ dollars per 100 percentage points}$$

$$\left.\frac{\partial A}{\partial r}\right|_{(P,r)=(14{,}000,\ 0.127)} = \$31{,}556 \text{ per 100 percentage points}$$

c. The slope of the line tangent to a graph of $A(14{,}000, r)$ at $r = 0.127$ is \$31,566 per 100 percentage points.

(*Note:* It is difficult to distinguish the graph of A from the tangent line in this figure because the graph of A appears nearly linear in this close-up view. This is an illustration of the principle of local linearity discussed in Chapter 1.)

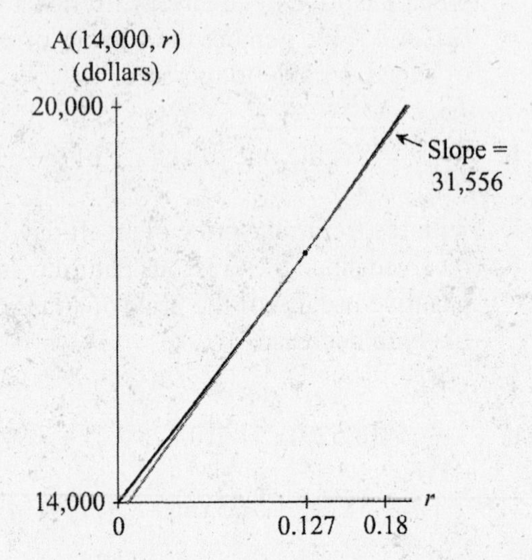

17. a. A cross-sectional model for the amount of UVA radiation with respect to the latitude for the month of March is $R(\text{March}, l) = -0.0073l^2 - 0.0346l + 49.7492$ watts/m²/month where l is the number of degrees of latitude north of the equator (negative values of l correspond to locations south of the equator).

An alternative model a sine model: $R(\text{March}, l) = 35.9104 \sin(-0.0232l + 1.5088) + 15.5463$ watts/m²/month where l is the number of degrees of latitude north of the equator.

A cross-sectional model for the amount of UVA radiation with respect to the month at 50° south of the equator is $R(m, -50) = 27.1052\sin(0.4846m + 1.7070) + 32.9106$ watts/m²/month where m is the number of months since the beginning of the year.

An alternative model is the piecewise continuous model

$$R(m,-50^\circ) = \begin{cases} 0.389m^3 - 3.321m^2 - 3.425m + 61.857 & \text{when } 0 \le m < 7 \\ \quad \text{watts per square meter per month} & \\ -0.565m^3 + 16.008m^2 - 137.689m + 381.183 & \text{when } 7 \le m \le 12 \\ \quad \text{watts per square meter per month} & \end{cases}$$

b. Quadratic model:

$$\left.\frac{\partial R}{\partial l}\right|_{m=\text{March}} = -0.0146l - 0.0342 \text{ watts/m}^2\text{/month/degree of latitude}$$

where l is the number of degrees of latitude north of the equator.

$$\left.\frac{\partial R}{\partial l}\right|_{(m,l)=(\text{March},-50)} = -0.0146(-50) - 0.0342 \approx 0.70 \text{ watt/m}^2\text{/month/degree of latitude}$$

Sine model:

$$\left.\frac{\partial R}{\partial l}\right|_{m=\text{March}} = 35.2303(-0.0236)\cos(-0.0236l + 1.5069) \text{ watts/m}^2\text{/month/degree of}$$

latitude where l is the number of degrees of latitude north of the equator.

$$\left.\frac{\partial R}{\partial l}\right|_{(m,l)=(\text{March},-50)} = 35.2303(-0.0236)\cos(-0.0236(-50) + 1.5069)$$

$$\approx 0.75 \text{ watt/m}^2\text{/month/degree of latitude}$$

c. Sine model:

$$\left.\frac{\partial R}{\partial m}\right|_{l=-50} = 27.1052(0.4846)\cos(0.4846m + 1.7070) \text{ watts/m}^2\text{/month/month where } m \text{ is the}$$

number of degrees of latitude north of the equator.

$$\left.\frac{\partial R}{\partial m}\right|_{(m,l)=(3,-50)} = 27.1052(0.4846)\cos(0.4846(3) + 1.7070)$$

$$\approx -13.13 \text{ watts/m}^2\text{/month/month}$$

Piecewise continuous model:

$$\left.\frac{\partial R}{\partial m}\right|_{l=-50} = \begin{cases} 3(0.389)m^2 - 2(3.321)m - 3.425 & \text{when } 0 \le m < 7 \\ \text{watts per square meter per month}^2 & \\ 3(-0.565)m^2 + 2(16.008)m - 137.689 & \text{when } 7 < m \le 12 \\ \text{watts per square meter per month}^2 & \end{cases}$$

$$\left.\frac{\partial R}{\partial m}\right|_{(m,l)=(3,-50)} = 3(0.389)3^2 - 2(3.321)3 - 3.425 \approx -12.85 \text{ watts/m}^2\text{/month/month}$$

d.

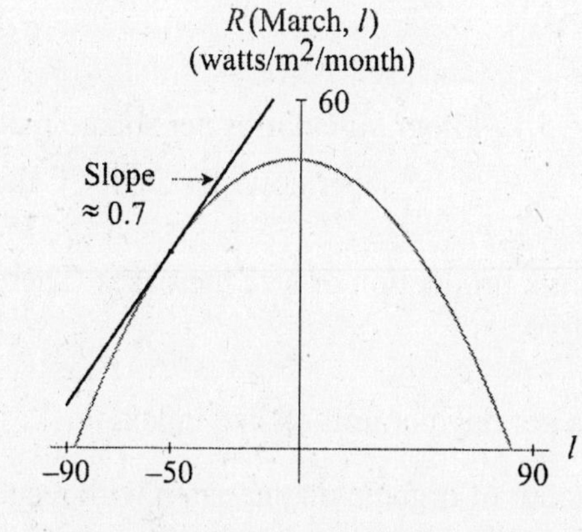

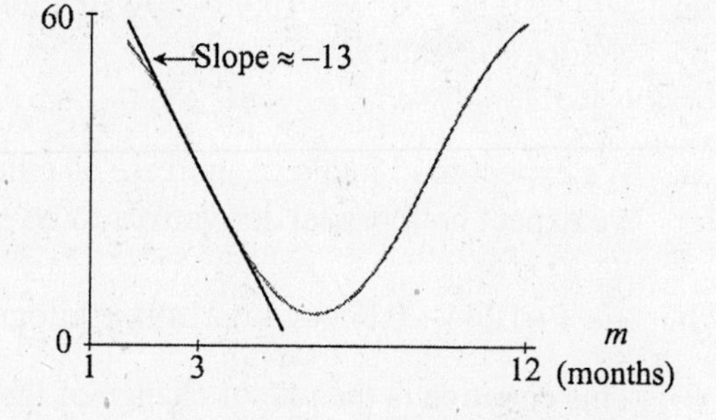

19. a. $\dfrac{7.6 - 6.2 \text{ pounds per person per year}}{4 - 2 \text{ thousand dollars}} = 0.7$ pound/person/year per thousand dollars of income

b. $\dfrac{6.9 - 7.1 \text{ pounds per person per year}}{0.3 - 0.1 \text{ dollar per pound}} = -1$ pound/person/year per dollar per pound

c. Answers vary. One possible answer follows.

Using a quadratic model to answer part *a* results in $\frac{\partial C}{\partial i} \approx 0.8$ pound/person/year per thousand dollars of income

Using a linear model to answer part *b* results in $\frac{\partial C}{\partial p} \approx -0.8$ pound/person/year per dollar per pound

d. Using the partial derivative $\frac{\partial C}{\partial i} = \frac{2}{i}$ with $i = 3$ gives $\frac{2}{3} \approx 0.7$ pound/person/year per thousand dollars of income. Using the partial derivative $\frac{\partial C}{\partial p} = -\ln 2.7183(2.7183)^{-p}$ with $p = 0.2$ gives approximately –0.8 pound/person/year per dollar per pound

e. Answers will vary.

21. a. $\frac{\partial H}{\partial t} = (10.45 + 10\sqrt{v} - v)(-1) = -10.45 - 10\sqrt{v} + v$ kilogram-calories per square meter per hour per degree Celsius

$\frac{\partial H}{\partial v} = (\frac{1}{2}v^{-1/2} - 1)(33 - t) = (33 - t)\left(\frac{5}{\sqrt{v}} - 1\right)$ kilogram-calories per square meter per hour per meter per second

b. $\frac{\partial H}{\partial v}$ should be positive because an increase in wind speed (when temperature is constant) should increase heat loss.

c. $\left.\frac{\partial H}{\partial v}\right|_{(v,\,t) = (20,\,12)} = (33 - 12)\left(\frac{5}{\sqrt{20}} - 1\right) \approx 2.48$ kilogram-calories per square meter per hour per meter per second

d. $\frac{\partial H}{\partial t}$ should be negative because an increase in temperature (when wind speed is constant) should decrease heat loss.

e. $\left.\frac{\partial H}{\partial t}\right|_{(v,\,t) = (20,\,12)} = -10.45 - 10\sqrt{20} + 20 \approx -35.17$ kilogram-calories per square meter per hour per degree Celsius

23. a. We expect food intake to increase as either milk production or size increases. Therefore, we expect both partial derivatives to be positive.

b. $\frac{\partial I}{\partial s} = -1.244 + 0.1794s + 0.21491m$ kilograms per day per unit of size index
This equation is the rate of change of the amount of organic matter eaten with respect to the size of the cow (when the amount of milk produced is constant).
$\frac{\partial I}{\partial m} = -0.20988 + 0.071894m + 0.214915s$ kilograms per day per kilogram of milk

This equation is the rate of change of the amount of organic matter eaten with respect to the amount of milk produced (when the size of the cow is constant).

c. $\left.\dfrac{\partial I}{\partial m}\right|_{(s,\, m)\, =\, (2,\, 6)} = -0.20988 + 0.071894(6) + 0.214915(2) \approx 0.65$ kilogram per day per kilogram of milk per day

d. $\left.\dfrac{\partial I}{\partial s}\right|_{(s,\, m)\, =\, (2,\, 6)} = -1.244 + 0.1794(2) + 0.21491(6) \approx 0.40$ kilogram per day per unit of size index

e. $\dfrac{\partial^2 I}{\partial s^2} = 0.1794$; $\dfrac{\partial^2 I}{\partial m^2} = 0.071894$; $\dfrac{\partial^2 I}{\partial m \partial s} = \dfrac{\partial^2 I}{\partial s \partial m} = 0.214915$

$$\begin{array}{c} \\ s \\ m \end{array} \begin{array}{c} \begin{array}{cc} s & m \end{array} \\ \begin{bmatrix} 0.1794 & 0.214915 \\ 0.214915 & 0.071894 \end{bmatrix} \end{array}$$

Because all second partials are positive, we know that the rates of change in the s and m directions increase as both s and m increase. This indicates that the surface is concave up in the s and m directions.

25. a. $\dfrac{\partial A}{\partial t} = 1000re^{rt}$ dollars per year and $\dfrac{\partial A}{\partial t} = 1000te^{rt}$ dollars per 100 percentage points

b. The second partials matrix is $\begin{array}{c} \\ t \\ r \end{array} \begin{array}{c} \begin{array}{cc} t & r \end{array} \\ \begin{bmatrix} 1000r^2e^{rt} & 1000e^{rt}(rt+1) \\ 1000e^{rt}(rt+1) & 1000t^2e^{rt} \end{bmatrix} \end{array}$. Note that we use the Product Rule to find the mixed partials. Evaluating this matrix at $t = 30$ and $r = 0.047$, we have $\begin{bmatrix} 9.05 & 9871.25 \\ 9871.25 & 3{,}686{,}359.86 \end{bmatrix}$.

When a $1000 investment has been earning 4.7% compounded continuously for 30 years,

(1) the rate at which the amount is growing with respect to time is increasing with respect to time by $9.05 per year per year.
(2) the rate at which the amount is growing with respect to time is increasing with respect to the interest rate by $9871.25 per year per 100 percentage points.
(3) the rate at which the amount is growing with respect to the interest rate is increasing with respect to time by $9871.25 per 100 percentage points per year.
(4) the rate at which the amount is growing with respect to the interest rate is increasing with respect to the rate by $3,686,359.86 per 100 percentage points per 100 percentage points.

27. a. $\dfrac{\partial A}{\partial t} = (1+r)^t \ln(1+r)$ million dollars per year

b. $\dfrac{\partial A}{\partial r} = t(1+r)^{t-1}$ million dollars per 100 percentage points

c. When $t = 5$ and $r = 0.15$, $\dfrac{\partial A}{\partial t} \approx 0.28$ million dollars per year

d.

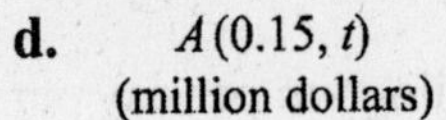

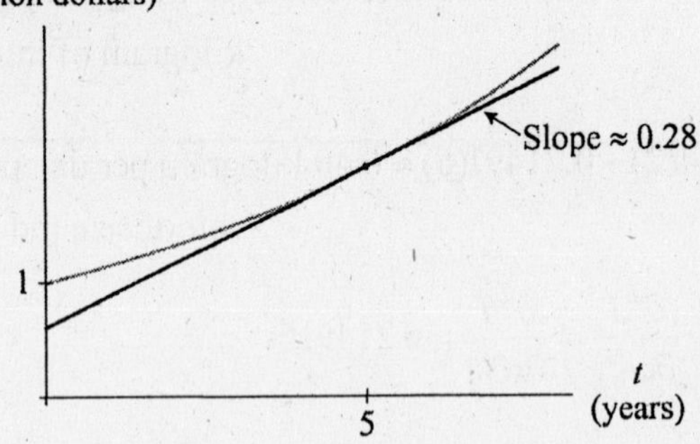

29.

$$\begin{array}{c|cc} & x & y \\ \hline x & \dfrac{-2y}{x^3} & \dfrac{-1}{y^2}+\dfrac{1}{x^2} \\ y & \dfrac{-1}{y^2}+\dfrac{1}{x^2} & \dfrac{2x}{y^3} \end{array}$$

31.

$$\begin{array}{c|cc} & x & y \\ \hline x & 4e^{2x-3y} & -6e^{2x-3y} \\ y & -6e^{2x-3y} & 9e^{2x-3y} \end{array}$$

33. Quill Activity

Section 9.4 Compensating for Change

1. $\dfrac{\partial f}{\partial x} = 30xy^3$, $\dfrac{\partial f}{\partial y} = 45x^2y^2$

$$\frac{dx}{dy} = \frac{-f_y}{f_x} = \frac{-\left(45x^2y^2\right)}{30xy^3} = \frac{-3x}{2y}$$

3. $\dfrac{\partial g}{\partial m} = \dfrac{59.372}{m} + 49.283n$, $\dfrac{\partial g}{\partial n} = 49.283m$

$$\frac{dm}{dn} = \frac{-g_n}{g_m} = \frac{-49.283m}{\dfrac{59.372}{m} + 49.283n}$$

5. $\dfrac{\partial g}{\partial x} = 1.0511^y$ $\qquad$ $\dfrac{\partial g}{\partial y} = x(\ln 1.0511)1.0511^y$

$$\frac{dx}{dy} = \frac{-g_y}{g_x} = -x \ln 1.0511$$

When $y = 5$ and $g(x, y) = 100$, $x = \dfrac{100}{1.0511^5} \approx 77.94$, so $\dfrac{dx}{dy} = -\left(\dfrac{100}{1.0511^5}\right) \ln 1.0511 \approx -3.88$.

Alternatively, $\dfrac{dy}{dx} \approx \dfrac{1}{-3.88} \approx -0.26$.

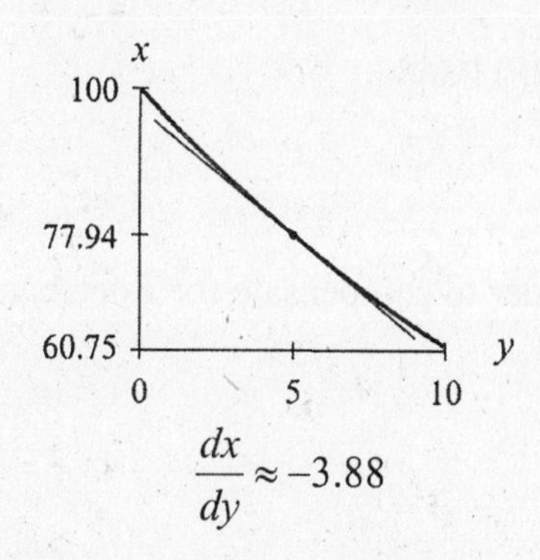

$\dfrac{dx}{dy} \approx -3.88$

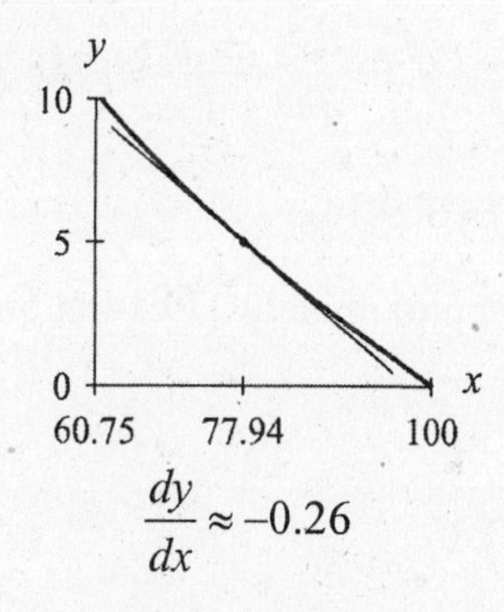

$\dfrac{dy}{dx} \approx -0.26$

7. $\dfrac{\partial f}{\partial a} = 5.692ab^3 - 3.668a$ $\qquad$ $\dfrac{\partial f}{\partial b} = 8.538a^2b^2 + 12.5$

$$\frac{da}{db} = \frac{-f_b}{f_a} = \frac{-(8.538a^2b^2 + 12.5)}{5.692ab^3 - 3.668a}$$

When $b = 0.9$ and $f(a, b) = 15$, $a \approx 3.9468$, so $\dfrac{da}{db} \approx -63.27$.

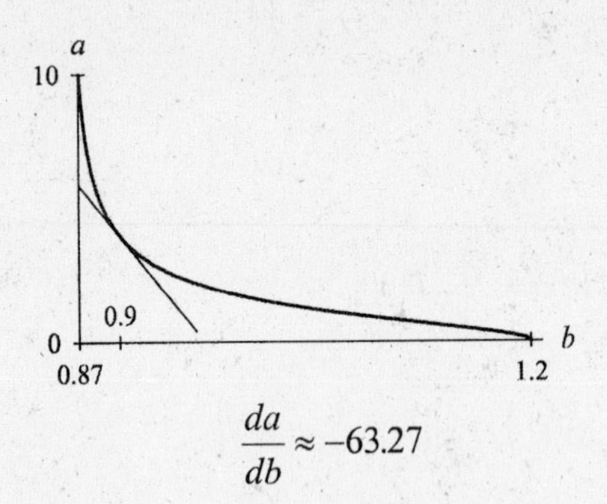

$$\frac{db}{da} \approx -0.016$$

$$\frac{da}{db} \approx -63.27$$

9. $f(2, 1) = 21$

$\frac{\partial f}{\partial m} = 6m + 2n$, $\frac{\partial f}{\partial n} = 2m + 10n$

$\frac{dm}{dn} = \frac{-f_m}{f_n} = \frac{-(2m+10n)}{6m+2n}$

When $m = 2$, $n = 1$, and $\Delta m = 0.2$, $\frac{dm}{dn} = \frac{-14}{14} = -1$ and $\Delta n \approx \frac{dn}{dm}\Delta m = (-1)(0.2) = -0.2$.

The value of n should decrease be approximately 0.2 in order to compensate for an increase of 0.2 in m.

11. $f(3.5, 1148) \approx 3.7217$

$\frac{\partial f}{\partial h} = 0.00091s[0.103(\ln 2.505)(2.505^h)]$, $\frac{\partial f}{\partial s} = 0.00091[0.103(2.505^h) + 1]$

When $h = 3.5$, $s = 1148$, and $\Delta h = -0.5$, $\frac{ds}{dh} = \frac{-f_h}{f_s} \approx -758.2811$ and

$\Delta s \approx \frac{ds}{dh}\Delta h \approx (-758.2811)(-0.5) \approx 379.14$.

The input s should increase by approximately 379.14 in order to compensate for a decrease of 0.05 in h.

13. a. $A(6, 250) \approx 7.16$

The average cost is about $7.16.

b. $\frac{\partial A}{\partial n} = (-0.02c^2 + 0.35c + 0.99)(\ln 0.99897)(0.99897^n)$

When $c = 6$ and $n = 250$,

$\frac{\partial A}{\partial n} = (-0.02c^2 + 0.35c + 0.99)(\ln 0.99897)(0.99897^n) \approx -0.001888$

the average cost is changing at a rate of about –$0.002 per shirt.

c. $\frac{\partial A}{\partial c} = (-0.04c + 0.35)(0.99897^n) + 0.46$

$\frac{dn}{dc} = \frac{-A_c}{A_n} = \frac{-[(-0.04c + 0.35)(0.99897^n) + 0.46]}{(-0.02c^2 + 0.35c + 0.99)(\ln 0.99897)(0.99897^n)}$ shirts per color

We expect $\frac{dn}{dc}$ to be positive because if the number of colors increases, the order size would also need to increase to keep average cost constant.

d. When $c = 4$ and $n = 500$, $\frac{dn}{dc} \approx 450$ shirts per color. For each additional color, the order size would need to increase by approximately 450 shirts. Similarly, if the number of colors decreases by 1, the order size could decrease by approximately 450 shirts and the average cost would remain constant.

15. a, b.

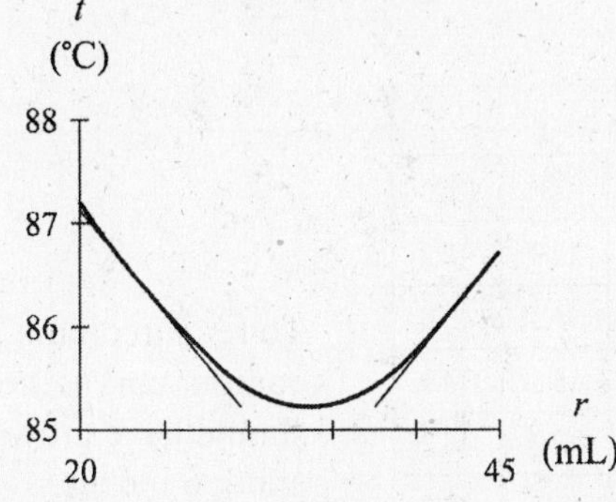

c. $\frac{\partial p}{\partial t} = -9.6544 + 0.14736t$, $\frac{\partial p}{\partial r} = 1.9836 - 0.05916r$

$\frac{dt}{dr} = \frac{-p_r}{p_t} = \frac{-(1.9836 - 0.05916r)}{-9.6544 + 0.14736t}$ °C per milliliter

d. We can solve for r in $p(86.5, r) = 53$ to get $r \approx 23.125$ or $r \approx 43.934$.

When $t = 86.5$ and $r = 23.125$, $\frac{dt}{dr} \approx -0.199$ °C per milliliter

When $t = 86.5$ and $r = 43.934$, $\frac{dt}{dr} \approx 0.199$ °C per milliliter

It is also possible to calculate the slope formula as

$\frac{dr}{dt} = \frac{-p_t}{p_r} = \frac{-(-9.6544 + 0.14736t)}{1.9836 - 0.05916r}$ milliliter per °C and the two specific slopes as follows:

When $t = 86.5$ and $r = 23.125$, $\frac{dr}{dt} \approx -5$ milliliter per °C

When $t = 86.5$ and $r = 43.934$, $\frac{dt}{dr} \approx 5$ milliliter per °C

To illustrate the tangent lines whose slopes these values represent, we must draw the parabola sideways with one tangent line on top and one on the bottom.

17. a. $B(67, 129) \approx 20.044$ points

Solving $B(h, w) = 20.044$ for w, we get $w = \frac{(20.044)(0.00064516)h^2}{0.45}$ pounds

b. $\frac{dw}{dh} = \frac{2(20.044)(0.00064516)h}{0.45}$

When $h = 67$, $\frac{dw}{dh} = \frac{2(20.044)(0.00064516)(67)}{0.45} \approx 3.85$ pounds per inch

c. The answer to part *b* agrees with the answer given in Example 1.

19. $\frac{\partial A}{\partial w} = 0.6416(0.425)w^{-0.575}h^{0.725}$, $\frac{\partial A}{\partial h} = 0.6416(0.725)w^{0.425}h^{-0.275}$

When $w = 130$, $h = 71$, and $\Delta h = 2$, $\Delta w = \frac{-\frac{\partial A}{\partial h}}{\frac{\partial A}{\partial w}} \Delta h \approx -6.25$ pounds.

The person must lose about 6.25 pounds in order for the skin surface area to remain constant if the person grows 2 inches.

21. a,b.

Weight (pounds)	Height (inches)						
	60	62	64	66	68	70	72
90	17.6	16.5	15.4	14.5	13.7	12.9	12.2
100	19.5	18.3	17.2	16.1	15.2	14.3	13.6
110	21.5	20.1	18.9	17.8	16.7	15.8	14.9
120	23.4	21.9	20.6	19.4	18.2	17.2	16.3
130	25.4	23.8	22.3	21.0	19.8	18.7	17.6
140	27.3	25.6	24.0	22.6	21.3	20.1	19.0
150	29.3	27.4	25.7	24.2	22.8	21.5	20.3
160	31.2	29.3	27.5	25.8	24.3	23.0	21.7
170	33.2	31.1	29.2	27.4	25.8	24.4	23.1

20-pound gain

6-inch growth

20

To remain on the 20 BMI contour curve, the girl should have grown by about 6 inches.

c. Using the cross-sectional models $f(110, h) = 0.012h^2 - 2.118h + 105.689$ and $f(w, 62) = 0.183w + 0.009$,where h is height in inches and w is weight in pounds, we estimate that the girl needs to grow by

$$\frac{-f_w}{f_h}\Delta w \approx \left(\frac{-0.813}{-0.642} \text{ inches per pound}\right)(20 \text{ pounds}) \approx 5.7 \text{ inches}$$

d. $\frac{-B_w}{B_h}\Delta w \approx \left(\frac{-0.812}{-0.643} \text{ inches per pound}\right)(20 \text{ pounds}) \approx 5.7 \text{ inches}$

e. Answer will vary.

23. a. From the table, we estimate that Coke would need to lower its prices by more than \$0.50 per can.

b. Using the equations $S(1.00, P) = 196P + 25$ cans of Coke products and $S(c, 1.25) = -50.286C^2 + 7.771C + 312.6$ cans of Coke products where \$$P$ is the price of Pepsi products and \$$C$ is the price of Coke products, we estimate the change in Coke prices as $\Delta C \approx \frac{-S_P}{S_C}\Delta P = \left(\frac{-196}{-92.8}\right)(-0.25) \approx -\$0.53$. Coke would need to lower its price from \$1.00 a can to about \$0.47 a can.

c. Rather than lower its prices so drastically, Coke should probably consider such alternatives as more advertising on campus.

25. Quill Activity

Chapter 9 Review Test

1. a.

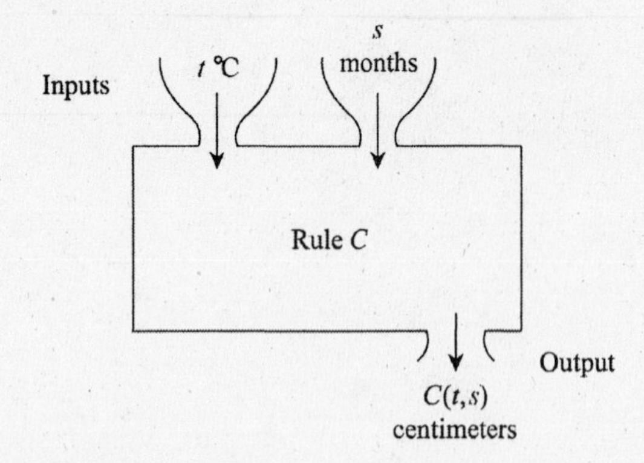

b. $C(3, 71) = 2.9$ cm
When apples are stored for 3 months and blanched at 71°C, the applesauce flows 2.9 cm down a vertical surface in 30 seconds.

c. Find a model for the consistometer value as function of storage time, using storage time as input and the values in the 35°C column as output. A piecewise continuous model is

$$C(s,35) = \begin{cases} -0.05s^2 + 0.35s + 3 \text{ cm} & \text{when } 0 \le s < 2 \\ -0.05s^2 + 0.15s + 3.4 \text{ cm} & \text{when } 2 \le s \le 4 \end{cases}$$

Assuming that 1 month is 4.3 weeks, 2 weeks corresponds to $s \approx \dfrac{2 \text{ wks}}{4.3 \text{ wks/month}} \approx 0.465$ month

$C(0.465, 35) \approx 3.2$ centimeters

d. $\dfrac{\partial C(4,t)}{\partial t}$ is the rate of change of the consistometer value with respect to the blanching temperature when the storage time is 4 months.

e. Find a model for the consistometer value as function of the blanching temperature, using temperature as the input and the values in the row corresponding to a 4-month storage time as the output. A quadratic model is $C(4,t) = (6.9444 \cdot 10^{-4})t^2 - 0.0869t + 5.4124$ centimeters for $35 \le t \le 83$ where t is the blanching temperature in degrees Celsius.

$$\frac{\partial C(4,t)}{\partial t} = 2(6.9444 \cdot 10^{-4})t - 0.0869 \text{ cm per °C}$$

$$\frac{\partial C(4,45)}{\partial t} = 2(6.9444 \cdot 10^{-4})(45) - 0.0869 \approx -0.0244 \text{ cm per °C}$$

When the storage time is a constant 4 months and the blanching temperature is 45°C, the consistometer value is decreasing by approximately 0.024 cm per °C. That is, if the blanching temperature is increased to 46°C, the consistometer value should decrease by approximately 0.024 cm.

f.

Storage time (months)	Temperature (Celsius)				
	35°	47°	59°	71°	83°
0	3.0	2.8	2.6	2.6	2.8
1	3.3	3.1	2.8	2.8	3.0
2	3.5	3.2	3.0	2.9	3.2
3	3.4	3.2	3.0	2.9	3.2
4	3.2	2.9	2.7	2.7	3.0

Contour labels: 3.0, 3.2, 2.8, 3.0, 3.2, 2.8

2. a. $E(24, 60) \approx 466$ eggs
A female insect kept at 24°C and 60% relative humidity will lay approximately 466 eggs in 30 days.

b. $\frac{\partial E}{\partial h} = 23.1412 - 0.1874h - 0.4023t$

When $t = 27$°C and $h = 77$%, $\frac{\partial E}{\partial h} = 23.1412 - 0.1874(77) - 0.4023(27) \approx 2.2$ eggs per percentage point of humidity. When the temperature is held constant at 27°C and the relative humidity is 77%, the number of eggs is decreasing by about 2.2 eggs per percentage point of relative humidity. That is, if the relative humidity were increased to 78%, the number of eggs would decrease by approximately 2.

c. $\frac{\partial E}{\partial t} = 299.7038 - 10.4420t - 0.4023h$

$\frac{dh}{dt} = \frac{-E_t}{E_h} = \frac{-(299.7038 - 10.4420t - 0.4023h)}{23.1412 - 0.1874h - 0.4023t}$ percentage points per °C

d. When $h = 63$% and $t = 25$°C, $\frac{dh}{dt} \approx -10.4$ percentage points per °C.

$\Delta h \approx \frac{dh}{dt}\Delta t \approx (-10.4)(-0.5) = 5.2$ percentage points
The humidity should increase by approximately 5.2%.

e. On the contour curve corresponding to the egg production for $t = 25$°C and $h = 63$% (the 490.1601 egg contour curve), the slope of the tangent line at that point is −10.4 percentage points per °C. The h value on the tangent line when $t = 24.5$°C is $h \approx 63 + 5.2 = 68.2$%. This is an approximation to the value of h that corresponds to $t = 24.5$°C on the 490.1601 egg contour curve.

3. a. Inputs are $t \approx 26$ °C and $h \approx 56$%. The output is approximately 485 eggs.

b. An increase in temperature by 3°C will result in a greater change in the number of eggs laid.

c. The number of eggs laid will decrease more rapidly when the temperature decreases than when the humidity decreases.

d. By sketching a tangent line on Figure 9.42 and calculating its slope, you should obtain an estimate of approximately 13 percentage points per °C.

e. $\frac{dh}{dt} = \frac{-E_t}{E_h} = \frac{-24.1532}{1.8672} \approx -12.9$ percentage points per °C

When the temperature is 24°C and the humidity is 62%, the change in humidity needed to compensate for a small change in temperature (so that the number of eggs remains constant) can be estimated as $\Delta h \approx (-12.9)\Delta t$ percentage points.

4. From Question 2 we have $\frac{\partial E}{\partial h} = 23.1412 - 0.1874h - 0.4023t$

$$\frac{\partial E}{\partial t} = 299.7038 - 10.4420t - 0.4023h$$

$E_{tt} = -10.4420$, $E_{th} = -0.4023$, $E_{hh} = -0.1874$, and $E_{ht} = -0.4023$

The second partials matrix for any values of h and t is $\begin{array}{c} \\ t \\ h \end{array}\begin{array}{c} \begin{array}{cc} t & h \end{array} \\ \begin{bmatrix} -10.442 & -0.4023 \\ -0.4023 & -0.1874 \end{bmatrix} \end{array}$.

Chapter 10

Section 10.1 Multivariable Critical Points

1. a. A relative maximum occurs when a table value is greater than all 8 values surrounding it.

b. A relative minimum occurs when a table value is less than all 8 values surrounding it.

c. If a table value appears to be a maximum in one direction but a minimum in another direction, then the value corresponds to a saddle point.

d,e. If all the edges of a table are terminal edges, then the absolute maximum and minimum are simply the largest and smallest values in the table. If all the edges are not terminal edges, then you must know whether any critical points exist outside the table in order to determine whether absolute extrema exist. If no critical points exist outside the table, then in determining absolute extrema, you must consider relative extrema, output values on terminal edges, and the behavior of the function beyond the edges of the table.
It is often helpful to sketch contour curves on a table when determining critical points and absolute extrema.

3. The point is a relative maximum point because the values of the contour curves decrease in all directions away from the point.

5. a.

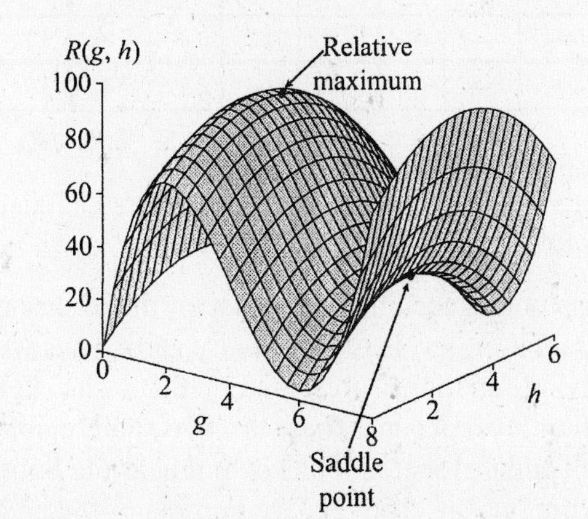

b. Relative maximum point: $(g, h, R) \approx (2, 3, 95$; Saddle point: $(g, h, R) \approx (6, 3, 30)$
(Answers may vary.)

7. Relative maximum point: (May, 1995, \$1.45 per pound)
Relative maximum point: (May, 1998, \$0.88 per pound)

9. The point is a saddle point because the values of the contour curves decrease in one direction and increase in other direction away from the point.

11. a. The table gives yearly averages, so it doesn't make sense to extend the columns. However, the choice of January as the first column is not mandatory. The best way to visualize this table is to wrap it around a cylinder so that the January and December columns are adjacent columns and there are no left or right edges on the table. The top and bottom rows are terminal edges.

b.

	Jan	Feb	Mar	Apr	May	Jun	Jul	Aug	Sep	Oct	Nov	Dec
North Pole	—	—	—	3.1	6.9	8.9	7.9	4.9	0.80	—	—	—
80°	—	—	0.77	3.4	6.9	8.8	7.8	4.8	1.55	0.13	—	—
70°	—	0.51	2.0	4.5	7.1	8.5	7.8	5.5	2.9	0.95	0.06	—
60°	0.55	1.53	3.4	5.7	7.7	8.8	8.2	6.5	4.3	2.2	0.81	0.34
50°	1.66	2.8	4.7	6.7	8.4	9.1	8.8	7.4	5.5	3.4	1.97	1.35
40°	3.0	4.2	5.9	7.5	8.8	9.3	9.0	8.1	6.5	4.8	3.4	2.6
30°	4.4	5.6	6.9	8.1	9.0	9.2	9.1	8.4	7.4	6.1	4.7	4.1
20°	5.8	6.7	7.8	8.5	8.8	8.9	8.8	8.6	8.0	7.1	6.0	5.5
10°	7.1	7.7	8.3	8.5	8.4	8.3	8.3	8.4	8.3	7.9	7.2	6.8
Equator	8.1	8.5	8.6	8.3	7.8	7.5	7.6	8.0	8.4	8.4	8.2	7.9
10°	8.9	8.8	8.4	7.7	6.9	6.4	6.5	7.2	8.1	8.6	8.8	8.8
20°	9.4	9.0	8.1	6.9	5.7	5.1	5.4	6.3	7.5	8.6	9.2	9.5
30°	9.6	8.8	7.4	5.8	4.4	3.8	4.1	5.2	6.7	8.2	9.3	9.8
40°	9.6	8.3	6.5	4.6	3.1	2.5	2.7	3.9	5.6	7.5	9.1	9.9
50°	9.3	7.6	5.4	3.3	1.84	1.25	1.49	2.6	4.5	6.6	8.7	9.7
60°	8.7	6.6	4.1	2.0	0.72	0.31	0.47	1.36	3.1	5.6	8.0	9.3
70°	8.2	5.5	2.8	0.84	—	—	—	0.38	1.78	4.3	7.2	9.1
80°	8.2	4.7	1.42	0.099	—	—	—	—	0.62	3.2	7.0	9.3
South Pole	8.1	4.6	0.60	—	—	—	—	—	—	2.9	7.0	9.4

c. There are three relative maximum points: (June, North Pole, 8.9 kW-h/m^2), (June, 40° North, 9.3 kW-h/m^2), and (December, 40° South, 9.9 kW-h/m^2).

It is difficult to estimate relative minima points accurately because of the dashes in the table, which we can interpret to mean radiation levels of essentially zero. Thus we conclude that the regions of the underlying function represented by the dashes in the table are those in which the minimum radiation level occurs. There are two such regions: one at and near the North Pole between March and October and one at and near the South Pole between April and September. If there are two specific relative minima of the underlying function, then we estimate that they occur at the end of December at the North Pole and in the middle of June at the South Pole.

There are four points that can be considered saddle points: (April, 10° North, 8.5 kW-h/m^2) (August, 10° North, 8.4 kW-h/m^2), (June, 70° North, 8.5 kW-h/m^2), and (December, 70° South, 9.1 kW-h/m^2). (Answers may vary.)

d. The greatest radiation level shown in the table is 9.9 kW-h/m^2 which occurs in December at 40° South. The smallest radiation level shown is 0.06 kW-h/m^2 which occurs in November at 80° North. If we consider the dashes to be zeros, then the smallest radiation level is zero and occurs many times in the table.

e. The absolute maximum value is 9.9 kW-h/m^2, and the absolute minimum is approximately zero. Because the table cannot extend in any direction, these answers do correspond to those in part *d*.

f. The largest and smallest values in the table will be the absolute maximum and minimum, respectively, if the table cannot be extended in any direction. That is, either the edges are terminal edges or the table "wraps around," as in this case.

13. a. The expected corn yield is 100% of the annual average yield. That is, there is no expected increase or decrease in yield from the average.

b. The expected corn yield is 40% of the annual average yield.

c.

T (°C) \ P (%)	-100	-90	-80	-70	-60	-50	-40	-30	-20	-10	0	10	20	30	40	50	60	70	80
6	0	7	14	18	23	27	32	40	48	56	64	67	69	72	74	75	75	75	76
	0	7	15	20	25	29	34	43	51	59	67	70	73	75	78	78	78	78	78
5	0	8	16	21	26	32	37	45	54	62	71	73	76	79	81	81	81	81	81
	0	8	17	22	28	34	39	48	57	65	74	77	79	82	85	85	84	84	84
4	0	9	18	24	30	36	42	51	60	68	77	80	83	86	88	88	88	87	87
	0	9	19	25	32	38	45	54	62	71	80	83	86	89	92	91	91	90	89
3	0	10	20	27	33	40	47	56	65	75	84	87	90	93	96	95	94	93	92
	0	10	21	28	35	42	49	59	69	78	87	91	94	96	98	97	95	94	93
2	0	11	21	29	37	44	52	63	74	82	90	95	99	99	100	98	97	96	94
	0	11	22	30	38	46	54	65	76	85	93	99	101	101	102	100	99	97	95
1	0	12	23	31	40	48	56	67	78	87	95	102	103	104	104	102	100	98	96
	0	12	24	33	41	50	58	69	80	90	98	104	105	106	106	104	102	100	97
0	0	12	25	34	43	51	60	71	83	92	100	107	107	108	109	106	103	101	98
	0	13	26	35	43	52	61	72	84	93	101	107	108	108	107	105	102	100	97
-1	0	13	27	35	44	53	61	73	85	94	103	108	109	108	106	104	101	99	96
	0	14	28	36	45	53	62	73	84	93	102	107	108	107	105	103	100	98	95
-2	0	14	29	37	46	54	63	73	84	93	101	105	107	106	104	102	99	97	94
	0	15	29	38	46	55	63	72	82	90	98	101	103	103	103	101	98	95	93
-3	0	15	30	39	47	55	64	72	80	88	96	97	99	101	102	100	97	94	92
	0	15	30	38	46	54	62	70	77	85	92	94	95	98	98	96	94	91	89
-4	0	15	31	38	46	53	61	68	74	81	88	90	91	93	95	92	90	88	85
	0	15	31	38	45	52	59	65	72	78	85	86	88	89	91	89	87	85	82
-5	0	15	31	38	44	51	58	63	69	75	81	82	84	85	87	85	83	81	79
	0	16	31	37	44	50	56	61	67	72	77	79	80	82	83	81	80	78	76
-6	0	16	31	37	43	49	55	59	64	69	74	75	76	78	79	78	76	75	73

d. The maximum percentage yield is 109%. This maximum occurs twice, at the points (40%, 0°C, 109%) and (20%, –1°C, 109%). This means that a yield of 109% above normal can be expected when temperatures are average (a change of 0°C) and there is 40% more precipitation than normal or when temperatures are 1°C below normal and precipitation is 20% above normal.

15. a. Answers will vary.

b. Relative maximum: 17.3 mm per day in June at a latitude between 40° and 42°

c. The table wraps around as in Activity 11; however, the rows can extend above the top row for degrees of latitude greater than 50.

d. The greatest amount of extraterrestrial radiation is approximately 17.3 mm per day, which occurs in June for latitudes between 40° and 42°. We cannot accurately estimate the least amount of radiation from the table, although we suspect that it is near zero and occurs in December at the North Pole.

17. a.

Storage time (months)	Temperature (°C)				
	35	47	59	71	83
0	3.0	2.8	2.6	2.6	2.8
1	3.3	3.1	2.8	2.8	3.0
2	3.5	3.2	3.0	2.9	3.2
3	3.4	3.2	3.0	2.9	3.2
4	3.2	2.9	2.7	2.7	3.0

b. Saddle point at (71 °C, 2–3 months, 2.9 cm)

c. Absolute maximum: 3.5 cm at 35°C and 2 months

Absolute minimum: 2.6 cm at 59°C and 71°C and 0 months

d. Thick applesauce is desirable, and small consistometer values correspond to think applesauce. Thus minimum values are of the greatest interest. From the table, we see that to produce the thickest applesauce possible, fresh apples should be blanched between 59°C and 71°C.

19. a. Maximum volume

b. The maximum occurs at approximately 4.3 grams of leavening and a baking time between 29 and 30 minutes.

c. The maximum volume index appears to be approximately 114. This probably means that the maximum volume possible is 114% of the volume of the cake batter.

20. a.

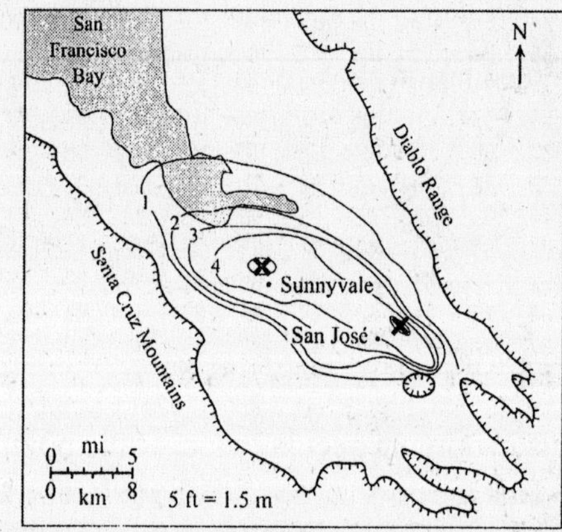

b. The point... points in... greatest l... land subs... more tha...

22.

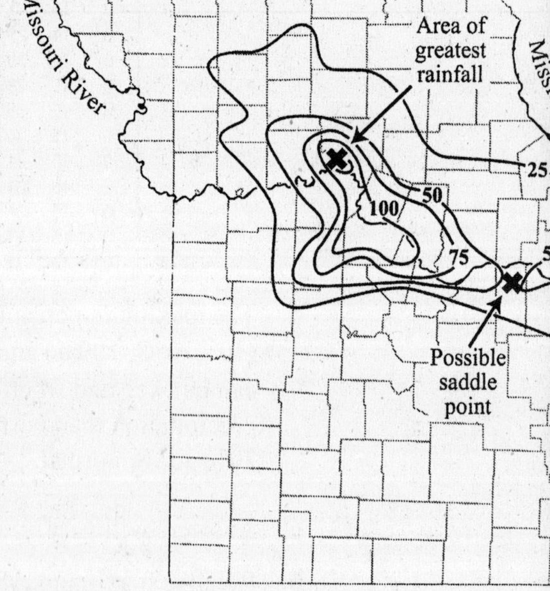

21. a.

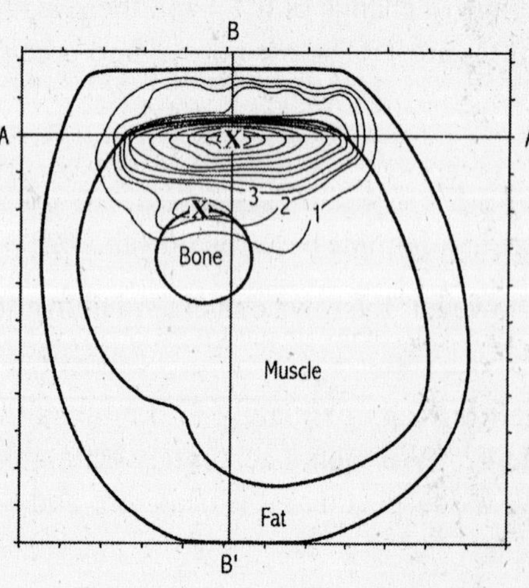

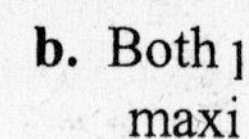

Section 10.2 Multivariable Optimization

1. Find the partial derivatives of R, and set them equal to zero.

$R_k = 6k - 2m - 20 = 0$

$R_m = -2k + 6m - 4 = 0$

Solving for k and m gives $k = 4$ and $m = 2$.

$R_{kk} = 6$, $R_{mm} = 6$, $R_{km} = R_{mk} = -2$

$$D(4,\ 2) = \begin{vmatrix} 6 & -2 \\ -2 & 6 \end{vmatrix} = 36 - 4 = 32 > 0$$

$R_{kk}(4,\ 2) = 6 > 0$

Because $D > 0$ and $R_{kk} > 0$, we know the critical point corresponds to a minimum. We conclude that a relative minimum of $R(4, 2) = 16$ is located at $k = 4$ and $m = 2$.

3. Find the partial derivatives of G, and set them equal to zero.

$G_t = pe^t = 0$

$G_p = e^t - 3 = 0$

To solve for t and p, note that in order for $pe^t = 0$, either $p = 0$ or $e^t = 0$. Because e^t can never be zero, we conclude that $p = 0$. The second equation gives $e^t = 3$ or $t = \ln 3 \approx 1.099$.

$G_{tt} = pe^t$, $G_{pp} = 0$, $G_{tp} = G_{pt} = e^t$

At $p = 0$ and $t = \ln 3$,

$G_{tt} = 0$, $G_{pp} = 0$, $G_{tp} = G_{pt} = e^{\ln 3} = 3$

$$D(\ln 3,\ 0) = \begin{vmatrix} 0 & 3 \\ 3 & 0 \end{vmatrix} = -9 < 0$$

Because $D < 0$, we know the point is a saddle point. The output at that point is $G(\ln 3, 0) = 0$. A saddle point is located at $(\ln 3, 0, 0)$.

5. Find the partial derivatives of h, and set them equal to zero.

$h_w = 1.2w - 4.7z = 0$

$h_z = 3.0z^2 - 4.7w = 0$

To solve this system of equations, solve the first equation for w: $w = \frac{4.7z}{1.2}$. Substitute this expression into the second equation:

$$3.0z^2 - 4.7\left(\frac{4.7z}{1.2}\right) = 0$$

$$z\left(3.0z - \frac{4.7^2}{1.2}\right) = 0$$

Solving for z gives two solutions: $z = 0$ and $z \approx 4.720$. The two critical points are $w = 0, z = 0, h = 0$ and

$w \approx 18.487, z \approx 4.720$, and $h \approx -68.353$

The second partials are

$h_{ww} = 1.2$, $h_{zz} = 6.0z$

$h_{zw} = h_{wz} = -4.7$

$D(0,\ 0) = \begin{vmatrix} 1.2 & -4.7 \\ -4.7 & 0 \end{vmatrix} = -20.89 < 0$, thus (0, 0, 0) is a saddle point.

$D(18.487,\ 4.720) = \begin{vmatrix} 1.2 & -4.7 \\ -4.7 & 28.32 \end{vmatrix} \approx 11.89 > 0$ and $h_{ww} > 0$, thus (18.487, 4.720, –68.353) is a relative minimum.

7. Solving $f_x = 6x - 3x^2 = 3x(2-x) = 0$ and $f_y = 24y - 24y^2 = 24y(1-y) = 0$ yields $x = 0$, $x = 2$, $y = 0$, $y = 1$. Thus we consider 4 points: (0, 0), (0, 1), (2, 0), (2, 1). The second partials are $f_{xx} = 6 - 6x$, $f_{yy} = 24 - 48y$, and $f_{xy} = f_{yx} = 0$.

For (0, 0), $D = \begin{vmatrix} 6 & 0 \\ 0 & 24 \end{vmatrix} = 144 > 0$ and $f_{xx} > 0$ thus (0, 0, 60) is a relative minimum.

For (0, 1), $D = \begin{vmatrix} 6 & 0 \\ 0 & -24 \end{vmatrix} = -144 < 0$, thus (0, 1, 64) is a saddle point.

For (2, 0), $D = \begin{vmatrix} -6 & 0 \\ 0 & 24 \end{vmatrix} = -144 < 0,$

thus (2, 0, 64) is a saddle point.

For (2, 1), $D = \begin{vmatrix} -6 & 0 \\ 0 & -24 \end{vmatrix} = 144 > 0$ and $f_{xx} < 0$, thus (2, 1, 68) is a relative maximum.

9. a. Solving $R_b = 14 - 6b - p = 0$ and $R_p = -b - 4p + 12 = 0$ yields

$b \approx 1.91$ and $p \approx 2.52$.

The manager should try to buy ground beef at \$1.91 a pound and sausage at \$2.52 a pound.

b. We verify that these inputs give a maximum revenue by finding the determinant of the second partials matrix: $R_{bb} = -6$, $R_{pp} = -4$, $R_{bp} = R_{pb} = -1$

$D = \begin{vmatrix} -6 & -1 \\ -1 & -4 \end{vmatrix} = 23 > 0$ and $R_{bb} < 0$

Thus we have found the prices that result in maximum quarterly revenue.

c. $R(1.91,\ 2.52) \approx \$28.5$ thousand

11. a. Find the partial derivatives of E, and set them equal to zero.

$E_T = 299.7038 - 10.4420T - 0.4023H = 0$

$E_H = 23.1412 - 0.1874H - 0.4023T = 0$

Solving for T and H gives

$T \approx 26.1032°C$ and $H \approx 67.4488\%$.
$E(26.1032, 67.4488) \approx 500.343$ eggs The critical point is approximately (26.1, 67.4, 500).

b. When exposed to approximately 26.1 °C and 67.4% relative humidity, a *C. grandis* female will lay approximately 500 eggs in 30 days.
$E_{TT} = -10.442$, $E_{RR} = -1.874$, $E_{TR} = E_{RT} = -0.4023$

$$D = \begin{vmatrix} -10.442 & -0.4023 \\ -0.4023 & -1.874 \end{vmatrix} \approx 19.4 > 0$$

Because $E_{TT} < 0$ and $D > 0$, the critical point is a maximum.

13. Find the partial derivatives of *R*, and set them equal to zero.

$$R_P = -1.544 + 9.810P - 3T = 0$$
$$R_T = -1.625 - 14.106T - 3P = 0$$

Solving for *P* and *T* gives $P \approx -0.1307$ and $T \approx -0.0874$.
$R(-0.1307, -0.0874) \approx 32.8$
The critical point is approximately (–0.13, –0.09, 32.8).
The second partials are
$R_{PP} = -9.810$, $R_{TT} = -14.106$, $R_{PT} = R_{TP} = -3$

$$D = \begin{vmatrix} -9.810 & -3 \\ -3 & -14.106 \end{vmatrix} \approx 129 > 0$$

Because $D > 0$ and $R_{PP} < 0$, the critical point is a maximum.

To maximize the rate, the pH is about $5.5 + 1.5(-0.13) \approx 5.3$ and the temperature is about $60 + 8(-0.09) \approx 59.3°C$.

15. a. From the graph, the maximum appears to be about 2.35 mg when pH is about 9 and the temperature is about 65 °C.

b. Find the partial derivatives of *P*, and set them equal to zero.

$$P_x = -0.26 - 0.46x - 0.25y = 0$$
$$P_y = -0.34 - 0.32y - 0.25x = 0$$

Solving for *x* and *y* gives $x \approx 0.0213$ and $y \approx -1.0791$.
The second partials are $P_{xx} = -0.46$, $P_{yy} = -0.32$, and $P_{xy} = P_{yx} = -0.25$

$$D = \begin{vmatrix} -0.46 & -0.25 \\ -0.25 & -0.32 \end{vmatrix} \approx 0.21 > 0$$

Because $P_{xx} < 0$ and $D > 0$, the critical point is a maximum.

A pH of about $9 + 0.0213 \approx 9.02$ and a temperature of about $70 + 5(-1.0791) \approx 64.6°F$ will result in the maximum production of $P(0.0213, -1.0791) \approx 2.32$ mg.

17. Find the partial derivatives of L, and set them equal to zero.

$L_w = 1.13 - 5.83s = 0$

$L_s = 1.04 - 5.83w = 0$

Solving for w and s gives $w \approx 0.18$ and $s \approx 0.19$. The second partials are

$L_{ww} = 0, L_{ss} = 0, L_{sw} = L_{ws} = -5.83$

$$D = \begin{vmatrix} 0 & -5.83 \\ -5.83 & 0 \end{vmatrix} \approx -34 < 0$$

Because $D < 0$, the critical point is a saddle point. The corresponding proportions are

whey protein: $w \approx 0.18$

skim milk powder: $s \approx 0.19$

sodium caseinate: $c \approx 1 - 0.18 - 0.19 = 0.63$

19. a. Find the partial derivatives of E, and set them equal to zero.

$E_e = -30.372e^2 + 42.694e - 13.972 = 0 \qquad E_n = -5n + 2.497 = 0$

Solving for n gives $n \approx 0.4994$ and solving for e gives $e \approx 0.5185$ or $e \approx 0.8872$. Thus there are two critical points:

$A \approx (0.5185, 0.4994, 799.91)$

$B \approx (0.8872, 0.4994, 800.16)$.

b. We use the contour graph to conclude that Point A is a saddle point and point B corresponds to a relative maximum.

c. $E_{ee} = -60.744e + 42.694$, $E_{nn} = -5$, $E_{ne} = E_{en} = 0$

When $e \approx 0.5185$ and $n \approx 0.4994$,

$D \approx -59.98$ so point A is confirmed to be a saddle point. When $e \approx 0.8872$ and $n \approx 0.4994$,

$D \approx 59.98$ and $e \approx -11.20$ so point B is confirmed to correspond to relative maximum.

21. a. Find the partial derivatives of A, and set them equal to zero.

$A_g = 4.26 - 0.1g = 0$

$A_m = 5.69 - 0.28m - 0.07h = 0$

$A_s = 0.67 - 0.06s = 0$

$A_h = 2.48 - 0.1h - 0.07m = 0$

Solving this system, we get

$g = 42.6\%$, $m \approx 17.1169\%$,

$s \approx 11.1667\%$, $h \approx 12.8182$ days

$A(42.6, 7.1169, 11.1667, 12.8182) \approx 7.29$

b. One method is to evaluate points close to the critical point. By doing so, it is possible to conjecture that the point is a relative maximum.

23. Quill Activity

Section 10.3 Optimization Under Constraints

1. **a.** The optimal point is (45, 45, 2025). The optimal value is 2025.

   b. The point is a constrained maximum.

   c. $f_a = b$, $f_b = a$, $g_a = 1$, $g_b = 1$
   We have the following system of equations:
   $$b = \lambda(1)$$
   $$a = \lambda(1)$$
   $$a + b = 90$$
   Solving this system, we get $a = 45$,
   $b = 45$, and $\lambda = 45$. Thus $f(45, 45) = 2025$ is a constrained optimal value.

3. **a.**

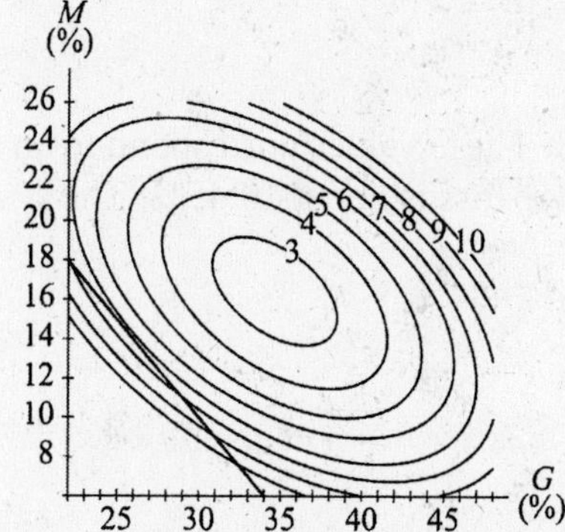

The constrained minimum is approximately 7, which occurs when $G \approx 26\%$ and $M \approx 15\%$. (Answers may vary.)

5. $f_r = 4r + p$, $f_h = r - 2p + 1$,
   $g_k = 2$, $g_h = 3$
   We have the following system of equations.
   $$4r + p = \lambda(2)$$
   $$r - 2p + 1 = \lambda(3)$$
   $$2r + 3p = 1$$
   Solving this system we get $r = \frac{-1}{16}$ and $p = \frac{3}{8}$. $f\left(\frac{-1}{16}, \frac{3}{8}\right) = \frac{7}{32}$ The constrained optimal point is $\left(\frac{-1}{16}, \frac{3}{8}, \frac{7}{32}\right)$.

a.

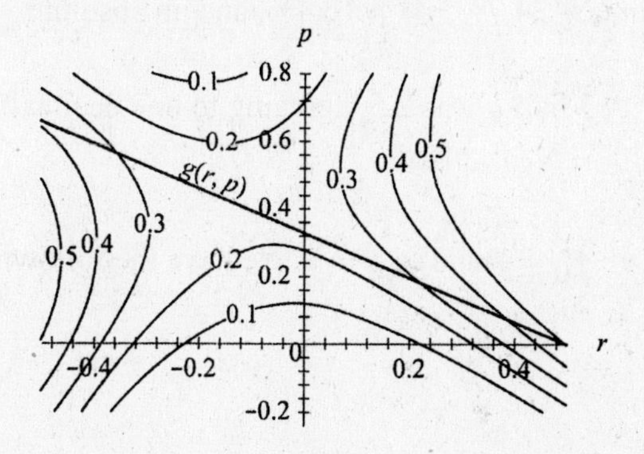

A contour graph confirms that the point is a constrained minimum, because the contour to which the constraint line is tangent is the smallest-valued contour that the constraint line touches.

b. We evaluate $f(r, p)$ for values of r near $\frac{-1}{16} = 0.0625$. We choose 0.6 and 0.65 (these are arbitrarily chosen). We find the corresponding p-values that lie on the constraint curve $2r + 3p = 1$, but substituting each r-value and solving for p. Thus we have the input pairs (0.065,0.29) and (0.06, 0.2933). Evaluating f at these inputs gives the outputs 0.2321 and 0.2332. Both of these values are greater than $f\left(\frac{-1}{16}, \frac{3}{8}\right) = \frac{7}{32} = 0.21875$. These calculations suggest that the point is the location of a constrained minimum.

7. We have the following system of equations (the first two are the same as in Example 2).

$$1.13 - 5.83s = \lambda$$
$$1.04 - 5.83w = \lambda$$
$$w + s = 0.9$$

Solving this system gives $s \approx 0.4577$ and $w \approx 0.4423$. These values correspond to an output $P(0.4577, 0.4423) \approx 10.45$, so the minimum percentage loss is about 10.45%. The approximation in Example 2 was 10.46%.

9. a. $g(r, n) = 12r + 6n = 504$

$A_r = 0.2rn$, $A_n = 0.1r^2$, $g_r = 12$, $g_n = 6$

We have the following system of equations.

$$0.2rn = \lambda(12)$$
$$0.1r^2 = \lambda(6)$$
$$12r + 6n = 504$$

Solving this system, we get $r = 28$, $n = 28$, and $\lambda \approx 13.07$ (or $r = 0$, $n = 0$, and $\lambda = 0$ which gives 0 responses.) The club should allocate (28 ads)($12 per ad) = $336 for 28 radio ads and (28 ads)($6 per ad) = $168 for 28 newspaper ads.

b. $A(28, 28) \approx 2195$ responses

c. The Lagrange multiplier is $\lambda = 13.07$ responses per dollar. The change in the number of responses can be approximated as

$\Delta A \approx$ (13.07 responses per dollar)($26) ≈ 340 additional responses.

11. a. We solve the system of equations: $C_G = -3.76 + 0.08G + 0.06M = 0$

$C_M = -4.71 + 0.16M + 0.06G = 0$ to obtain $G = 34.7\%$, $M \approx 16.4\%$, and an absolute minimum cohesiveness of honey of

$C(34.7, 16.4) \approx 2.5$. Our estimate in Activity 3 part *a* was 2.5, agreeing to one decimal place with the actual minimum.

b. $g(G, M) = G + M = 40$

Using the partial derivatives from part *a* and $g_G = 1$ and $g_M = 1$, we have the following system of equations:

$$-3.76 + 0.08G + 0.06M = \lambda(1)$$
$$-4.71 + 0.16M + 0.06G = \lambda(1)$$
$$G + M = 40$$

Solving this system, we get $G \approx 25.4$, $M \approx 14.6$, and $\lambda \approx -0.85$.
The minimum measure of cohesiveness possible is about $C(25.4, 14.6) \approx 7.2$.

c. Figure 10.3.3 verifies that the point is a minimum since the contour to which the constraint line is tangent is the smallest-valued contour that the constraint line touches.

d. The estimate in Activity 3 part *b* of 7.5 is slightly higher than 7.2, the actual constrained minimum found in part *b* of this activity.

13. a. Worker expenditure:

$$\frac{(\$7.50/\text{hour})(100L \text{ hours})}{1000}$$

$= 0.75L$ thousand dollars
The constraint is
$g(L, K) = 0.75L + K$
The partial derivatives are
$f_L = 3.16389L^{-0.7}K^{0.5}$ $f_K = 5.27315L^{0.3}K^{-0.5}$
$g_L = 7.5$, $g_K = 1$
We solve the following system of equations

$$3.16389L^{-0.7}K^{0.5} = 0.75\lambda$$
$$5.27315L^{0.3}K^{-0.5} = \lambda$$
$$0.75L + K = 15$$

to obtain $L = 7.5$, $K = 9.375$, $\lambda \approx 3.152$.
Maximum production will be achieved by using 750 labor-hours and \$9375 in capital expenditures.

b. To verify that the value in part *a* is a maximum, evaluate the production at close points on the constraint curve, or examine the constraint curve graphed on a contour graph of the production function.

c. The marginal productivity of money is $\lambda \approx 3.152$ radios per thousand dollars. An increase in the budget of \$1000 will result in an increase in output of about 3 radios.

15. a. From Activity 11, $\lambda \approx -0.85$, so $\frac{dC}{dk} \approx -0.85$ cohesiveness unit per percentage point.

b. The minimum cohesiveness measure should decrease by about (0.85 unit per percentage point)(2 percentage points) = 1.7.

c. Find the partial derivatives of *C*, and set them equal to zero.
$C_G = -3.76 + 0.08G + 0.06M = 0$ $C_M = -4.71 + 0.16M + 0.06G = 0$
Solving for *G* and *M* gives
$G \approx 34.6739$ and $M \approx 16.4348$.

The second partials are
$C_{GG} = 0.08$, $C_{MM} = 0.16$, $C_{GM} = C_{MG} = 0.06$

$$D = \begin{vmatrix} 0.08 & 0.06 \\ 0.06 & 0.16 \end{vmatrix} = 0.0092 > 0$$

Because $C_{GG} > 0$ and $D > 0$, the critical point is a minimum.

$C(34.6739, 16.4348) \approx 3.08$

The relative minimum when there are no constraints is approximately 3.1 which is obtained when the percentage of glucose and maltose is approximately 34.7% and the percentage of moisture is approximately 16.4 %.

17. a. From Activity 13, $\lambda \approx 3.152$, so $\frac{dP}{dc} \approx 3.152$ radios/thousand dollars.

b. $\Delta P \approx$ (3.152 radios per thousand dollars)(1.5 thousand dollars)
≈ 4.7 radios

c. $\Delta P \approx$ (3.152 radios per thousand dollars)(–1 thousand dollars)
≈ -3.2 radios

$P(7.5, 9.375) \approx 30$ radios

We estimate the maximum production to be about 30 – 3 = 27 radios.

19. a. $S(r,h) = 2\pi rh + \pi r^2 + \pi\left(r + \frac{9}{8}\right)^2$

square inches when the radius is *r* inches and the height is *h* inches

b. $V(r,h) = \pi r^2 h = 808.5$ cubic inches

c. We solve the equations

$$2\pi h + 2\pi r + 2\pi\left(r + \frac{9}{8}\right) = \lambda 2\pi rh$$

$$2\pi r = \lambda \pi r^2$$

$$\pi r^2 h = 808.5$$

by isolating λ in the second equation and h in the third equation to obtain $\lambda = \frac{2\pi r}{\pi r^2} = \frac{2}{r}$ and $h = \frac{808.5}{\pi r^2}$. Substituting these expressions into the first equation gives

$$\frac{1617}{r^2} + 4\pi r + \frac{9\pi}{4} = \frac{3234}{r^2}$$

Solving for *r* gives $r \approx 4.87$ inches, $h \approx \frac{808.5}{\pi(4.87^2)} \approx 10.86$ inches, and

$S(r,h) \approx 519.5$ square inches.

d. The answers in part *c* agree with those in Section 5.3.

e. Answers will vary. The solution method in Section 5.3 involves using the constraint to write the multivariable function as a single-variable function and then finding where the derivative is zero. The method of Lagrange multipliers involves finding three partial derivatives and solving a system of three equations.

21. Quill Activity

Section 10.4 Least-Squares Optimization

1. a. $f(a,b) = (7-a-b)^2 + (11-6a-b)^2 + (19-12a-b)^2$

b. $\frac{\partial f}{\partial a} = 2(7-a-b)(-1) + 2(11-6a-b)(-6) + 2(19-12a-b)(-12) = 362a + 38b - 602$

$\frac{\partial f}{\partial b} = 2(7-a-b)(-1) + 2(11-6a-b)(-1) + 2(19-12a-b)(-1) = 38a + 6b - 74$

The second partials are $f_{aa} = 362$, $f_{bb} = 6$, $f_{ab} = f_{ba} = 38$

$$D = \begin{vmatrix} 362 & 38 \\ 38 & 6 \end{vmatrix} = 728$$

c. Set $\frac{\partial f}{\partial a} = 0$ and $\frac{\partial f}{\partial b} = 0$, and solve the resulting system of equations.

The solution is $a \approx 1.0989$ and $b \approx 5.3736$ corresponding to an output of $f(1.0989, 5.3736) \approx 1.4066$. This is a minimum because $D > 0$ and $f_{aa} > 0$.

d. The linear model that best fits the data is $y = 1.099x + 5.374$.

3. a. $f(a,b) = (3-b)^2 + (2-10a-b)^2 + (1-20a-b)^2$

b. $\frac{\partial f}{\partial a} = 2(2-10a-b)(-10) + 2(1-20a-b)(-20) = 1000a + 60b - 80$

$\frac{\partial f}{\partial a} = 2(3-b)(-1) + 2(2-10a-b)(-1) + 2(1-20a-b)(-2) = 60a + 6b - 12$

The second partials are $f_{aa} = 1000$, $f_{bb} = 6$, $f_{ab} = f_{ba} = 60$.

$$D = \begin{vmatrix} 1000 & 60 \\ 60 & 6 \end{vmatrix} = 2400$$

Set $\frac{\partial f}{\partial a} = 0$ and $\frac{\partial f}{\partial b} = 0$, and solve the resulting system of equations.

The solution is $a = -0.1$ and $b = 3$, corresponding to an output of $f(-0.1, 3) = 0$.

This is a minimum because $D > 0$ and $f_{aa} > 0$.

Because the minimum SSE is zero, we know that the line of best fit is a perfect line—that is, all of the data points lie on the line.

c. The linear model that best fits the data is $y = -0.1x + 3$ percent, where x is the number of years since 1970.

d. Answers will vary depending on the year.

5. a.

b. Using technology, $y = 1.176x + 1.880$ dollars to make x cases of ball bearings. The vertical intercept is the fixed cost per case. The slope is the cost to produce one case.

c.

x	$y(x)$	Data value	Deviation: data − $y(x)$	Squared deviations
1	3.0561	3.10	0.044	0.00192
2	4.2324	4.25	0.018	0.00031
6	8.9375	8.95	0.013	0.00016
9	12.4663	12.29	−0.176	0.03108
14	18.3477	18.45	0.102	0.01047
			Sum of squared deviations ≈ 0.044	

d. To find the best-fitting line, first construct the function f with inputs a and b, which represents the sum of the squared errors of the data points from the line $y = ax + b$. Find the partial derivatives of f with respect to a and b. Simplify the partials, and find the point (a, b) where the partials are simultaneously zero. These are the coefficients of the model given in part b. The function f evaluated at (a, b) gives the value of SSE shown in part c.

7. $f(a,b) = (5.5 - b)^2 + (5 - 5a - b)^2 + (4.8 - 8a - b)^2 + (4.6 - 10a - b)^2$

$$\frac{\partial f}{\partial a} = 2(5 - 5a - b)(-5) + 2(4.8 - 8a - b)(-8) + 2(4.6 - 10a - b)(-10) = 378a + 46b - 218.8$$

$$\frac{\partial f}{\partial b} = 2(5.5 - b)(-1) + 2(5 - 5a - b)(-1) + 2(4.8 - 8a - b)(-1) + 2(4.6 - 10a - b)(-1)$$

$$= 46a + 8b - 39.8$$

Set $\frac{\partial f}{\partial a} = 0$ and $\frac{\partial f}{\partial b} = 0$, and solve the resulting system of equations. The solution is $a \approx -0.0885$ and $b \approx 5.4841$.

The linear model that best fits the data is $y = -0.0885x + 5.4841$ million experiments, where x is the number of years since 1970.

9. a.

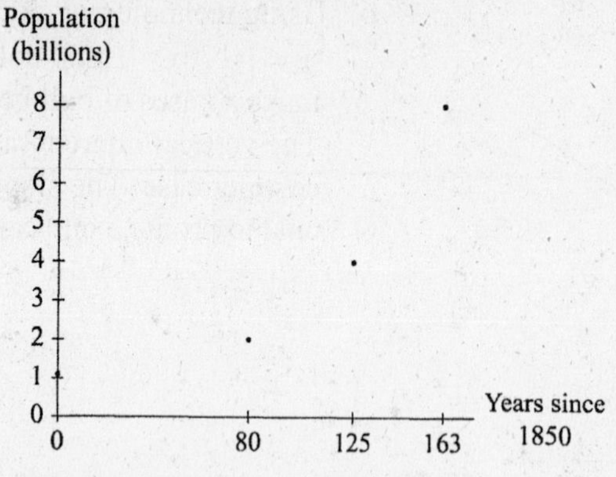

The plot of the data points appears to be concave up.

b. Plot the data points (0, ln 1.1), (80, ln 2.0), (125, ln 4.0), and (163, ln 8.0). The plot of the data points also appears to be concave up, but less so than the plot in part *a*.

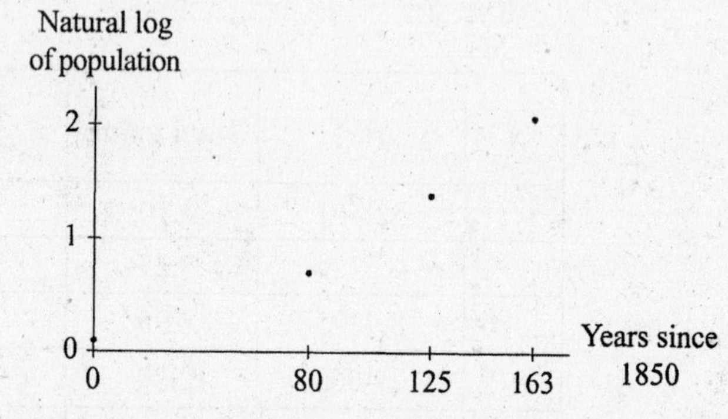

c. Using the least-squares technique, we begin with the function describing the sum of the squared errors:

$$f(a,b) = (\ln 1.1 - b)^2 + (\ln 2.0 - 80a - b)^2 + (\ln 4.0 - 125a - b)^2 + (\ln 8.0 - 163a - b)^2$$

Next we find the first partial derivatives and set them equal to zero:

$$\frac{\partial f}{\partial a} = 2(\ln 2.0 - 80a - b)(-80) + 2(\ln 4.0 - 125a - b)(-125) + 2(\ln 8.0 - 163a - b)(-163)$$

$$= 97{,}188a + 736b - (160\ln 2 + 250\ln 4 + 326\ln 8) = 0$$

$$\frac{\partial f}{\partial b} = 2(\ln 1.1 - b)(-1) + 2(\ln 2.0 - 80a - b)(-1) + 2(\ln 4.0 - 125a - b)(-1) + 2(\ln 8.0 - 163a - b)(-1)$$

$$= 736a + 8b - 2(\ln 1.1 + \ln 2 + \ln 4 + \ln 8) = 0$$

The solution to this system of linear equations is $a \approx 0.012$ and $b \approx -0.037$.

$y = 0.012x - 0.037$ whose output is the natural log of the population in billions x years after 1850

d.

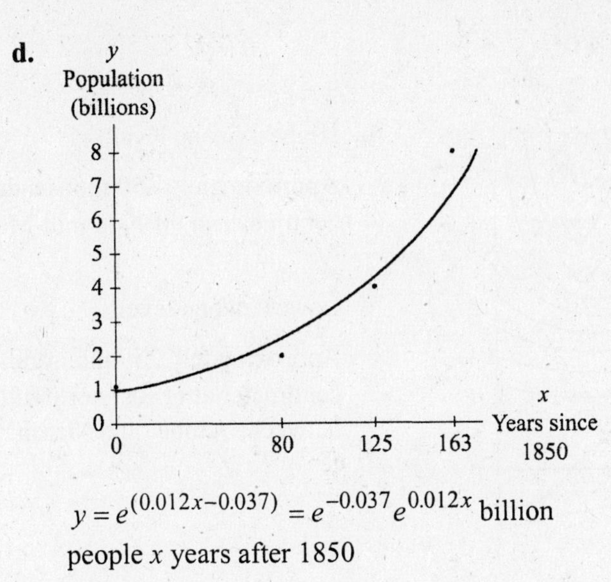

$y = e^{(0.012x - 0.037)} = e^{-0.037}e^{0.012x}$ billion people x years after 1850

e. Using technology, we get $y = 0.964(1.012^x)$ billion people x years after 1850. This confirms our result in part *d* because $0.964 \approx e^{-0.037}$ and $1.012 \approx e^{0.012}$.

11. Quill Activity

Chapter 10 Review Test

1. a.

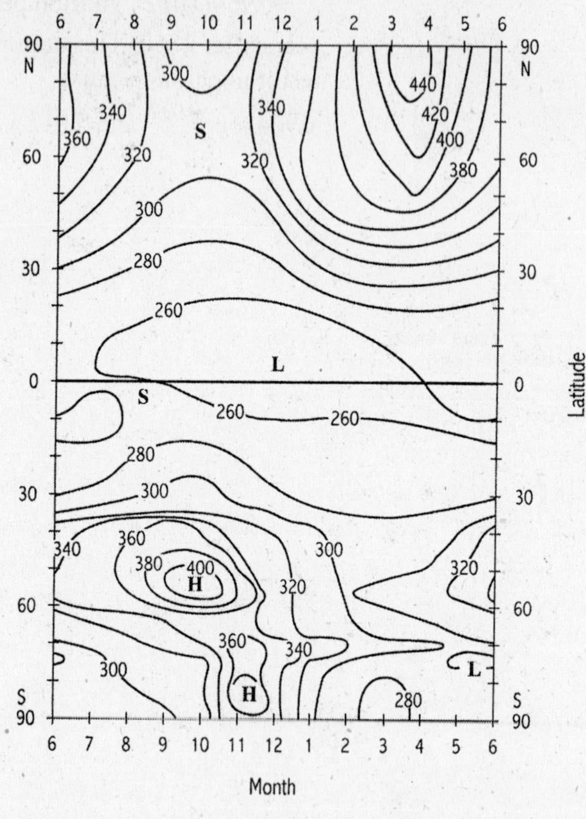

b. Highest ozone level:

Approximately 450 thousandths of a centimeter at 90°N in mid-March.

Lowest ozone level:

Approximately 250 thousandths of a centimeter at or just north of 0° (the equator) between October and March.

2. a.

Blanching temperature (°C)	Blanching time (minutes)		
	2	15	30
50	4.2	6.8	4.2
60	4.5	4.7	7.1
70	8.2	11.4	8.6
80	7.3	6.9	3.9

b. The maximum crispness is approximately 11.4, occurring for a blanching time of 15 minutes at 70°C.

c. Find the partial derivatives of C and set them equal to zero.

$C_x = -10.4x + 299.7 - 0.4y = 0$

$C_y = -0.2y + 23.1 - 0.4x = 0$

Solving for x and y gives $x \approx 26.4$ minutes and $y \approx 62.7$ °C.
The crispness index is $C(26.4, 62.7) \approx 10$.

d. $C_{xx} = -10.4$, $C_{yy} = -0.2$, $C_{xy} = C_{yx} = -0.4$

$D = \begin{vmatrix} -10.4 & -0.4 \\ -0.4 & -0.2 \end{vmatrix} \approx 1.92 > 0$ and $C_{xx} < 0$, so the point is a maximum. Because the point corresponds to the only critical value and the function decreases in every direction away from the maximum, it is an absolute maximum.

3. a. The maximum profit is approximately \$202,500 for about 52,000 shirts and 9000 hats sold. (Answers may vary.)

b. $4s + 1.25h = 150$

c.

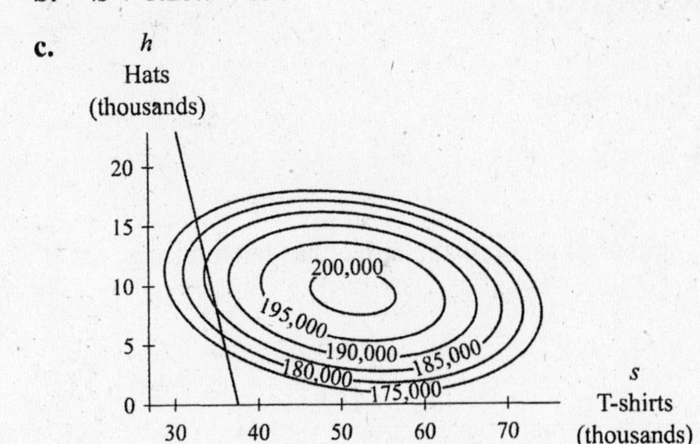

The constrained maximum revenue is approximately $185,000 for about 34,000 shirts and 10,000 hats sold. (Answers may vary.)

4. a.

$$-104.4s + 5935.5 - 59.1h = 4\lambda$$

$$-59.1s + 10,299.3 - 768.6h = 1.25\lambda$$

$$4 + 1.25h = 150$$

b. Solving this system of linear equations, we get $s \approx 34.3611$, $h \approx 10.0446$, and $\lambda \approx 438.64$. The corresponding profit is about $P(34.3611, 10.0446) \approx \$186,599$.

c. $\lambda \approx \$438.64$ of profit per thousand dollars spent, which means that for each additional thousand dollars budgeted, profit will increase by approximately $439.

d. The change in budget is 2 thousand dollars, so expect profit to increase by about ($438.64 of profit per thousand dollars budgeted)($2 thousand) $\approx \$877$.

5. a. $f(a,b) = (29.9 - 10a - b)^2 + (33.4 - 15a - b)^2 + (37.5 - 20a - b)^2$

b.

$$\frac{\partial f}{\partial a} = 2(29.9 - 10a - b)(-10) + 2(33.4 - 15a - b)(-15) + 2(37.5 - 20a - b)(-20)$$

$$= 1450a + 90b - 3100$$

$$\frac{\partial f}{\partial a} = 2(29.9 - 10a - b)(-1) + 2(33.4 - 15a - b)(-1) + 2(37.5 - 20a - b)(-1)$$

$$= 90a + 6b - 201.6$$

Set $\frac{\partial f}{\partial a} = 0$ and $\frac{\partial f}{\partial b} = 0$, and solve the resulting system of equations. The solution is

$a = 0.76$ and $b = 22.2$, corresponding to $f(0.76, 22.2) = 0.06$.

$f_{aa} = 1450$, $f_{bb} = 6$, $f_{ab} = f_{ba} = 90$

$$D = \begin{vmatrix} 1450 & 90 \\ 90 & 6 \end{vmatrix} = 600$$

This is a minimum because $D > 0$ and $f_{aa} > 0$.

c. The linear model that best fits the data is $y = 0.76x + 22.2$ kilograms, where x is the body temperature in °C. The sum of the squared deviations from this line is 0.06, and this is the smallest possible sum.

Chapter 11

Section 11.1 Differential Equations and Slope Fields

1. $c = kg$

3. Let v be the velocity and let t be the number of seconds the object has been falling. $v = kt$

5. $\dfrac{dp}{da} = ka$

7. a.

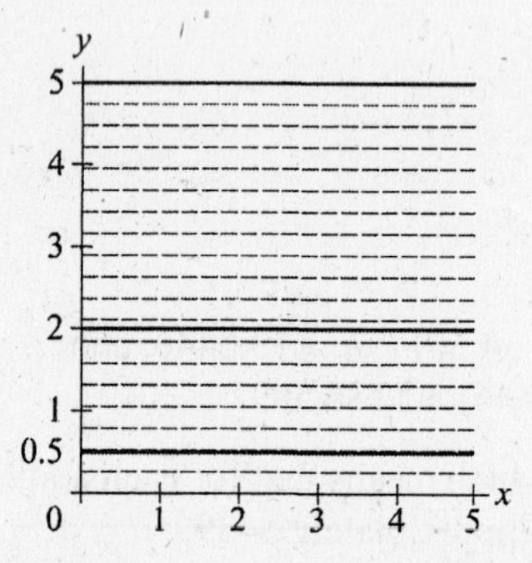

The displayed particular solutions go through the points (0, 0.5), (2, 2) and (4, 5). Answers will vary.

b. All particular solutions are horizontal lines, each passing through the chosen initial condition.

c. $y = C$, where C is a constant

9. a.

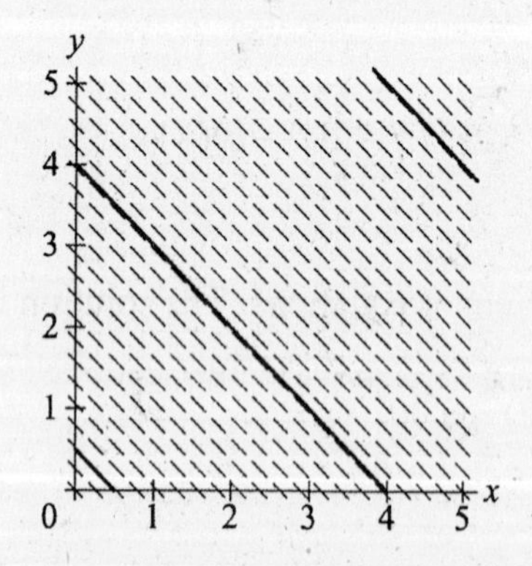

The displayed particular solutions go through the points (0, 0.5), (2, 2) and (4, 5). Answers will vary.

b. All particular solutions are parallel lines with slope –1 and differing vertical shifts.

c. $y = -x + C$, where C is a constant

11. a.

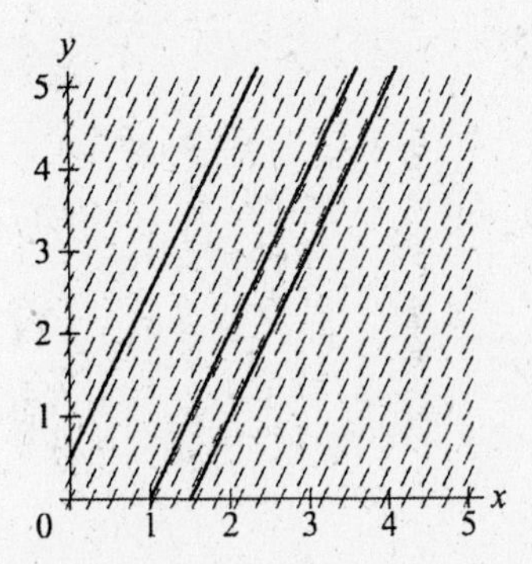

The displayed particular solutions go through the points (0, 0.5), (2, 2) and (4, 5). Answers will vary.

b. All particular solutions are parallel lines with slope 2 and differing vertical shifts.

c. $y = 2x + C$, where C is a constant

12. a.

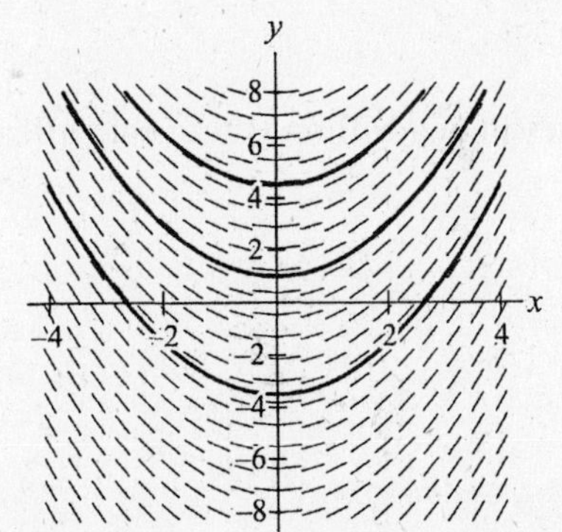

The displayed particular solutions go through the points (0, 1), (1,–3) and (2, 6.5). Answers will vary.

b. All particular solutions are parabolas, each passing through the chosen initial condition.

c. $y = \frac{x^2}{2} + C$, where C is a constant

13. a.

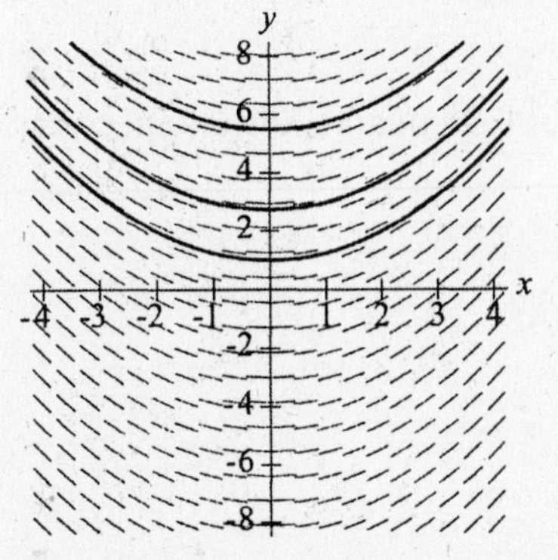

The displayed particular solutions go through the points (0, 1), (1, 3) and (2, 6.5). Answers will vary.

b. All particular solutions are concave-up parabolas with minimum points on the vertical axis.

c. $y = 0.25x^2 + C$, where C is a constant

15. a.

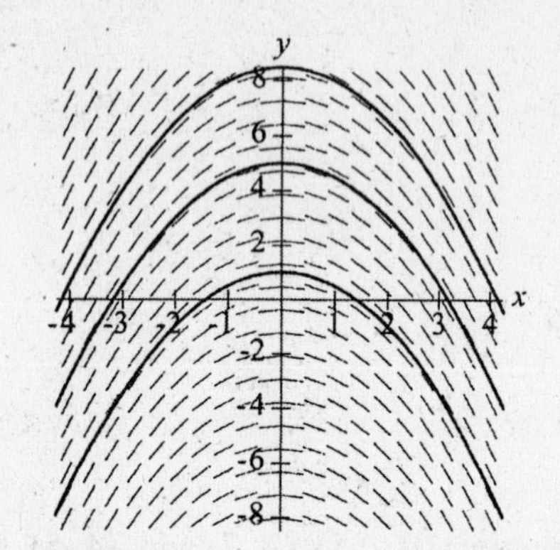

The displayed particular solutions go through the points (0, 1), (2, 3) and (2, 6.5). Answers will vary.

b. All particular solutions are concave-down parabolas with maximum points on the vertical axis.

c. $y = -0.5x^2 + C$, where C is a constant

19. a. Let p be the energy production in the United States in quadrillion Btu t years after 1975.

$\frac{dp}{dt} = 0.98$ quadrillion Btu / year

t years after 1975

b. $p(t) = 0.98t + C$ quadrillion Btu
t years after 1975

c. $p(5) = 0.98(5) + C = 64.8$, so
$C = 59.9$
$p(t) = 0.98t + 59.9$ quadrillion Btu
t years after 1975

d. $p(0) = 59.9$ and $\frac{dp}{dt} = 0.98$ when

$t = 0$.
In 1975 the production was
59.9 quadrillion Btu and was increasing at a rate of 0.98 quadrillion Btu per year.

e. Particular solution with $p(0)$=59.9

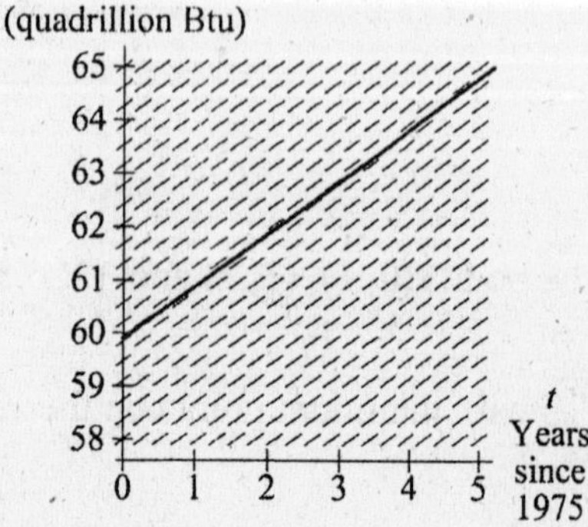

21. a. Let c be the amount of arable and permanent cropland in millions of square kilometers t years after 1970.

$\frac{dc}{dt} = 0.0342$ million square kilometers per year t years after 1970

b. $c(t) = 0.0342t + C$ million square kilometers t years after 1970

c. $c(10) = 0.0342(10) + C = 14.17$, so $C = 13.828$
$c(t) = 0.0342t + 13.828$ million square kilometers t years after 1970

d. $c(0) = 13.828$, $c(20) = 14.512$ and $\frac{dc}{dt} = 0.0342$ when $t = 0$ and $t = 20$.
Cropland was increasing at a rate of 0.0342 million square kilometers per year in both 1970 and 1990. In 1970 there were 13.828 million square kilometers of cropland, and in 1990, there were 14.512 million square kilometers of cropland.

23. a. $v(t) = -32t$ feet per second t seconds after the object is dropped

b. Let s be distance.

$\frac{ds}{dt} = -32t$ feet per second t seconds after the object is dropped

c. $s(t) = -16t^2 + C$ feet t seconds after the object is dropped

d. $s(0) = -16(0^2) + C = 35$, so $C = 35$

$s(t) = -16t^2 + 35$

When $s(t) = 0$, $t \approx \pm 1.479$, where only the positive answer makes sense in this context. It takes approximately 1.5 seconds after the object is dropped for the object to hit the ground.
$v(1.479) \approx -47.3$
The object has a terminal velocity of approximately –47.3 feet per second. (The negative sign on the velocity indicates downward motion.)

25. a. $\frac{df}{dx} = kx$

b. $f(x) = \frac{k}{2}x^2 + C$

c. Taking the derivative of f, we get $\frac{d}{dx}\left(\frac{k}{2}x^2 + C\right) = \frac{k}{2}(2x) + 0 = kx$ Thus we have the identity $kx = kx$, and our solution is verified.

27. a. $\frac{dh}{dt} = \frac{k}{t}$ feet per year after t years

b. $h(t) = k \ln|t| + C$ feet in t years

$h(2) = k \ln 2 + C = 4$ and

$h(7) = k \ln 7 + C = 30$

Solving this system of equations, we get $k \approx 20.75$ and $C \approx -10.39$.

$h(t) = 20.75 \ln t - 10.39$ feet in t years

c. $h(15) \approx 45.8$ feet

Over time, the tree will continue to grow, but the rate of increase will be smaller and smaller.

29. a.

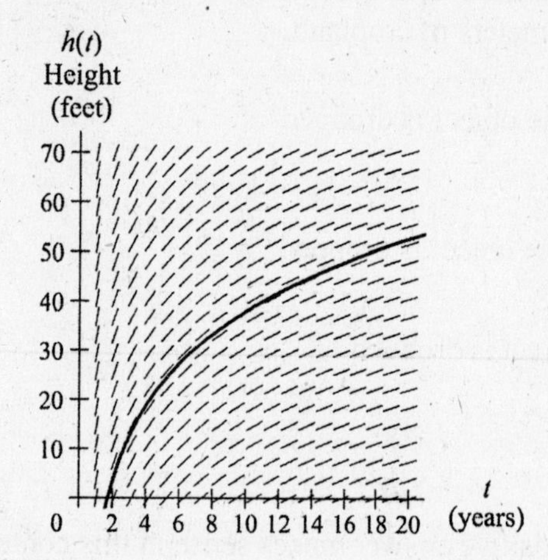

b. The output value on the graph corresponding to an input of $t = 15$ is $h(15) \approx 46$ feet.

31. a.

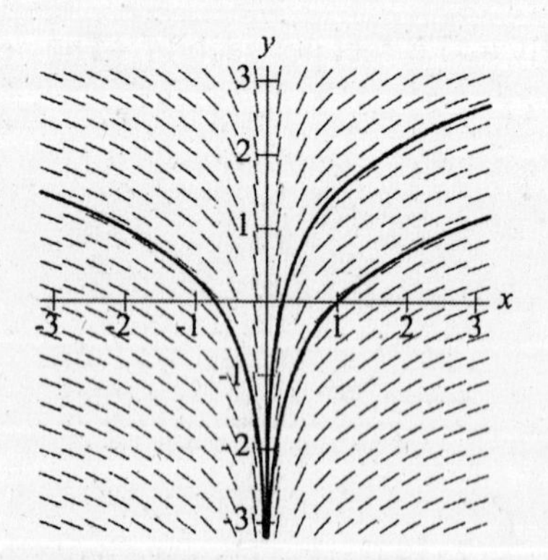

i. The particular solutions shown go through (1, 1.5), (–2, 1), and (1, 0). Answer will vary.

ii. When $x > 0$, the graph of a particular solution rises as x gets larger. When $x < 0$, the solution graph rises as x gets smaller. The particular solution graphs are concave down.

iii. The family of solutions appears to increase rapidly as x moves away from the origin (in both directions), and then the increase slows down. The line $x = 0$ (lying on the y-axis) appears to be a vertical asymptote for the family.

b.

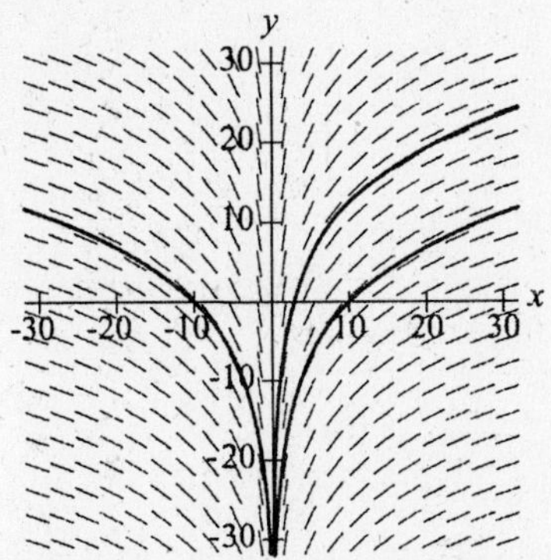

i. The particular solutions shown go through (10, 0), (–10, 0), and (5, 5). Answer will vary.

ii. When $x > 0$, the graph of a particular solution rises as x gets larger. When $x < 0$, the solution graph rises as x gets smaller. The particular solution graphs are concave down.

iii. The family of solutions appears to behave the same as that in part *a*, but the slope at each point on a particular solution graph is 10 times the slope at the corresponding point on a particular solution graph in part *a*. Again, the line $x = 0$ appears to be a vertical asymptote for the family.

c.

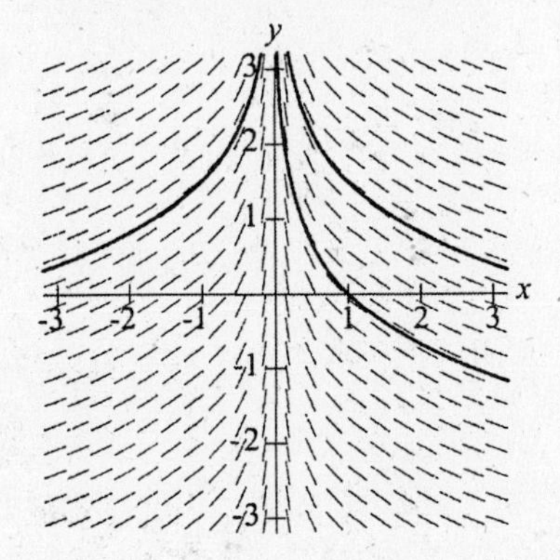

i. The particular solutions shown go through (1, 1.5), (–2, 1), and (1, 0). Answer will vary.

ii. When $x > 0$, the graph of a particular solution falls as x gets larger. When $x < 0$, the solution graph falls as x gets smaller. The particular solution graphs are concave up.

iii. The slope at each point on a particular solution graph is the negative of the slope at a corresponding point on a particular solution graph in part *a*. The family of solutions appears to decrease rapidly as x moves away from the origin (in both directions), and then the decrease levels off. The line $x = 0$ appears to be a vertical asymptote for the family.

d.

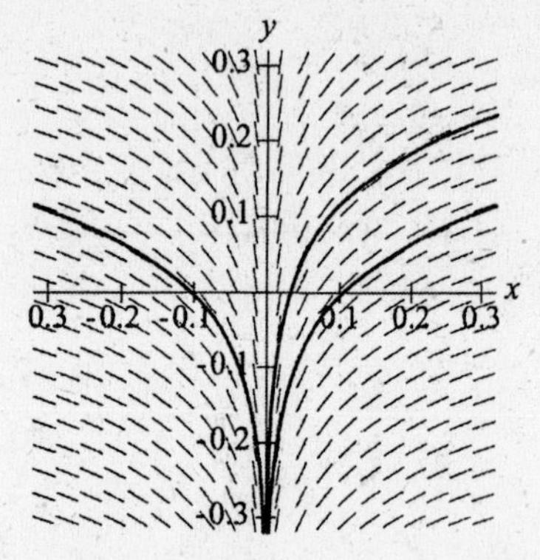

i. The particular solutions shown go through (0.1, 0), (–0.1, 0), and (0.05, 0.05). Answer will vary.

ii. When $x > 0$, the graph of a particular solution rises as x gets larger. When $x < 0$, the solution graph rises as x gets smaller. The particular solution graphs are concave down.

iii. The family of solutions appears to behave the same as that in part *a*, but the slope at each point on a particular solution graph is $\frac{1}{10}$ times the slope at the corresponding point on a particular solution graph in part *a*. Again, the line $x = 0$ appears to be a vertical asymptote for the family.

33.

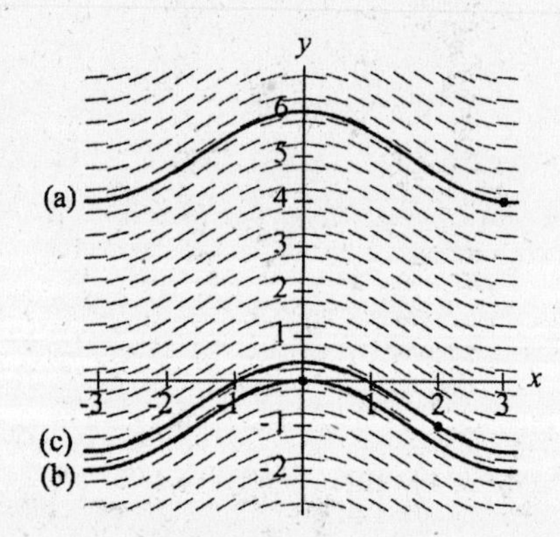

35.

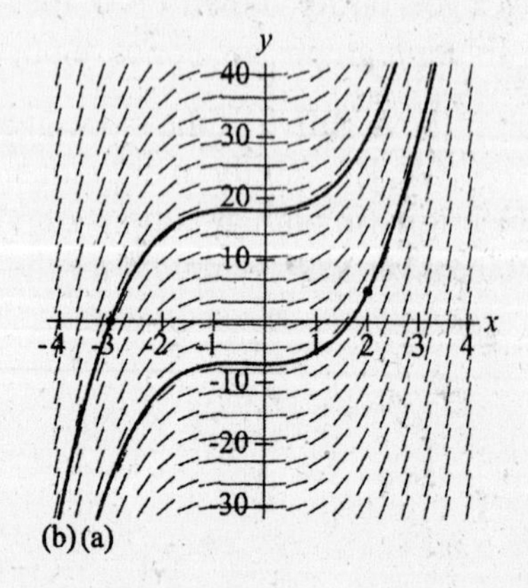

Section 11.2 Separable Differential Equations

1. $\frac{dT}{dt} = \frac{k}{T}$

This is a separable differential equation.

$$TdT = kdt$$
$$\int TdT = \int kdt$$
$$\tfrac{1}{2}T^2 + c_1 = kt + c_2$$
$$T^2 = 2kt + C$$
$$T = \pm\sqrt{2kt + C}$$

Because thickness can't be negative,
$T(t) = \sqrt{2kt + C}$.

3. $\frac{dA}{dt} = kA$

This a separable differential equation.

$$\frac{1}{A}dA = kdt$$
$$\int \frac{1}{A}dA = \int kdt$$
$$\ln A + c_1 = kt + c_2$$
$$\ln A = kt + C$$
$$A = e^{kt+C}$$
$$A = ae^{kt}$$

(Note that A is positive, so we omitted the absolute value signs from the natural logarithms, and a is a positive constant.)

$$A(t) = ae^{kt}$$

5. $\frac{dx}{dt} = kx(N - x)$

This differential equation has a logistic function as its solution.

$$x(t) = \frac{N}{1 + Ae^{-Nkt}}$$

7. Flow-in rate $= \frac{k}{\sqrt{D}}$ where s is a constant

Flow-out rate $= hD$ where h is a constant

$$\frac{dD}{dt} = \frac{k}{\sqrt{D}} - hD$$

9.

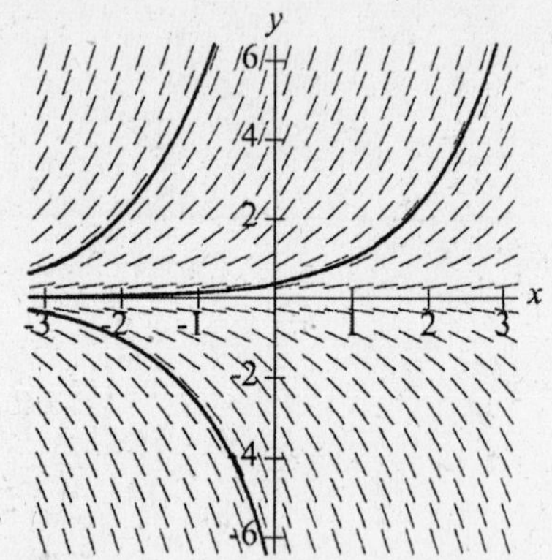

The particular solutions shown go through (–2, 2), (–2, –1), and (1, 1). Answers will vary.

11.

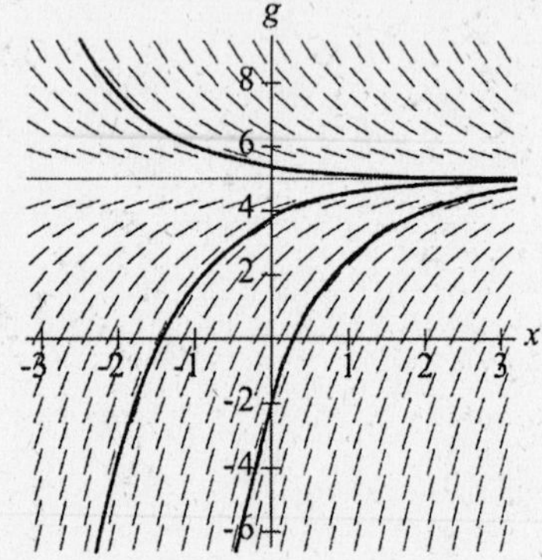

The particular solutions shown go through (0, –2), (–2, –4), and (–1, 6). Answers will vary.

13.

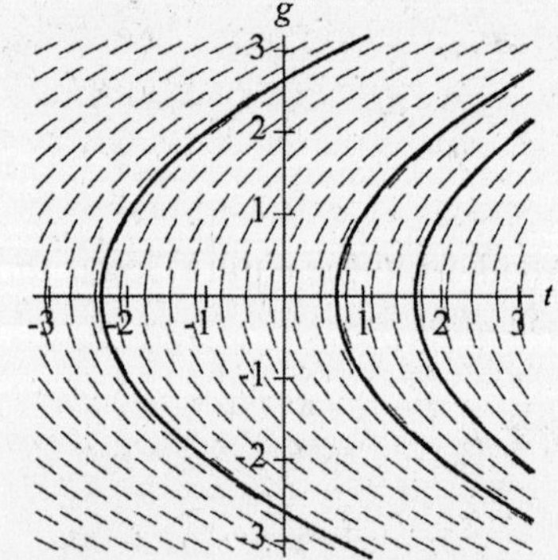

The particular solutions shown go through (–2, 1), (2, 1), and (1, –1). Answers will vary.

15.

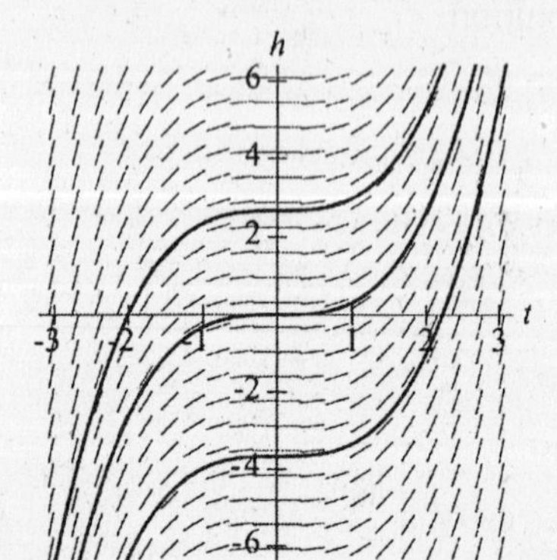

The particular solutions shown go through (0, 0), (–1, –4), and (1, 3). Answers will vary.

17.

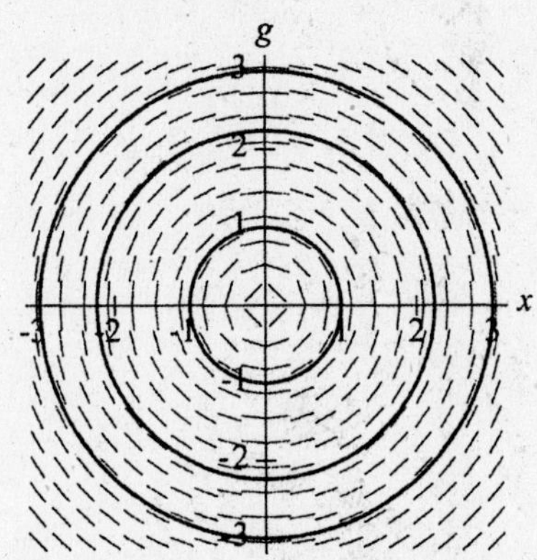

The particular solutions shown go through (0, 3), (1, 0), and (2, 1). Answers will vary.

19. Solve by separation of variables.

$$\frac{1}{y}dy = kdx$$

$$\int \frac{1}{y}dy = \int kdx$$

$$\ln|y| + c_1 = kx + c_2$$

$$\ln|y| = kx + C$$

$$y = \pm e^{kx+C}$$

$$y(x) = \pm ae^{kx}$$

21. Solve by antidifferentiation.

$y(x) = k\ln|x| + C$

23. Solve by separation of variables.

$$\frac{1}{y}dy = \frac{k}{x}dx$$

$$\int \frac{1}{y}dy = \int \frac{k}{x}dx$$

$$\ln|y| + c_1 = k\ln|x| + c_2$$

$$\ln|y| = k\ln|x| + C$$

$$|y| = e^C e^{\ln|x|^k}$$

$$|y| = a|x|^k$$

$$y(x) = \pm ax^k$$

25. a. $\frac{dq}{dt} = kq$ milligrams per hour

b. Solve by separation of variables.

$$\frac{1}{q}dq = kdt$$
$$\int \frac{1}{q}dq = \int kdt$$
$$\ln|q| + c_1 = kt + c_2$$
$$\ln|q| = kt + C$$
$$q = e^{kt+C}$$
$$q(t) = ae^{kt}$$

When $t = 0$, $q = 200$, and when $t = 2$, $q = 100$. Thus, we have

$q(0) = a = 200$ and $q(2) = ae^{2k} = 100$

$$200e^{2k} = 100$$
$$e^{2k} = \frac{1}{2}$$
$$2k = \ln\frac{1}{2}$$
$$k = \frac{1}{2}\ln\frac{1}{2}$$

Thus we have $a = 200$ and $k = \frac{1}{2}\ln\frac{1}{2} \approx -0.346574$.

$q(t) = 200e^{-0.346574t}$ milligrams after t hours

c. After 4 hours, $q(4) \approx 50$ milligrams will remain. After 8 hours, $q(8) \approx 12.5$ milligrams will remain.

27. a. $\frac{da}{dt} = ka$ units per day

b. Solve by separation of variables.

$$\frac{1}{a}da = kdt$$
$$\int \frac{1}{a}da = \int kdt$$
$$\ln|a| + c_1 = kt + c_2$$
$$\ln|a| = kt + C$$
$$a = e^{kt+C}$$
$$a(t) = ce^{kt}$$

$a(0) = c =$ the initial amount.

When $t = 3.824$,

a = half of the initial amount $= \frac{c}{2}$. Thus, we have

$$\frac{c}{2} = ce^{3.824k}$$
$$\frac{1}{2} = e^{3.824k}$$
$$\ln\frac{1}{2} = 3.824k$$
$$k = \frac{1}{3.824}\ln\frac{1}{2}$$
$$\approx -0.181262$$

$a(t) = ce^{-0.181262t}$ units after t days

c. Here, $c = 1$ gram.
After 12 hours = 0.5 days,
$a(0.5) \approx 0.91$ gram will remain. After 4 days, $a(4) \approx 0.48$ gram will remain. After 9 days, $a(9) \approx 0.20$ gram will remain. After 30 days, $a(30) \approx 0.004$ grams will remain.

29. a. Let N be number of countries that issued stamps.

$$\frac{dN}{dt} = 0.0049N(37 - N) \text{ countries per year}$$

b. This differential equation has a logistic function as its solution.

$$N(t) = \frac{37}{1 + Ae^{-0.0049(37)t}}$$
$$= \frac{37}{1 + Ae^{-0.1813t}} \text{ countries}$$

t years after 1800

c. Because $N(55) = 16$, we have the equation $N(55) = \frac{37}{1 + Ae^{-0.1813(55)}} = 16$. Solving this equation for A, we get

$$\frac{37}{1 + Ae^{-0.1813(55)}} = 16$$
$$37 = 16 + 16Ae^{-9.9715}$$
$$A = \frac{21}{16e^{-9.9715}}$$
$$\approx 28{,}097.439$$

$$N(t) = \frac{37}{1 + 28{,}097.439e^{-0.1813t}} \text{ countries } t \text{ years after 1800}$$

d. $N(40) \approx 2$ and $N(60) \approx 24$; There were 2 countries in 1840 and 24 countries in 1860.

e. The upper asymptote is $N(t)=37$, and the lower asymptote is $N(t) = 0$.

f.

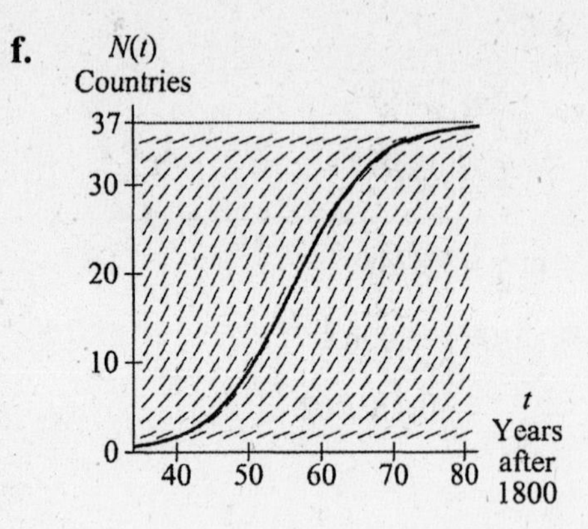

31. a. $\frac{df}{dx} = kf(L - f)$

b. This differential equation has a logistic function as its solution: $f(x) = \frac{L}{1 + Ae^{-Lkx}}$

c. Taking the derivative in part *b*, we get

$$\frac{df}{dx} = -L(1 + Ae^{-Lkx})^{-2} Ae^{-Lkx}(-Lk)$$

$$= \frac{-L}{(1 + Ae^{-Lkx})^2}(-LkAe^{-Lkx}) = \frac{L^2kAe^{-Lkx}}{(1 + Ae^{-Lkx})^2}$$

Substituting $f(x) = \frac{L}{1 + Ae^{-Lkx}}$ into $kf(L - f)$ and simplifying gives

$$kf(L - f) = k\left(\frac{L}{1 + Ae^{-Lkx}}\right)\left(L - \frac{L}{1 + Ae^{-Lkx}}\right)$$

$$= k\left(\frac{L}{1 + Ae^{-Lkx}}\right)\left(\frac{L(1 + Ae^{-Lkx})}{1 + Ae^{-Lkx}} - \frac{L}{1 + Ae^{-Lkx}}\right)$$

$$= k\left(\frac{L}{1 + Ae^{-Lkx}}\right)\left(\frac{L + LAe^{-Lkx} - L}{1 + Ae^{-Lkx}}\right) = \frac{kL^2Ae^{-Lkx}}{\left(1 + Ae^{-Lkx}\right)^2}$$

Thus we have the identity $\frac{kL^2Ae^{-Lkx}}{\left(1 + Ae^{-Lkx}\right)^2} = \frac{kL^2Ae^{-Lkx}}{\left(1 + Ae^{-Lkx}\right)^2}$, and our solution is verified.

Section 11.3 Numerically Estimating by Using Differential Equations: Euler's Method

1. a. $\frac{dy}{dx}=\frac{1}{2}$, initial condition $(0, 0)$, step size $\frac{4-0}{2\text{ steps}}=2$ units per step

x	Estimate of $y(x)$	Slope at x
0	0	0.5
2	$0+2(0.5)=1$	0.5
4	$1+2(0.5)=2$	

Thus we estimate $y(4)\approx 2$.

b. $\frac{dy}{dx}=2x$, initial condition $(1, 4)$, step size $\frac{7-1}{2\text{ steps}}=3$ units per step

x	Estimate of $y(x)$	Slope at x
1	4	$2(1)=2$
4	$4+3(2)=10$	$2(4)=8$
7	$10+3(8)=34$	

Thus we estimate $y(7)\approx 34$.

3. a. $\frac{dy}{dx}=\frac{5}{y}$, initial condition (1, 1), step size is $\frac{5-1}{2}=2$ units per step

x	Estimate of $y(x)$	Slope at x
1	1	$\frac{5}{1}=5$
3	$1+2(5)=11$	$\frac{5}{11}\approx 0.4545$
5	$11+2\left(\frac{5}{11}\right)\approx 11.91$	

We estimate that $y(5)\approx 11.91$.

b. $\frac{dy}{dx}=\frac{5}{x}$, initial condition (2, 2), step size is $\frac{8-2}{2}=3$ units per step

x	Estimate of $y(x)$	Slope at x
2	2	$\frac{5}{2}=2.5$
5	$2+3\left(\frac{5}{2}\right)=9.5$	$\frac{5}{5}=1$
8	$9.5+3(1)=12.5$	

We estimate that $y(8)\approx 12.5$.

5. a. $\frac{dw}{dt} = \frac{33.67885}{t}$ pounds per month after t months

b. Initial condition (1, 6), step size 0.25 month

t (months)	Estimate of $w(t)$ (pounds)	Slope at t (pounds per month)	t (months)	Estimate of $w(t)$ (pounds)	Slope at t (pounds per month)
1	6.000	33.679	3.25	48.768	10.363
1.25	14.420	26.943	3.5	51.359	9.623
1.5	21.155	22.453	3.75	53.764	8.981
1.75	26.769	19.245	4	56.010	8.420
2	31.580	16.839	4.25	58.115	7.924
2.25	35.790	14.968	4.5	60.100	7.484
2.5	39.532	13.472	4.75	61.967	7.090
2.75	42.900	12.247	5	63.739	6.736
3	**45.961**	11.226	5.5	67.027	6.123
3.25	48.768	10.363	5.75	68.558	5.857
3.5	51.359	9.623	**6**	**70.022**	

The Euler estimates using a step size of 0.25 month are $w(3) \approx 46.0$ lbs; $w(6) \approx 70.0$ lbs

c. Initial condition (1, 6), step size 1 month $w(3) \approx 56.52$ pounds; $w(6) \approx 82.90$ pounds

t (months)	Estimate of $w(t)$ (pounds)	Slope at t (pounds per month)	t (months)	Estimate of $w(t)$ (pounds)	Slope at t (pounds per month)
1	6	33.679	4	67.745	8.420
2	39.679	16.839	5	76.164	6.736
3	**56.518**	11.226	**6**	**82.90**	

The Euler estimates using a step size of 1 month are $w(3) \approx 56.5$ lbs; $w(6) \approx 82.9$ lbs

d. The answer to part *b* should be more accurate because it uses a smaller step size.

7. a. $\frac{dp}{dt} = 3.935t^{3.55}e^{-1.35135t}$ thousand barrels per year t years after production begins, where $p(t)$ is the total amount of oil produced after t years

b. Initial condition: (0, 0); Step size: 0.5 year

t	**Estimate of $p(t)$**	$p'(t)$	t	**Estimate of $p(t)$**	$p'(t)$
0	0	0	3	4.9680	3.3734
0.5	0	0.1709	3.5	6.6547	2.9668
1	0.0855	1.0187	4	8.1381	2.4251
1.5	0.5948	2.1865	4.5	9.3507	1.8744
2	1.6881	3.0891	**5**	**10.2879**	
2.5	3.2326	3.4707			

After 5 years of production, the well has produced about 10.3 thousand barrels.

c.

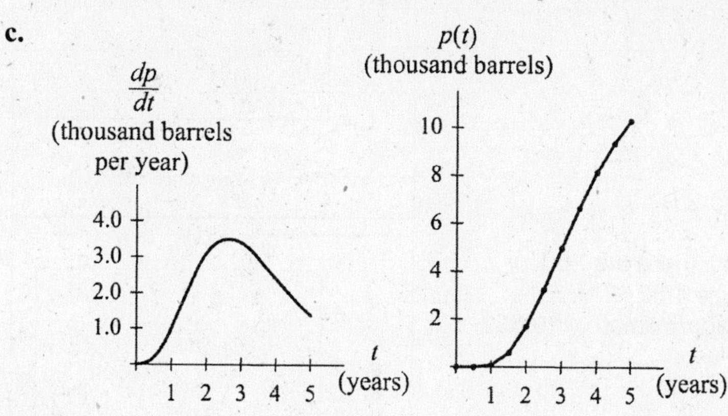

The graph of the differential equation is the slope graph for the graph of the Euler estimates. Similarly, the graph of the Euler estimates is an approximation to the accumulation graph of the differential equation graph.

9. a. $\frac{dT}{dt} = k(T - A)$ °F per minute after t minutes.

b. Solve the equation $k(98 - 70) = -1.8$ to get $k \approx -0.064$.

c.

t	**Estimate of $p(t)$**	$p'(t)$	t	**Estimate of $p(t)$**	$p'(t)$
0	98	−1.8	8	86.4552	−1.05784
1	96.2	−1.68429	9	85.3974	−0.989833
2	94.5157	−1.57601	10	84.4076	−0.926201
3	92.9397	−1.47470	11	83.4814	−0.866659
4	91.465	−1.37989	12	82.6147	−0.810945
5	90.0851	−1.29119	13	81.8038	−0.758813
6	88.7939	−1.20818	14	81.0449	−0.710032
7	87.5857	−1.13051	**15**	**80.3349**	

After 15 minutes, the temperature of the object is approximately 80.3°F.

11. **Quill Activity**
Euler's method uses tangent-line approximations. Tangent lines generally lie close to a curve near the point of tangency and deviate more and more as you move farther and farther away from that point. Thus, smaller steps generally result in better approximations.

13.a-b. With monthly intervals, there are 60 intervals with step size 1/12. The Euler estimate for $p(5)$ is approximately 10.594 thousand barrels.

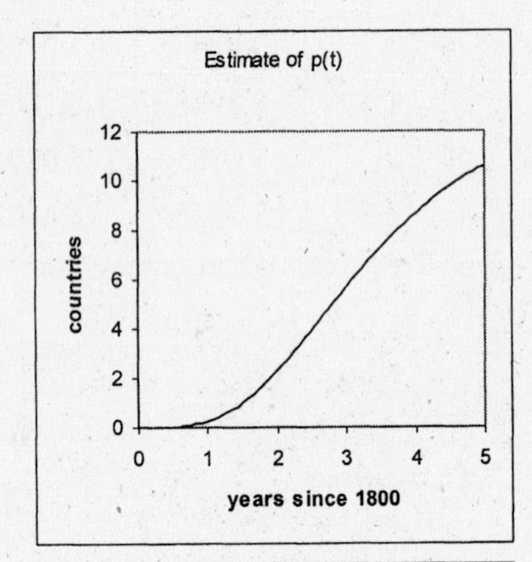

With weekly intervals, there are 260 intervals with step size 1/52. The Euler estimate for $p(5)$ is approximately 10.639 thousand barrels.

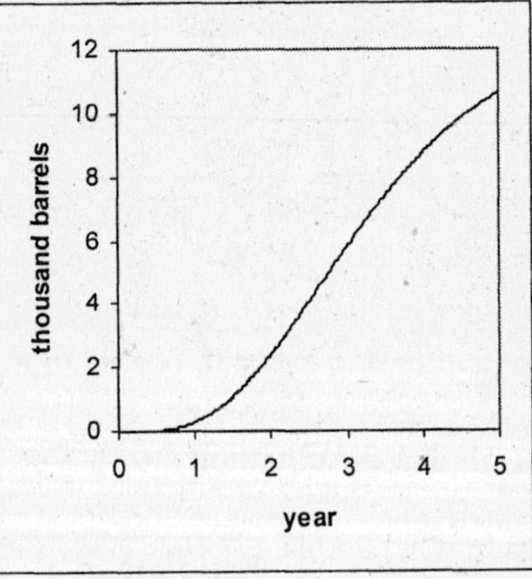

With daily intervals, there are 1825 intervals with step size 1/365. The Euler estimate for $p(5)$ is approximately 10.650 thousand barrels.

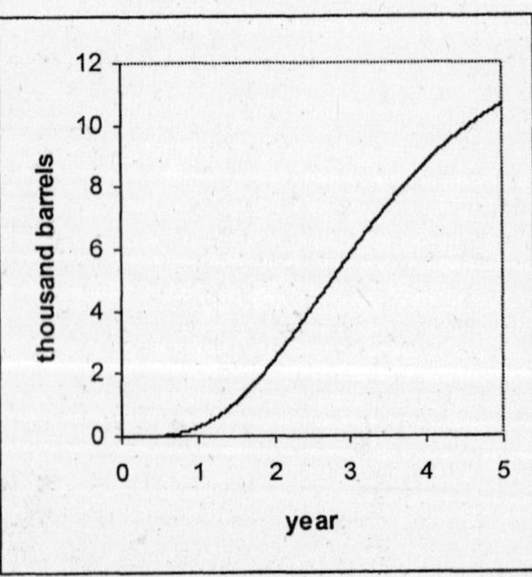

c. Answers will vary.

15. a-b. There are 790 one-second intervals with step size 1/60. The Euler estimate for $T(15)$ is approximately 80.6696349 or 80.7 °F.

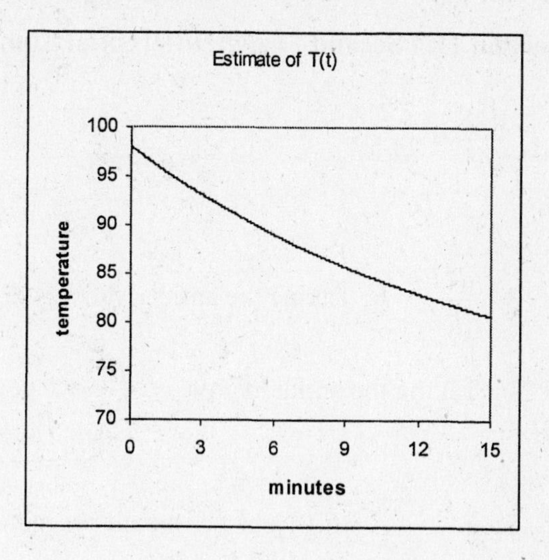

c. The estimate using 790 intervals instead is 0.4 degrees higher than the estimate using 15 intervals. The estimate using 790 intervals is probably more accurate than the estimate using 15 intervals since the function has no really steep areas of descent and no change of curvature.

Section 11.4 Second-Order Differential Equations

1. $\frac{d^2S}{dt^2} = \frac{k}{S^2}$

3. $\frac{d^2P}{dy^2} = k$; Taking the antiderivative, we get $\frac{dP}{dy} = ky + C$.

Taking the antiderivative of $\frac{dP}{dy}$, we get $P(y) = \frac{k}{2}y^2 + Cy + D$.

5. a. $\frac{d^2R}{dt^2} = 6.14$ jobs per month per month in the tth month of the year

b. Taking the antiderivative of $\frac{d^2R}{dt^2}$, we get $\frac{dR}{dt} = 6.14t + C$.

When $t = 1$, $\frac{dR}{dt} = -0.87$. Solving for C, we get $-0.87 = 6.14(1) + C$
$C = -7.01$

$\frac{dR}{dt} = 6.14t - 7.01$ jobs per month in the tth month of the year

Taking the antiderivative of $\frac{dR}{dt}$, we get $R(t) = 3.07t^2 - 7.01t + C$.

When $t = 2$, $R = 14$. Solving for C, we get $14 = 3.07(2^2) - 7.01(2) + C$ The particular
$C = 15.94$

solution is $R(t) = 3.07t^2 - 7.01t + 15.94$ jobs in the tth month of the year

c. $R(8) \approx 156$ and $R(11) \approx 310$
We estimate the number of jobs in August to be about 156 and the number in November to be 310.

7. a. $\frac{d^2A}{dt^2} = -2009$ cases per year per year, where t is the number of years since 1988

b. Taking the antiderivative of $\frac{d^2A}{dt^2}$, we get $\frac{dA}{dt} = -2099t + C$. When $t = 0$,

$\frac{dA}{dt} = 5988.7$, so $C = 5988.7$. The particular solution is $\frac{dA}{dt} = -2099t + 5988.7$ cases per year, where t is the number of years since 1988.

Taking the antiderivative of $\frac{dA}{dt}$, we get $A(t) = -1049.5t^2 + 5988.7t + C$. When $t = 0$,
$A = 33{,}590$, so $C = 33{,}590$. The particular solution is
$A(t) = -1049.5t^2 + 5988.7t + 33{,}590$ cases, where t is the number of years since 1988

c. When $t = 3$, $\frac{dA}{dt} = -308.3$ and $A(3) \approx 42{,}111$.

We estimate that in 1991 there were 42,111 AIDS cases and the number of cases was decreasing at rate of 308.3 cases per year.

9. a. $\frac{d^2 f}{dx^2} = k$

b. Taking the antiderivative, we have $\frac{df}{dx} = kx + C$.

Taking the antiderivative of the result, we get $f(x) = \frac{k}{2}x^2 + Cx + D$.

c. Taking the derivative in part *b*, we get $\frac{d}{dx}\left(\frac{k}{2}x^2 + Cx + D\right) = \frac{k}{2}(2x) + C + 0 = kx + C$

Taking the derivative of this result, we get $\frac{d}{dx}(kx + C) = k$ so we have the identity $k = k$, and our solution is verified.

11. a. $\frac{d^2 E}{dt^2} = -0.212531E$ mm per day per month per month, where $E(t)$ is the amount of radiation in mm per day and t is measured in months

b. The general solution to the equation is of the form $E(t) = a\sin(\sqrt{k}t + c)$ with $\sqrt{k} = \sqrt{0.212531} \approx 0.461011$.

In June, the amount of radiation is 4.5 mm above the expected value, and in December, the amount of radiation is 4.7 mm below the expected value. Thus we have $4.5 = a\sin(0.461011(6) + c)$ and $-4.7 = a\sin(0.461011(12) + c)$. We solve for a in the first equation $a = \frac{4.5}{\sin(0.461011(6) + c)}$ and substitute this into the second equation:

$$-4.7 = \frac{4.5}{\sin(0.461011(6) + c)} \sin(0.461011(12) + c).$$

Using technology, we find $c \approx 2.24801$ and $a \approx -4.71284$.

$E(t) = -4.71284\sin(0.461011t + 2.24801)$ mm per day, where t is the month of the year.

c. $E(t) = -4.71284\sin(0.461011t + 2.24801) + 12.5$ mm/day where t is the month of the year

d. In March the amount is $f(3) \approx 14.7$ mm per day, and in September the amount is $f(6) \approx 12.0$ mm per day.

Chapter 11 Review Test

1. **a.** The relative risk of having a car accident is changing with respect to blood alcohol level at a rate that is proportional to the risk of having a car accident at certain blood alcohol level.

   b. Solve by separation of variables.

$$\frac{1}{R}dR = kdb$$

$$\int \frac{1}{R}dR = \int kdb$$

$$\ln R + c_1 = kb + c_2 \quad \text{(Because the risk is always positive, we omit the absolute value.)}$$

$$\ln R = kb + C$$

$$R = e^{kb+C}$$

$$R = ae^{kb}$$

$$R(b) = ae^{kb}$$

   c. $R(0) = a = 1$ and $R(0.14) = ae^{0.14k} = 20$. Solving these equations, we get $a = 1$ and $k = \frac{\ln 20}{0.14} \approx 21.398$.

   $R(b) = e^{21.398b}$ percent, where b is the proportion of alcohol in the blood stream

   d. A certain occurrence corresponds to a relative risk of 100%, so $R = 100$. We solve $e^{21.398b} = 100$ for b:

$$e^{21.398b} = 100$$

$$21.398b = \ln 100$$

$$b = \frac{\ln 100}{21.398}$$

$$\approx 0.215$$

   Thus a crash is certain to occur when the blood alcohol level is about 21.5%.

   e.

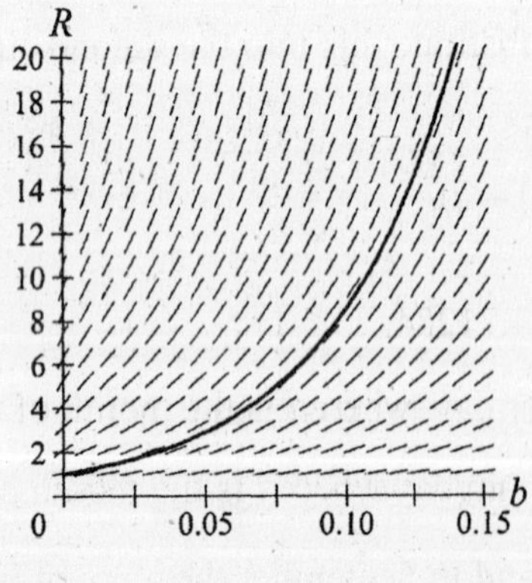

2. **a.** $\frac{dP}{dx} = 0.001175P(16.396 - P)$ million people per year x years after 1800

   b. This differential equation has a logistic function as its solution.

   $P(x) = \frac{16.396}{1 + Ae^{-0.001175(16.396)x}} = \frac{16.396}{1 + Ae^{-0.019265x}}$ million people x years after 1800

   c. We use the fact that $P(-20) = 4.0$, to solve for A:

$$\frac{16.396}{1+Ae^{0.385306}}=4$$

$$16.396=4(1+Ae^{0.385306})$$

$$12.396=4Ae^{0.385306}$$

$$A=\frac{12.396}{4e^{0.385306}}\approx 2.108$$

Thus $P(x)=\dfrac{16.396}{1+2.108e^{-0.019265x}}$ million people x years after 1800.

d. $P(40)\approx 8.3$ and $P(50)\approx 9.1$
We estimate that there were 8.3 million people in 1840 and 9.1 million people in 1850.

3. a.

x (years after 1800)	Estimate of $P(x)$ (million people)	Slope at x (million people per year)
–20	4	0.058
–10	4.583	0.064
0	5.219	0.069
10	5.904	0.073
20	6.632	0.076
30	7.393	0.078
40	**8.175**	0.079
50	**8.965**	

We estimate that $P(40)\approx 8.18$ million people; $P(50)\approx 8.97$ million people.

b.

x	Estimate of $P(x)$	Slope at x	x	Estimate of $P(x)$	Slope at x
–20	4	0.058	20	6.684	0.076
–15	4.291	0.061	25	7.066	0.077
–10	4.596	0.064	30	7.453	0.078
–5	4.915	0.066	35	7.844	0.079
0	5.247	0.069	**40**	**8.239**	0.079
5	5.590	0.071	45	8.633	0.079
10	5.945	0.073	**50**	**9.027**	
15	6.310	0.075			

We estimate that $P(40)\approx 8.24$ million people and $P(50)\approx 9.03$ million people.

c. The 1940 estimates in this question are 0.12 and 0.06 million people less than population found in the Question 2. The 1950 estimates are 0.13 and 0.07 million people less than the Question 2 answer.

4. a. $\frac{dQ}{dx} = -0.008307Q(7.154 - Q)$ million people per year, where the population is $P(x) = Q(x) + 4.4$ million people and x is the number of years since 1800

b. This differential equation has a logistic function as its solution.

$Q(x) = \frac{7.154}{1 + Ae^{0.008307(7.154)x}} = \frac{7.154}{1 + Ae^{0.059428x}}$ million people, where x is the number of years since 1800

Because $Q(100) = 4.5$, we have the equation $\frac{7.154}{1 + Ae^{0.59428}} = 0.1$. Solving this equation for A (as illustrated in Question 2 part *c*), we get $A = \frac{7.054}{0.1e^{5.9428278}} \approx 0.185$.

$Q(x) = \frac{7.154}{1 + 0.185e^{0.059428x}}$ million people, where x is the number of years since 1800

c. $P(x) = \begin{cases} \frac{16.396}{1 + 2.108e^{-0.019265x}} \text{ million people} & \text{when } x \le 40 \\ \frac{7.154}{1 + 0.185e^{0.059428x}} + 4.4 \text{ million people} & \text{when } x \ge 50 \end{cases}$

where x is the number of years since 1800

d. Using the model in part *c*, $P(50) \approx 6.0$
We estimate that there were 6.0 million people in 1850. This answer is significantly smaller than the one found in part *d* of Question 2.